Design of Caspase Inhibitors as Potential Clinical Agents

T0100476

CRC Enzyme Inhibitors Series

Series Editors

H. John Smith and Claire Simons
Cardiff University
Cardiff, UK

Carbonic Anhydrase: Its Inhibitors and Activators
Edited by Claudiu T. Supuran, Andrea Scozzafava and Janet Conway

Design of Caspase Inhibitors as Potential Clinical Agents
Edited by Tom O'Brien and Steven D. Linton

Enzymes and Their Inhibition: Drug Development
Edited by H. John Smith and Claire Simons

Inhibitors of Cyclin-dependent Kinases as Anti-tumor Agents
Edited by Paul J. Smith and Eddy W. Yue

Protein Misfolding in Neurodegenerative Diseases: Mechanisms and Therapeutic Strategies
Edited by H. John Smith, Claire Simons, and Robert D. E. Sewell

CRC Enzyme Inhibitors Series

Design of Caspase Inhibitors as Potential Clinical Agents

Edited by
Tom O'Brien
Steven D. Linton

CRC Press
Taylor & Francis Group
Boca Raton London New York

CRC Press is an imprint of the
Taylor & Francis Group, an **informa** business

CRC Press
Taylor & Francis Group
6000 Broken Sound Parkway NW, Suite 300
Boca Raton, FL 33487-2742

First issued in paperback 2019

ISBN-13: 978-1-4200-4540-6 (hbk)
ISBN-13: 978-0-367-38657-3 (pbk)

Library of Congress Cataloging-in-Publication Data

Design of caspase inhibitors as potential clinical agents / editor(s), Tom O'Brien,
Steven D. Linton.
p. ; cm. -- (CRC enzyme inhibitors series)
Includes bibliographical references and index.
ISBN 978-1-4200-4540-6 (hardback : alk. paper)
1. Cysteine proteinases. 2. Cysteine proteinases--Inhibitors. I. O'Brien, Tom. II. Linton, Steven D. III. Series.
[DNLM: 1. Caspases--antagonists & inhibitors. 2. Apoptosis. 3. Caspases--physiology. 4. Caspases--therapeutic use. QU 136 D457 2009]

QP609.C94D47 2009
572'.76--dc22
2008025506

Visit the Taylor & Francis Web site at
http://www.taylorandfrancis.com

and the CRC Press Web site at
http://www.crcpress.com

Contents

Preface

In 1992 the first report was published describing the identification of caspase-1 or Interleukin-1β Converting Enzyme (ICE), the founding member of a new class of cysteine based proteases. Since then, there has been an enormous effort to discover small molecule inhibitors of this therapeutically important class of enzymes. During this period, a large volume of literature has emerged describing the identification of a diverse range of inhibitors, ranging from substrate-like peptidic to non-peptidic heterocycles. Despite this significant effort, only four different inhibitors have initiated clinical trials: Pralnacasan (VX740) for rheumatoid arthritis (subsequently withdrawn during phase II trials); VX765, a second generation caspase-1 inhibitor (entered Phase II trials for psoriasis, but no current status is available); Emricasan (PF-03491390/IDN-6556), an anti-fibrotic agent for the treatment of chronic liver disease (apparently discontinued after completing Phase II trials); and LB-84451, for undisclosed indications, but possibly focusing on its anti-fibrotic activity (currently in Phase II trials). This emphasizes the difficulty in progressing a caspase small molecule inhibitor from discovery into the clinic, and, in this book, we intend to outline these efforts and highlight the complex issues that have been encountered. We also will present the current status of clinical trials and the future potential for caspase inhibitors as therapeutic agents.

The first three chapters outline what is currently known about the inflammatory and apoptotic caspase pathways. Logue and Martin (Chapter 1) present a comprehensive overview of the key caspase proteolytic pathways, both apoptotic and inflammatory, and the importance of these pathways in normal cellular activities. Chapters by Kersse et. al. (Chapter 2) and Matsui (Chapter 3) provide an in-depth coverage of the potential therapeutic value of an inflammatory or apoptotic (respectively) caspase inhibitor. In each of these chapters, the authors present an overview of the relevant pathways, along with *in vitro* and *in vivo* evidence supporting the use of a caspase inhibitor for each indication. In Chapter 4, Nalley reviews the catalytic properties of caspases, how to use these properties in designing inhibitors, and the potential difficulties involved therein. In particular, one of the key outstanding questions in caspase small molecule discovery is whether a specific or a pan-caspase inhibitor is preferred. As different indications will likely require a caspase inhibitor with a unique specificity profile, Nalley illustrates that by understanding the architecture of each caspase active site and its known substrate specificities, it should be possible to design inhibitors with different profiles.

We then turn to applying this knowledge to the design of caspase inhibitors and, in subsequent chapters, review the progress already made toward discovering small molecule inhibitors. Ivachtchenko et al (Chapter 5) outline progress towards discovering non-peptide inhibitors and give a very comprehensive overview spanning a large number of different chemical classes. This is followed by a chapter by Wos and Demuth (Chapter 6) that summarizes the current status of the discovery of inflammatory caspase inhibitors and by Ullman (Chapter 7) that gives a perspective of the discovery of apoptotic caspase inhibitors. Chapter 8 presents an example of the preclinical approach undertaken to identify an apoptotic caspase inhibitor. In this case, Holgen presents a case-study of the discovery and characterization of Emricasan, a compound that completed Phase II clinical trials, but is now reported as discontinued from further development.

As described in the previous chapters, caspase inhibitors have been predominantly identified either by functional screens, structure-based design, or by computational modeling. In most cases, a combination of all three approaches has been used. However, as new technologies have emerged so have the approaches taken to inhibitor discovery, and some of these novel approaches are presented in Chapter 9 (Scheer and Romanowski). The advantage of using novel approaches has been validated by the discovery of an allosteric regulatory site that lies at the dimeric interface between the caspase large and small subunits. The presence of an allosteric site presents an opportunity to inhibit catalytic activity that avoids the limitations associated with designing a molecule that binds to the active site.

In the final chapter (Chapter 10), Eda gives an overview of the current status of ongoing clinical trials with caspase inhibitors. A survey of the literature indicates that potent caspase inhibitors can be discovered; however, advancing these compounds into the clinic has been challenging. Some of the key issues that are discussed revolve around questions such as what is the desired selectivity profile, whether reversible or irreversible inhibition is more relevant to the indication, and what impact the mode of inhibition has upon the toxicity profile. The answers will likely depend upon the indication being pursued and whether the caspase being targeted is apoptotic or inflammatory.

Despite the difficulties involved in caspase inhibitor discovery, considerable progress has been made and early clinical studies have shown promise. However, of the four compounds that have entered clinical trials, two have been discontinued and the fate of the remaining two compounds remains unclear. Nevertheless, the potential therapeutic benefits are tremendous, and there appears to be a renewed enthusiasm for caspase small molecule drug discovery. If one of the current compounds shows clinical benefit and makes it to market as a "first-in-class" drug, we have no doubt that this will fuel an enhanced discovery effort for additional compounds that could be "best-in class."

Editors

Dr. Steve Linton has been involved with San Diego biotech for over 15 years, not only focusing on modulators of apoptosis while at Idun, but also working in several therapeutic areas, including inflammation and oncology, as well as metabolic and CNS disorders. Dr. Linton graduated with a BS degree in chemistry from Texas Christian University and received his PhD in the area of natural product synthesis from Rice University under the direction of Dr. Tohru Fukuyama. After beginning his medicinal chemistry career at Gensia Pharmaceuticals in 1992, Dr. Linton joined Idun Pharmaceuticals in 1995. He contributed to the development of Idun's caspase inhibitor drug discovery platform as well as investigated other modulators of apoptosis and is widely published in this area. Dr. Linton was promoted to medicinal chemistry section head and was part of the core team that presented Idun technology to potential investors. Idun was acquired by Pfizer in 2005, and its flagship caspase inhibitor, Emricasan, is currently in late-stage clinical development. Dr. Linton has also participated in other start-ups, such as Synstar, Inc., a custom synthesis contract research organization (Hangzhou, San Diego), as well as Novasite Pharmaceuticals, where he served as director of chemistry. He is currently director of project management at Halozyme Therapeutics.

Dr. Tom O'Brien graduated from Trinity College, Dublin, Ireland, with a BA (Mod) degree in genetics. Dr. O'Brien completed his PhD degree at Cornell University, Ithaca, New York, in the laboratory of Dr. John Lis, and subsequently moved to the University of California at Berkeley where he pursued postdoctoral research studies in the laboratory of Dr. Robert Tjian, where his research focused on dissecting the biochemical and molecular regulation of eukaryotic transcription. In 1999 Dr. O'Brien moved to the newly formed company Sunesis Pharmaceuticals, where his work focused on the discovery of novel small-molecule caspase inhibitors. As the biology project leader for their caspase small-molecule programs, Dr. O'Brien was one of the key people that helped optimize and validate their fragment-based approaches to small-molecule drug discovery. During his time at Sunesis, Dr. O'Brien was also the lead biologist for a number of additional programs, one of which recently entered clinical trials. In 2006 Dr. O'Brien joined Genentech, Inc., in their newly formed Department of Cell Regulation.

Contributors

Konstantin V. Balakin
ChemDiv, Inc.
San Diego, California

Wim Declercq
Department of Molecular Biology
Ghent University
and
Department for Molecular Biomedical
  Research
VIB
Ghent, Belgium

Thomas P. Demuth Jr.
Clinical and Regulatory Affairs
Procter & Gamble Pharmaceuticals
Mason, Ohio

Hiroyuki Eda
Cell Biology and Enzymology
Global Research and Development
  Research
St. Louis Laboratories
Pfizer, Inc.
Chesterfield, Missouri

Niel C. Hoglen
Aires Pharmaceutical, Inc.
San Diego, California

Alexandre V. Ivachtchenko
ChemDiv, Inc.
San Diego, California

Yan A. Ivanenkov
ChemDiv, Inc.
San Diego, California

Kristof Kersse
Department of Molecular Biology
Ghent University
and
Department for Molecular Biomedical
  Research
VIB
Ghent, Belgium

Alex S. Kiselyov
ChemDiv, Inc.
San Diego, California

Saskia Lippens
Department of Molecular Biology
Ghent University
and
Department for Molecular Biomedical
  Research
VIB
Ghent, Belgium

Susan E. Logue
Molecular Cell Biology Laboratory
Department of Genetics
The Smurfit Institute
Trinity College
Dublin, Ireland

Seamus J. Martin
Molecular Cell Biology Laboratory
Department of Genetics
The Smurfit Institute
Trinity College
Dublin, Ireland

Takashi Matsui
Cardiovascular Research
Cardiovascular Division
Beth Israel Deaconess Medical Center
Harvard Medical School
Boston, Massachusetts

Kip A. Nalley
Laboratory of Receptor Biology and
  Gene Expression
Center for Cancer Research
National Cancer Institute
National Institutes of Health
Bethesda, Maryland

Ilya Okun
ChemDiv, Inc.
San Diego, California

Michael J. Romanowski
Department of Protein Sciences and
  Structural Biology
Sunesis Pharmaceuticals, Inc.
South San Francisco, California

Justin M. Scheer
Department of Protein Chemistry
  MS 63
Genentech, Inc.
South San Francisco, California

Sergey E. Tkachenko
ChemDiv, Inc.
San Diego, California

Brett R. Ullman
Arena Pharmaceuticals, Inc.
San Diego, California

Peter Vandenabeele
Department of Molecular Biology
Ghent University
and
Department for Molecular Biomedical
  Research
VIB
Ghent, Belgium

Tom Vanden Berghe
Department of Molecular Biology
Ghent University
and
Department for Molecular Biomedical
  Research
VIB
Ghent, Belgium

John A. Wos
Global New Business & Technology
  Development
Procter & Gamble Pharmaceuticals
Mason, Ohio

1 Mammalian Caspase Activation Pathways in Apoptosis and Inflammation

Susan E. Logue and Seamus J. Martin

CONTENTS

1.1 INTRODUCTION

Members of the caspase family of cysteine proteases play key roles in signal transduction cascades in apoptosis (programmed cell death) and inflammation. Caspases are normally expressed as inactive precursor enzymes (zymogens), a subset of which become activated during apoptosis and coordinate the demolition of the cell from within. To date, three major apoptosis-associated pathways to caspase activation have been elucidated. Certain caspases, such as caspases-1, -4, and -5, also play key roles in signaling pathways associated with immune responses to microbial pathogens.

1

In this situation, caspase activation results in the maturation of pro-inflammatory cytokines, such as IL-1β and IL-18. Here we discuss the current understanding of how caspases are activated during apoptosis and inflammation and the roles these proteases play in either context.

1.2 MAMMALIAN CASPASES

Early studies directed toward identifying genes involved in the regulation of programmed cell death (PCD) were conducted in the nematode worm *Caenorhabditis elegans* and led to the identification of the cell death defective gene-3 (*ced-3*).[1] Worms defective for *ced-3* failed to eliminate many of the 131 cells that normally undergo PCD during worm development and implicated this gene as a major regulator of PCD in this organism. Ensuing searches for human homologues of *ced-3* resulted in the publication of a landmark paper by Horvitz and colleagues in 1993 describing interleukin-1β converting enzyme (ICE) as a human homologue of CED-3.[1] ICE, or caspase-1, as it is now commonly known, became the founding member of the family of aspartic acid–specific proteases, called caspases. To date, twelve members of the human caspase family have been identified (caspases-1, -2, -3, -4, -5, -6, -7, -8, -9, -10, -12, -14). Based upon the functional data available, these caspases fall into two distinct groups; apoptotic caspases (caspases-2, -3, -6, -7, -8, -9, -10) and inflammatory caspases (caspases-1, -4, -5, -12) with the role of caspase-14 somewhat poorly defined at present (Figure 1.1). Irrespective of their function, all members of this protease family are thought to cleave their substrates following an aspartate (Asp) residue.[2] Caspases recognize the Asp residues they cleave within a tetrapeptide motif, P4-P3-P2-P1, with substrate cleavage occurring at the peptidyl bond distal to the P1 residue. Depending upon the caspase in question, residues P2 to P4 can vary; however, position P1 has a near absolute requirement for Asp.[2]

Caspases are highly active proteases that are initially expressed as largely inactive precursors (pro-caspases) that require further proteolytic processing to achieve their active forms. Pro-caspases are comprised of three distinct domains: an N-terminal pro-domain, a large subunit containing the active site cysteine within a conserved QACXG motif, and a small C-terminal subunit (Figure 1.1). An Asp cleavage site frequently demarcates the N-terminal pro-domain from the large subunit. Similarly, a linker domain, containing one or two Asp cleavage sites, divides the large and small subunits.[2] Receipt of an activation signal initiates proteolytic processing of pro-caspases via a two-step process. Initial proteolytic cleavage at the Asp residues within the linker domain separates the large and small subunits. The caspase pro-domain is frequently, but not always, removed by proteolytic cleavage at the Asp residue located between this domain and the large subunit.[3] This series of proteolytic events results in the formation of active heterotetramers, comprised of two large subunits, two small subunits, and two active sites.[4–6] The substrate specificity of active caspases for Asp residues, combined with their own requirement for cleavage at specific Asp resides, suggested that caspase activation occurred either by autoproteolytic means or via cleavage by other caspases.

Apoptotic caspases can be further subdivided on the basis of their domain structures (Figure 1.1). Initiator caspases (caspases-2, -8, -9, -10) possess long pro-domains

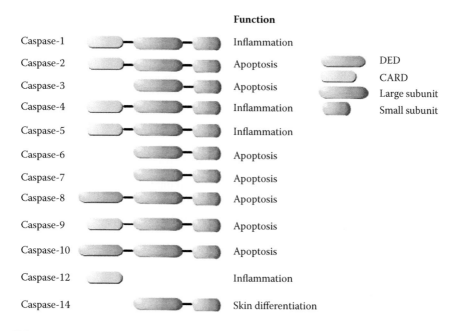

	Function		
Caspase-1	Inflammation		DED
Caspase-2	Apoptosis		CARD
Caspase-3	Apoptosis		Large subunit
Caspase-4	Inflammation		Small subunit
Caspase-5	Inflammation		
Caspase-6	Apoptosis		
Caspase-7	Apoptosis		
Caspase-8	Apoptosis		
Caspase-9	Apoptosis		
Caspase-10	Apoptosis		
Caspase-12	Inflammation		
Caspase-14	Skin differentiation		

FIGURE 1.1 (See color insert.) Domain structures of the human caspase family.

with protein-protein interaction motifs, such as caspase recruitment domains (CARDs) or death effector domains (DEDs).[2] These motifs enable initiator caspase clustering upon scaffold molecules following receipt of activation signals. The clustering of multiple initiator caspases into close proximity induces dimerization followed by auto-processing, a mechanism referred to as the induced-proximity model.[7] Conversely, effector caspases (caspases-3, -6, -7, -14) have short pro-domains, lacking protein interaction motifs, and are dependent upon upstream initiator caspases for processing.[2,8] Therefore, caspase activation occurs in a hierarchical manner with initiator caspase activation both preceding and facilitating downstream effector caspase activation.

Apoptotic cell death is characterized by a specific morphology, which includes blebbing of the plasma membrane, nuclear condensation, and fragmentation.[9] This characteristic morphology is a consequence of effector caspase-mediated cleavage of numerous cellular substrates, the precise details of which remain obscure. To date, over 400 effector caspase substrates have been identified.[10] However, the cleavage of only a small subset of these substrates has been definitively linked to specific features of apoptosis. The serine/threonine kinase rho-associated kinase I (ROCK I), structural proteins vimentin, Gas2, and plectin, and the nuclear protein ICAD have all been linked to the morphological changes associated with apoptosis. For example, caspase-3-mediated cleavage of the inhibitor of caspase-activated DNase (ICAD), breaks the inhibitory association of ICAD with caspase-activated DNase (CAD), allowing CAD to initiate DNA fragmentation.[11] Targeting of cytoskeletal proteins vimentin,[12,13] Gas2,[14] and plectin[15] by caspases contributes to changes in cell shape, while proteolysis of ROCK I has been associated with nuclear fragmentation and plasma membrane blebbing.[16,17]

1.3   CASPASE ACTIVATION PATHWAYS

Caspase activation pathways have been the focus of intense research over the past 10 years. Presently, the most studied and accepted pathways leading to caspase activation are the mitochondrial pathway, the death receptor pathway, and the granzyme B–initiated pathway. Other, less well-defined caspase activation pathways, such as the inflammasome and endoplasmic reticulum stress-induced caspase activation pathways, have also been described.

1.3.1   THE INTRINSIC PATHWAY TO CASPASE ACTIVATION

Early studies examining cell death initiated by cell damage, such as cytotoxic drugs or in ionizing radiation, found that overexpression of Bcl-2, a protein localized to mitochondria, blocked cell death.[18–20] These observations suggested the mitochondria, in addition to acting as the "powerhouse" of the cell, may be involved in cell death signaling.

Bcl-2 is the founding member of a large family of proteins important in the regulation of cellular life and death decisions.[21] Over the past years our knowledge of this protein family has expanded dramatically, and we now know that the Bcl-2 family is comprised of twenty-two members, some of which promote apoptosis, while others suppress this form of cell death. Although functionally distinct, each member of this family possesses at least one Bcl-2 homology (BH) domain. Pro-survival members of this family, Bcl-2, Bcl-xL, Bcl-w, Bcl-b, Mcl-1, and A1, contain three or four BH domains, while pro-apoptotic members contain between one and three BH domains. The pro-apoptotic members of the Bcl-2 family can be divided into two distinct groups, those that contain Bcl-2 homology (BH) domains 1–3 (Bax, Bak, and Bok) and those containing only a single BH-3 domain, referred to as BH3-only proteins (Noxa, PUMA, Bad, Bim, Bid, Bmf, HRK, Bik, and BLK).[22]

Intense research over the past 10 years, investigating the mechanisms by which Bcl-2 proteins regulate cell death decisions, has revealed a complex network of interactions between family members in which the ratio of pro- to anti-apoptotic Bcl-2 family members controls the release of cytochrome c from mitochondria. Pro-apoptotic members Bax and Bak have essential functions in regulating cytochrome c release. Normally, Bax is localized to the cytoplasm where it is maintained in an inactive conformation, possibly via interactions with pro-survival proteins Bcl-2, Bcl-xL, and Mcl-1.[23,24] Similarily, Bak, an integral membrane protein localized to the outer mitochondrial membrane, is restrained through binding to anti-apoptotic Bcl-2 proteins. Following receipt of pro-apoptotic signals, levels of active BH3-only proteins increase by a mixture of transcriptional upregulation (PUMA, Noxa) and posttranslation modification (Bim, Bad, Bid), depending upon the initiating stimulus.[25] Activation of the BH3-only cohort of proteins shifts the balance in favor of apoptosis by relieving the inhibition placed upon Bak and Bax. As a consequence, Bax and Bak undergo conformational changes that permit oligomerization of these proteins within the mitochondrial outer membrane and release of intermembrane space proteins, the most important of which is cytochrome c.

BH3-mediated repression of pro-survival Bcl-2 family members was thought, until recently, to be a relatively nonselective process. The use of peptides mimicking the α-helical BH3 domain permitted studies examining interactions between BH3-only proteins and other members of this family and found that the BH3-only subfamily can be divided into direct activators and de-repressors. Direct activators, such as Bid and Bim, have the ability to directly target and activate Bax and Bak.[26] Other BH3-only proteins, such as Bad, Bik, and PUMA, while not directly activating Bax or Bak, do so indirectly by neutralizing pro-survival Bcl-2 proteins. Furthermore, there is significant selectivity among the interaction of de-repressors with pro-survival Bcl-2 proteins. For instance, Bad has been demonstrated to interact with Bcl-2 and Bcl-xL, but not Mcl-1, while Noxa interacts with Mcl-1 but not with Bcl-2 or Bcl-xL.[27]

Ultimately, the balance of pro- and anti-apoptotic Bcl-2 family proteins controls permeabilization of the outer mitochondrial membrane and release of intermembrane space proteins. Cytochrome c resides in the mitochondrial intermembrane space and is released in response to diverse stress signals. *In vitro* systems, artificially reconstituting the intrinsic pathway of caspase activation, identified three apoptotic protease activating factors (Apafs) required for caspase activation. Further analysis identified these as Apaf-1, a homologue of *Caenorhabditis elegans* CED-4, caspase-9, and cytochrome c.[28–30]

1.3.1.1 Apoptosome Formation

Apaf-1 is comprised of an N-terminal CARD motif, a nucleotide binding and oligomerization domain, and thirteen WD40 repeats. Upon release from the mitochondrial intermembrane space, cytochrome c binds to the WD40 repeats on Apaf-1 initiating a conformational change and unmasking the CARD motif. Pro-caspase-9, like Apaf-1, contains a CARD motif within its pro-domain permitting association with Apaf-1.[31,32] ATP binding to the Apaf-1/pro-caspase-9/cytochrome c complex triggers further conformational changes culminating in the formation of a heptameric wheel-shaped complex called the apoptosome. The CARD domains of Apaf-1 and pro-caspase-9 are located at the center of the complex, while the WD40 domains form the "spokes" of the wheel.[33] Pro-caspase-9 undergoes autoprocessing within the apoptosome from which it triggers downstream caspase processing (Figure 1.2).

Elegant analysis of cytochrome c-initiated caspase activation cascades established that all intrinsic caspase activation is dependent on caspase-9.[34] Following activation within the apoptosome caspase-9 targets and simultaneously processes pro-caspases-3 and -7.[34,35] Caspase-3, in turn, cleaves pro-caspases-2 and -6, followed by caspase-6-mediated processing of caspases-8 and -10.[34] Pro-caspase-9 is also cleaved by caspase-3 generating a positive feedback loop between the initiator and effector caspase[34] (Figure 1.2). Knockout mouse studies have confirmed the importance of caspase-9 and Apaf-1 in the intrinsic pathway to caspase activation. Cells derived from *CASP-9* null animals demonstrated resistance to internal stress agents, such as cytotoxic drugs and radiation.[36,37] A similar resistance to apoptotic stimuli was also evident in *APAF-1* knockout animals, reinforcing the importance of apoptosome formation to intrinsic caspase activation.[38]

Apart from cytochrome c, other mitochondrial matrix proteins, including Smac/DIABLO, OMI/Htra2, and endonuclease G, are released from the mitochondrial intermembrane space as a result of outer membrane permeabilization during apoptosis.[39] Endonuclease G translocates to the nucleus and may contribute to oligonucleosomal

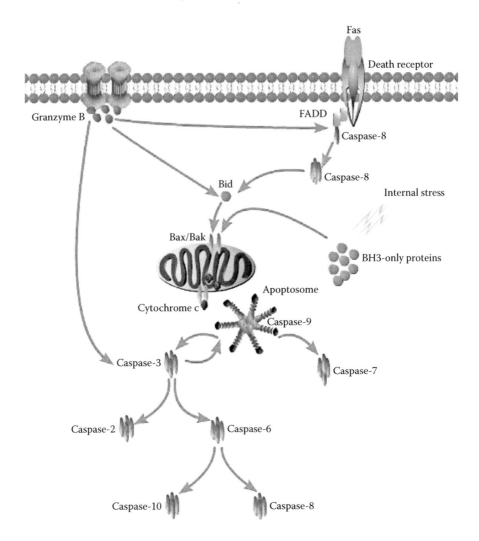

FIGURE 1.2 (See color insert.) The major routes to apoptosis-associated caspase activation. See main text for further details.

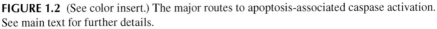

DNA fragmentation.[40] Both Omi/Htra2 and Smac/Diablo interact with inhibitor of apoptosis proteins (IAPs). Such interactions are thought to dissociate IAPs from effector caspases, effectively freeing these caspases for activation by upstream initiator caspases such as caspase-9.[41–43]

1.3.2 EXTRINSIC OR DEATH RECEPTOR–INITIATED CASPASE ACTIVATION PATHWAYS

Engagement of death receptors, present on the cell surface, can also induce apoptotic cell death. For example, activated cytotoxic T cells recognize virally infected cells, leading to caspase activation and cell death via death receptors present on the target cell surface, such as tumor necrosis factor alpha (TNFα) or Fas/CD95.[44]

Such cytotoxic T cells can also kill via the granule-dependent granzyme B pathway, as described later. Structurally, death receptors are comprised of an extracellular cysteine-rich domain (ligand binding region) and an intracellular region containing a death domain (DD). Ligand binding to death receptors initiates receptor trimerization and recruitment of adapter proteins, such as FADD and TRADD, forming the death-inducing signaling complex (DISC).[45] Similar to the apoptosome, the DISC functions as a caspase activation platform and recruits, via death effector domain (DED) interactions, multiple pro-caspase-8 or pro-caspase-10 molecules, leading to their activation.[46] Active caspase-8 transmits and propagates the apoptotic signal by cleaving and activating pro-caspase-3, which in turn cleaves pro-caspases-6 and -2 (Figure 1.2).

Two distinct pathways of Fas-mediated apoptosis have been identified, leading to the classification of cells as either type-I or -II.[47] Type-I cells generate sufficient quantities of active caspase-8 at the DISC leading to pro-caspase-3 activation and cellular demolition. In contrast, type-II cells fail to produce sufficient active caspase-8 at the DISC (for reasons that remain obscure) and require cross talk with the mitochondrial pathway to trigger downstream caspase cascades. In type-II cells, DISC-activated caspase-8 cleaves the BH3-only protein, Bid, generating truncated Bid (tBid). The amino terminus of tBid is subsequently myristolated, targeting it to mitochondria where it induces cytochrome c release via BAX or BAK.[48–51] Death in type-II cells is thus routed through the apoptosome pathway.

Both type-I and type-II Fas-induced apoptosis can be blocked by the cellular-FADD-like inhibitory protein (c-FLIP).[52] Like caspase-8, c-FLIP contains DEDs, allowing it to compete with the latter for recruitment to the DISC. However, c-FLIP lacks protease activity and its incorporation within the DISC inhibits progression of the signaling cascade.[53]

1.4 GRANZYME B-MEDIATED CASPASE ACTIVATION

Virally infected and transformed cells are removed from the body through natural killer (NK) cells and cytotoxic T lymphocytes (CTLs). Initial secretion of TNFα and IFN$_9$ by NK cells limits viral replication and spread. CTLs form the second wave of attack by seeking out and specifically targeting virally infected cells. Engagement of virally infected cells by a CTL results in the initiation of apoptosis by delivery of cytotoxic granules.[54] Perforin, a pore-forming protein, present in cytotoxic granules generates pores in the target cell that facilitate delivery of cytotoxic granzymes into the target (Figure 1.2). The precise mechanism by which perforin mediates delivery of granule components has not been resolved. However, the importance of perforin to this process is clearly illustrated by perforin-deficient mice, which display impaired clearance of viral pathogens.[55]

Granzyme B, a serine protease, is a key component of cytotoxic granules. Artificial systems have demonstrated that addition of perforin and granzyme B alone are sufficient for the induction of apoptosis.[56] Granzyme B, similar to the caspases, cleaves its substrates after aspartate residues,[57,58] suggesting that this protease has the ability to directly activate members of the caspase family. Indeed, caspase-3 was the first substrate identified for granzyme B.[59,60] Experiments utilizing Jurkat cell-free extracts demonstrated that addition of granzyme B resulted in processing of pro-caspase-3 and multiple caspase substrates.[59] The cohort and activation cascade of caspases activated in the granzyme B–mediated pathway has recently been elucidated.[61] Caspases-3, -7, -8, and -10 are directly processed by granzyme B, whereas caspases-2, -6, and -9 are processed in a second, caspase-3-dependent wave of processing.[61] Because granzyme B directly cleaves and activates caspases, addition of caspase-specific inhibitors might be expected to inhibit granzyme B–mediated cell death. However, although preincubation of Jurkat cells with caspase-3-like specific inhibitor DEVD-fmk or the broad range caspase inhibitor zVAD-fmk reduced DNA fragmentation, in response to granzyme B and perforin, no long-term protection was evident, indicating granzyme B has other noncaspase targets.[62] This is not entirely surprising when we consider that many viruses encode caspase inhibitors, such as Crm A or p35, as a means to aid their survival and replication in host cells. Therefore, by incorporating a caspase-independent route to cell death, granzyme B has the ability to overcome these viral defenses.

Subsequent studies demonstrated that overexpression of Bcl-2 clonogenically protected cells against granzyme B and perforin-mediated apoptosis.[62–64] Protection by Bcl-2 implied that granzyme B may also act upstream of mitochondria. Indeed, the BH3-only Bcl-2 family member Bid was subsequently identified as a target of granzyme B.[65] Cleavage of Bid by granzyme B generates a truncated fragment that translocates to mitochondria, initiating release of intermembrane space proteins such as cytochrome *c* (Figure 1.2). Both granzyme B and caspase-8 target and cleave Bid, although at distinct sites. Overexpression studies with a mutant, granzyme B–resistant Bid (D75E) demonstrated inhibition of granzyme B and perforin-induced apoptotic features, while overexpression of D59E Bid (mutated at the caspase-8 site) failed to abolish the apoptotic phenotype.[66] The latter result illustrates that granzyme B directly cleaves Bid rather than doing so indirectly by caspase-8-mediated cleavage.

Several studies now indicate that Bid, rather than caspases, is the preferential substrate for human granzyme B, with Bid cleavage being evident within minutes of granzyme B entry to the target cell. Upon entry into the target, granzyme B rapidly induces mitochondrial permeabilization, via Bid, and release of intermembrane space proteins leading to apoptosome assembly and caspase activation (Figure 1.2). Concurrently, granzyme B can also directly target and cleave caspases, thereby amplifying the level of caspase activation.

1.5   EMERGING CASPASE ACTIVATION PATHWAYS

The endoplasmic reticulum and inflammatory-mediated pathways of caspase activation are relatively poorly understood, compared to the death receptor (extrinsic) and mitochondrial-mediated (intrinsic) pathways. Our current knowledge of each pathway and the mechanisms utilized to achieve caspase activation is outlined below.

1.5.1   ENDOPLASMIC RETICULUM STRESS-INDUCED CASPASE ACTIVATION

The endoplasmic reticulum (ER) is responsible for the synthesis, folding, and maturation of proteins within the cell. Stresses negatively regulating energy availability or intracellular calcium levels, such as ischemia, can have a detrimental affect on protein folding, leading to the accumulation of unfolded proteins, a condition referred to as ER stress. Cells have evolved response mechanisms aimed at reducing levels of unfolded proteins and restoring cellular homeostasis. However, in certain situations, these survival mechanisms are insufficient and cell death ensues.[67] Unlike the death receptor or mitochondrial-mediated pathways the cohort of caspases and their activation mechanisms have not been firmly established in ER stress-induced apoptosis and are still a matter of debate. Processing of pro-caspases-12, -3, -6, -7, -8, and -9 in response to ER stress has been reported.[68–70] However, as yet, the order in which these caspases are activated is unknown, as is the apical caspase in this context.

Early work proposed caspase-12 as the initiator caspase in ER stress-induced apoptosis.[68] Mouse embryonic fibroblasts from caspase-12-deficient animals displayed partial resistance to the ER stress-inducing agents, brefeldin A, and tunicamycin.[68] However, recent studies using mouse caspase-12 knockout cells, from a different source, have cast doubt upon this claim. Saleh and colleagues reported that caspase-12, rather than being implicated in apoptosis, functions in pro-inflammatory responses as a negative regulator of IL-1β processing.[71] To date, no substrates for mouse caspase-12, aside from caspase-12 itself, have been reported. Indeed, the importance of caspase-12 for ER stress-induced apoptosis has been further undermined by the discovery that the majority of humans lack full-length caspase-12. A frameshift mutation, producing a premature stop codon, is present in most humans, resulting in the production of a short CARD-only protein.[72] Certain individuals of African descent lack this mutation and are able to produce full-length and presumably active caspase-12.[73] Studies examining the outcome of caspase-12 expression in this subset of the population determined that expression of full-length caspase-12 correlated with increased susceptibility to severe sepsis.[73] Collectively, these data argue that caspase-12 processing does not trigger ER stress-induced caspase activation but is involved as a negative regulator of inflammatory responses.

Additional data indicate that ER stress-induced apoptosis is dependent upon mitochondrial-mediated processes to promote caspase activation. Mitochondrial translocation of Bax and cytochrome c release has been observed in cells treated with tunicamycin or thapsigargin. Moreover, studies using cells devoid of a functional mitochondrial pathway (*BAX/BAK* null or *APAF-1* null mouse embryonic fibroblasts), or overexpressing anti-apoptotic Bcl-2, fail to activate caspases in response to ER stress signals.[74,75] These observations indicate that ER stress-induced caspase activation is dependent upon the intrinsic or mitochondrial pathway. As yet, the signaling pathways employed by the ER to trigger cytochrome c release have not been delineated but most likely occur by regulation of Bcl-2 family members.

1.5.2 INFLAMMATORY-MEDIATED CASPASE ACTIVATION

As previously described, caspases can be subdivided based upon their sequence identity and chromosome location. Based upon these criteria, caspases-1, -4, -5, -11, and -12 form the inflammatory branch of the caspase family. Caspase-1 is the most intensively studied member of this group and is the protease responsible for the cleavage and maturation of IL-1β, permitting its secretion from monocytes and macrophages in response to pathogens and other pro-inflammatory stimuli.[1] Caspase-1 activity is also required for the maturation of IL-18, a cytokine involved in IFNγ secretion, and more recently this caspase has also been implicated in the cleavage of IL-33, a cytokine involved in T helper cell type 2 (Th2) polarization.[76–78] Although early studies identified caspase-1 as the first homologue of *Caenorhabditis elegans* CED-3, suggesting it played a role in cell death, the generation of *CASP-1* null mice failed to support this observation and suggested an inflammatory rather than a cell death role for this protease.[79–81]

Other members of the inflammatory caspase subfamily include caspases-4, -5, -11, and -12. *CASP-11* null mice, similar to *CASP-1* null mice, do not process IL-1β and, as a direct consequence of this, are resistant to lipopolysaccharide (LPS)-induced endotoxic shock.[82] Unlike caspase-1, caspase-11 expression is inducible by inflammatory activators, such as LPS, via NFkB and STAT-1 signaling.[83] No direct homologue of murine caspase-11 has been identified in humans. However, human caspases-4 and -5 share a high degree of homology with murine caspase-11 and are thought to have arisen from a gene duplication of mouse caspase-11.[84] Like caspase-11, caspase-5 is inducible by LPS and has been implicated in IL-1β processing.[84] At present, little data concerning caspase-4 are available. It has been suggested to function as the human equivalent of mouse caspase-12[85] and has been implicated in ER stress-induced apoptosis, but as yet no convincing data are available to support this hypothesis.

Although caspase-1 was the founding member of the caspase family, relatively little is known about its activation mechanisms and substrates compared to other members of the caspase family. Structurally, owing to the presence of a long pro-domain containing a CARD motif, caspase-1 is a member of the initiator caspase family alongside caspases-8 and -9. Therefore, it is likely that caspase-1 requries a scaffold protein or complex to facilitate its activation. Unlike caspases-8 and -9, the activation platform for caspase-1 (dubbed the inflammasome) has not been fully resolved. Indeed, it is only in the past few years that a model for caspase-1 activation has been proposed.

1.5.2.1 Inflammatory Caspase Activation Pathways

The Nod-like receptor (NLR) family of cytoplasmic adaptor proteins has been implicated in the processing of inflammatory caspases. Members of this family are composed of an N-terminal pyrin, CARD or BIR (baculovirus inhibitor of apoptosis repeat) domain, a central NAIP, CIITA, HET-E TP-1 (NACHT) domain, and C-terminal leucine-rich repeats (LRRs). The N-terminal domain is believed to be important for protein-protein interactions, while the C-terminal LRRs are essential for pathogen detection. The central NACHT region, which is related to the NB-ARC domain of Apaf-1, is thought to induce oligomerization following LRR stimulation. Similar to Apaf-1, oligomerization of some members of the NLR family has been reported following LRR activation.[86] Furthermore, constitutive activation following removal of the LRRs has been reported, analogous to the constitutive activation observed by removal of the WD40 repeats from Apaf-1.[86] Based upon phylogenetic analyses, NLRs can be divided into four subfamilies, class II transactivator (CIITA),

Nods, ICE protease activating factor (IPAF), and NACHT leucine-rich repeat and pyrin domain-containing proteins (NALPs). Presently, only the NALP and IPAF subfamilies have been implicated in inflammatory caspase activation. The putative activation platforms mediating inflammatory caspase activation are described below. However, it should be emphasized that these activation platforms have been postulated based upon overexpression studies and as yet have not been purified as native complexes.

1.5.2.1.1 The NALP Subfamily

The NACHT-, LLR-, and PYD-containing protein (NALP) subfamily is the largest subfamily within the NLR family, encompassing fourteen members.[87] With the exception of NALP1, all members of the NALP family are composed of an N-terminal pyrin domain and C-terminal LRRs. The C terminus of NALP1 differs from that of all other members because it contains both FIIND and CARD interaction motifs in addition to the LRRs.[87] The NALP family has been implicated in the processing and activation of inflammatory caspases via formation of activation platforms referred to as inflammasomes (Figure 1.3). Although each member of the NALP family could potentially form inflammasomes, to date, the NALP3 and NALP1 inflammasomes are the best characterized.

1.5.2.1.2 NALP3 Inflammasomes

NALP3 is the most well studied of all the proposed inflammasome platforms and has been implicated in caspase-1 activation stimulated by ATP, monosodium sodium urate, and the bacteria *Listeria monocytogenes*.[88–90] NALP3 lacks a CARD motif and therefore is unable to directly bind and recruit caspases. To enable caspase recruitment, an adapter molecule, apoptosis-associated Speck-like protein containing a CARD (ASC), is recruited to NALP3.[91] ASC is a bipartite protein comprised of both a pyrin motif facilitating binding to NALP1 and a CARD region, enabling recruitment of caspase-1 to the complex. A second caspase-1 molecule may be recruited by the inclusion of a second adapter molecule, CARDINAL, within the complex (Figure 1.3). Structurally, CARDINAL resembles the FIIND and CARD interaction motifs, which are present on the C terminus of NALP1 but missing in all other NALP family members.[92–94] Recruitment of CARDINAL to the NALP3 inflammasome can recruit a second caspase-1 molecule to the activation platform via CARD-CARD interactions.[94,95] Assembly of the NALP3 inflammasome is thought to promote activation of caspase-1. The importance of NALP3 and the adapter molecule, ASC, in caspase-1 activation is clearly illustrated in cells lacking either molecule. Cells deficient in NALP3 or ASC display impaired activation of caspase-1 and release of cytokines in response to LPS.[96] Consequently, *ASC* null mice are resistant to endotoxic shock induced by LPS injection.[91] Conversely, individuals suffering from Muckle-Wells syndrome, an autoinflammatory disorder, express a mutated and constitutively active NALP3 resulting in unrestrained caspase-1 activation and cytokine release.[97]

1.5.2.1.3 NALP1 Inflammasomes

By virtue of its C-terminal CARD motif, NALP1 can directly recruit caspases (Figure 1.3). Studies have suggested that caspase-5 is recruited to the CARD region

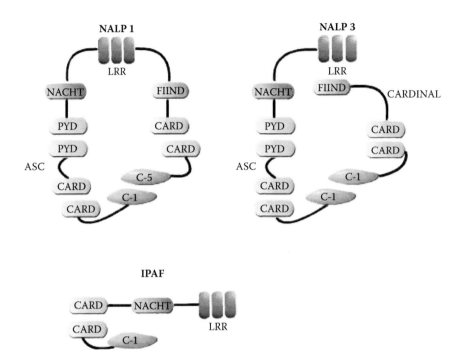

FIGURE 1.3 (See color insert.) Proposed composition of the NALP1, NALP3, and IPAF inflammasomes. See main text for further details.

of NALP1, while binding of caspase-1 to the complex is dependent upon the adapter molecule ASC[95] (Figure 1.3). NALP1 inflammasome formation therefore results in the activation of both caspases-1 and -5. The ligands capable of specifically triggering assembly of the NALP1 inflammasome have not been extensively determined. However, anthrax toxin has been reported to trigger assembly of NALP1 inflammasomes and inflammatory caspase activation.[98]

1.5.2.2   IPAF Subfamily

Members of the IPAF subfamily are composed of an N-terminal CARD, a central NACHT domain, and C-terminal LRRs. Caspase-1 activation, via assembly of the IPAF inflammasome, occurs by direct recruitment to IPAF.[86] Unlike the NALP1 or NALP3 models of caspase-1 activation, IPAF does not require adapter molecules such as ASC or CARDINAL, but rather, directly recruits caspase-1 via the N-terminal CARD domain. The C-terminal LRR in IPAF has autoactivation properties, as loss of the LRRs results in constitutive activation of IPAF.[86] Currently, flagellin is the only reported activator of IPAF-dependent caspase-1 processing.[99,100] Neuronal apoptosis inhibitor protein (NIAP), due to its high-sequence homology with the NACHT and LRR regions of IPAF, has been classified as an IPAF subfamily member.[87] Unlike IPAF, NIAP possesses a baculovirus inhibitor of apoptosis repeats (BIR), a motif associated with caspase inhibitors such as XIAP, at the N terminus rather than a

CARD domain. It has been suggested that NIAP and IPAF may interact within the same caspase-1 activation platform,[101] but the precise mechanism of interaction or the functional consequence of this interaction have not been determined.

1.6   CONCLUSIONS

Apoptotic cell death is required for the removal of damaged, infected, or aged cells from the body. Unlike other forms of cell death, such as necrosis, apoptosis is a highly ordered and regulated process characterized by the activation of caspases-2, -3, -6, -7, -8, -9, and -10. The caspase family consists of initiator and effector caspases. Initiator caspases are activated by scaffold molecules such as the DISC, the apoptosome, or the recently described inflammasome. Upon activation, initiator caspases proteolytically process the effector caspases, resulting in their activation. Effector caspases process hundreds of cellular substrates, some of which contribute to the classical morphology associated with apoptosis. Over the past 10 years, a huge amount of work has deciphered the major caspase activation pathways and has resulted in the defining of the apoptosome, death receptor, and granzyme B–mediated pathways to caspase activation. Deciphering less well-established pathways, such as inflammatory caspase activation mechanisms, is the challenge that currently preoccupies workers in this fast-moving field.

ACKNOWLEDGMENTS

We thank Science Foundation Ireland (PI.1/B038) for support of some of the work discussed in this review. S. L. is supported by a postdoctoral fellowship from the Irish Research Council for Science and Engineering Technologies (IRCSET).

REFERENCES

1. Yuan, J., et al. 1993. The *C. elegans* cell death gene ced-3 encodes a protein similar to mammalian interleukin-1 beta-converting enzyme. *Cell* 75:641–52.
2. Nicholson, D. W., and N. A. Thornberry. 1997. Caspases: Killer proteases. *Trends Biochem Sci* 22:299–306.
3. Wolf, B. B., and D. R. Green. 1999. Suicidal tendencies: Apoptotic cell death by caspase family proteinases. *J Biol Chem* 274:20049–52.
4. Walker, N. P., et al. 1994. Crystal structure of the cysteine protease interleukin-1 beta-converting enzyme: A (p20/p10)2 homodimer. *Cell* 78:343–52.
5. Wilson, K. P., et al. 1994. Structure and mechanism of interleukin-1 beta converting enzyme. *Nature* 370:270–75.
6. Rotonda, J., et al. 1996, The three-dimensional structure of apopain/CPP32, a key mediator of apoptosis. *Nat Struct Biol* 3:619–25.
7. Salvesen, G. S., and V. M. Dixit. 1999. Caspase activation: The induced-proximity model. *Proc Natl Acad Sci USA* 96:10964–67.
8. Hu, S., et al. 1998. Caspase-14 is a novel developmentally regulated protease. *J Biol Chem* 273:29648–53.
9. Kerr, J. F., A. H. Wyllie, and A. R. Currie. 1972. Apoptosis: A basic biological phenomenon with wide-ranging implications in tissue kinetics. *Br J Cancer* 26:239–57.

10. Luthi, A. U., and S. J. Martin. 2007. The CASBAH: A searchable database of caspase substrates. *Cell Death Differ* 14:641–50.

11. Enari, M., et al. 1998. A caspase-activated DNase that degrades DNA during apoptosis, and its inhibitor ICAD. *Nature* 391:43–50.

12. Byun, Y., et al. 2001. Caspase cleavage of vimentin disrupts intermediate filaments and promotes apoptosis. *Cell Death Differ* 8:443–50.

13. Nakanishi, K., et al. 2001. Identification of a caspase-9 substrate and detection of its cleavage in programmed cell death during mouse development. *J Biol Chem* 276:41237–44.

14. Brancolini, C., M. Benedetti, and C. Schneider. 1995. Microfilament reorganization during apoptosis: The role of Gas2, a possible substrate for ICE-like proteases. *Embo J* 14:5179–90.

15. Stegh, A. H., et al. 2000. Identification of the cytolinker plectin as a major early in vivo substrate for caspase 8 during CD95- and tumor necrosis factor receptor-mediated apoptosis. *Mol Cell Biol* 20:5665–79.

16. Coleman, M. L., et al. 2001. Membrane blebbing during apoptosis results from caspase-mediated activation of ROCK I. *Nat Cell Biol* 3:339–45.

17. Croft, D. R., et al. 2005. Actin-myosin-based contraction is responsible for apoptotic nuclear disintegration. *J Cell Biol* 168:245–55.

18. Vaux, D. L., S. Cory, and J. M. Adams. 1988. Bcl-2 gene promotes haemopoietic cell survival and cooperates with c-myc to immortalize pre-B cells. *Nature* 335:440–42.

19. Nunez, G., et al. 1989. Growth- and tumor-promoting effects of deregulated BCL2 in human B-lymphoblastoid cells. *Proc Natl Acad Sci USA* 86:4589–93.

20. Hockenbery, D., et al. 1990. Bcl-2 is an inner mitochondrial membrane protein that blocks programmed cell death. *Nature* 348:334–36.

21. Cory, S., and J. M. Adams. 2002. The Bcl2 family: Regulators of the cellular life-or-death switch. *Nat Rev Cancer* 2:647–56.

22. Petros, A. M., E. T. Olejniczak, and S. W. Fesik. 2004. Structural biology of the Bcl-2 family of proteins. *Biochim Biophys Acta* 1644:83–94.

23. Oltvai, Z. N., C. L. Milliman, and S. J. Korsmeyer. 1993. Bcl-2 heterodimerizes in vivo with a conserved homolog, Bax, that accelerates programmed cell death. *Cell* 74:609–19.

24. Sedlak, T. W., et al. 1995. Multiple Bcl-2 family members demonstrate selective dimerizations with Bax. *Proc Natl Acad Sci USA* 92:7834–38.

25. Puthalakath, H., and A. Strasser. 2002. Keeping killers on a tight leash: Transcriptional and post-translational control of the pro-apoptotic activity of BH3-only proteins. *Cell Death Differ* 9:505–12.

26. Kuwana, T., et al. 2005. BH3 domains of BH3-only proteins differentially regulate Bax-mediated mitochondrial membrane permeabilization both directly and indirectly. *Mol Cell* 17:525–35.

27. Chen, L., et al. 2005. Differential targeting of prosurvival Bcl-2 proteins by their BH3-only ligands allows complementary apoptotic function. *Mol Cell* 17:393–403.

28. Liu, X., et al. 1996. Induction of apoptotic program in cell-free extracts: Requirement for dATP and cytochrome c. *Cell* 86:147–57.

29. Li, P., et al. 1997. Cytochrome c and dATP-dependent formation of Apaf-1/caspase-9 complex initiates an apoptotic protease cascade. *Cell* 91:479–89.

30. Zou, H., et al. 1997. Apaf-1, a human protein homologous to C. *elegans* CED-4, participates in cytochrome c-dependent activation of caspase-3. *Cell* 90:405–13.

31. Hu, Y., et al. 1998. WD-40 repeat region regulates Apaf-1 self-association and procaspase-9 activation. *J Biol Chem* 273:33489–94.

32. Zou, H., et al. 1999. An APAF-1 cytochrome c multimeric complex is a functional apoptosome that activates procaspase-9. *J Biol Chem* 274:11549–56.

33. Acehan, D., et al. 2002. Three-dimensional structure of the apoptosome: Implications for assembly, procaspase-9 binding, and activation. *Mol Cell* 9:423–32.

34. Slee, E. A., et al. 1999. Ordering the cytochrome c-initiated caspase cascade: Hierarchical activation of caspases-2, -3, -6, -7, -8, and -10 in a caspase-9-dependent manner. *J Cell Biol* 144:281–92.

35. Pan, G., E. W. Humke, and V. M. Dixit. 1998. Activation of caspases triggered by cytochrome c in vitro. *FEBS Lett* 426:151–54.

36. Hakem, R., et al. 1998. Differential requirement for caspase 9 in apoptotic pathways in vivo. *Cell* 94:339–52.

37. Kuida, K., et al. 1998. Reduced apoptosis and cytochrome c-mediated caspase activation in mice lacking caspase 9. *Cell* 94:325–37.

38. Cecconi, F., et al. 1998. Apaf-1 (CED-4 homolog) regulates programmed cell death in mammalian development. *Cell* 94:727–37.

39. Loeffler, M., and G. Kroemer. 2000. The mitochondrion in cell death control: Certainties and incognita. *Exp Cell Res* 256:19–26.

40. Li, L. Y., X. Luo, and X. Wang. 2001. Endonuclease G is an apoptotic DNase when released from mitochondria. *Nature* 412:95–99.

41. Hegde, R., et al. 2002. Identification of Omi/HtrA2 as a mitochondrial apoptotic serine protease that disrupts inhibitor of apoptosis protein-caspase interaction. *J Biol Chem* 277:432–38.

42. Martins, L. M. 2002. The serine protease Omi/HtrA2: A second mammalian protein with a Reaper-like function. *Cell Death Differ* 9:699–701.

43. Du, C., et al. 2000. Smac, a mitochondrial protein that promotes cytochrome c-dependent caspase activation by eliminating IAP inhibition. *Cell* 102:33–42.

44. Smith, C. A., et al. 1989. Antibodies to CD3/T-cell receptor complex induce death by apoptosis in immature T cells in thymic cultures. *Nature* 337:181–84.

45. Chinnaiyan, A. M., et al. 1995. FADD, a novel death domain-containing protein, interacts with the death domain of Fas and initiates apoptosis. *Cell* 81:505–12.

46. Eberstadt, M., et al. 1998. NMR structure and mutagenesis of the FADD (Mort1) death-effector domain. *Nature* 392:941–45.

47. Scaffidi, C., et al. 1998. Two CD95 (APO-1/Fas) signaling pathways. *EMBO J* 17:1675–87.

48. Zha, J., et al. 2000. Posttranslational N-myristoylation of BID as a molecular switch for targeting mitochondria and apoptosis. *Science* 290:1761–65.

49. Li, H., et al. 1998. Cleavage of BID by caspase 8 mediates the mitochondrial damage in the Fas pathway of apoptosis. *Cell* 94:491–501.

50. Luo, X., et al. 1998. Bid, a Bcl2 interacting protein, mediates cytochrome c release from mitochondria in response to activation of cell surface death receptors. *Cell* 94:481–90.

51. Eskes, R., et al. 1998. Bax-induced cytochrome C release from mitochondria is independent of the permeability transition pore but highly dependent on Mg2+ ions. *J Cell Biol* 143:217–24.

52. Irmler, M., et al. 1997. Inhibition of death receptor signals by cellular FLIP. *Nature* 388:190–95.

53. Tschopp, J., M. Irmler, and M. Thome. 1998. Inhibition of fas death signals by FLIPs. *Curr Opin Immunol* 10:552–58.

54. Froelich, C. J., V. M. Dixit, and X. Yang. 1998. Lymphocyte granule-mediated apoptosis: Matters of viral mimicry and deadly proteases. *Immunol Today* 19:30–36.

55. Kagi, D., et al. 1994. Cytotoxicity mediated by T cells and natural killer cells is greatly impaired in perforin-deficient mice. *Nature* 369:31–37.

56. Waterhouse, N. J., and J. A. Trapani. 2002. CTL: Caspases terminate life, but that's not the whole story. *Tissue Antigens* 59:175–83.

57. Odake, S., et al. 1991. Human and murine cytotoxic T lymphocyte serine proteases: Subsite mapping with peptide thioester substrates and inhibition of enzyme activity and cytolysis by isocoumarins. *Biochemistry* 30:2217–27.

58. Poe, M., et al. 1992. Human cytotoxic lymphocyte granzyme B. Its purification from granules and the characterization of substrate and inhibitor specificity. *J Biol Chem* 266:98–103.

59. Martin, S. J., et al. 1996. The cytotoxic cell protease granzyme B initiates apoptosis in a cell-free system by proteolytic processing and activation of the ICE/CED-3 family protease, CPP32, via a novel two-step mechanism. *EMBO J* **15**:2407–16.

60. Darmon, A. J., et al. 1996. Cleavage of CPP32 by granzyme B represents a critical role for granzyme B in the induction of target cell DNA fragmentation. *J Biol Chem* 271:21709–12.

61. Adrain, C., B. M. Murphy, and S. J. Martin. 2005. Molecular ordering of the caspase activation cascade initiated by the cytotoxic T lymphocyte/natural killer (CTL/NK) protease granzyme B. *J Biol Chem* 280:4663–73.

62. Sutton, V. R., D. L. Vaux, and J. A. Trapani. 1997. Bcl-2 prevents apoptosis induced by perforin and granzyme B, but not that mediated by whole cytotoxic lymphocytes. *J Immunol* 158:5783–90.

63. Chiu, V. K., et al. 1995. Bcl-2 blocks degranulation but not fas-based cell-mediated cytotoxicity. *J Immunol* 154:2023–32.

64. Schroter, M., et al. 1995. Regulation of Fas(Apo-1/CD95)- and perforin-mediated lytic pathways of primary cytotoxic T lymphocytes by the protooncogene bcl-2. *Eur J Immunol* 25:3509–13.

65. Barry, M., et al. 2000. Granzyme B short-circuits the need for caspase 8 activity during granule-mediated cytotoxic T-lymphocyte killing by directly cleaving Bid. *Mol Cell Biol* 20:3781–94.

66. Sutton, V. R., et al. 2000. Initiation of apoptosis by granzyme B requires direct cleavage of Bid, but not direct granzyme B-mediated caspase activation. *J Exp Med* 192:1403–14.

67. Szegezdi, E., et al. 2006. Mediators of endoplasmic reticulum stress-induced apoptosis. *EMBO Rep* 7:880–85.

68. Nakagawa, T., et al. 2000. Caspase-12 mediates endoplasmic-reticulum-specific apoptosis and cytotoxicity by amyloid-beta. *Nature* 403:98–103.

69. Jimbo, A., et al. 2003. ER stress induces caspase-8 activation, stimulating cytochrome c release and caspase-9 activation. *Exp Cell Res* 283:156–66.

70. Morishima, N., et al. 2002. An endoplasmic reticulum stress-specific caspase cascade in apoptosis. Cytochrome c-independent activation of caspase-9 by caspase-12. *J Biol Chem* 277:34287–94.

71. Saleh, M., et al. 2006. Enhanced bacterial clearance and sepsis resistance in caspase-12-deficient mice. *Nature* 440:1064–68.

72. Fischer, H., et al. 2002. Human caspase 12 has acquired deleterious mutations. *Biochem Biophys Res Commun* 293:722–26.

73. Saleh, M., et al. 2004. Differential modulation of endotoxin responsiveness by human caspase-12 polymorphisms. *Nature* 429:75–79.

74. Distelhorst, C. W., and T. S. McCormick. 1996. Bcl-2 acts subsequent to and independent of Ca2+ fluxes to inhibit apoptosis in thapsigargin- and glucocorticoid-treated mouse lymphoma cells. *Cell Calcium* 19:473–83.

75. Wei, M. C., et al. 2001. Proapoptotic BAX and BAK: A requisite gateway to mitochondrial dysfunction and death. *Science* 292:727–30.

76. Gu, Y., et al. 1997. Activation of interferon-gamma inducing factor mediated by interleukin-1beta converting enzyme. *Science* 275:206–9.

77. Ghayur, T., et al. 1997. Caspase-1 processes IFN-gamma-inducing factor and regulates LPS-induced IFN-gamma production. *Nature* 386:619–23.

78. Schmitz, J., et al. 2005. IL-33, an interleukin-1-like cytokine that signals via the IL-1 receptor-related protein ST2 and induces T helper type 2-associated cytokines. *Immunity* 23:479–90.

79. Miura, M., et al. 1993. Induction of apoptosis in fibroblasts by IL-1 beta-converting enzyme, a mammalian homolog of the *C. elegans* cell death gene ced-3. *Cell* 75:653–60.

80. Li, P., et al. 1995. Mice deficient in IL-1 beta-converting enzyme are defective in production of mature IL-1 beta and resistant to endotoxic shock. *Cell* 80:401–11.

81. Kuida, K., et al. 1995. Altered cytokine export and apoptosis in mice deficient in interleukin-1 beta converting enzyme. *Science* 267:2000–3.

82. Wang, S., et al. 1998. Murine caspase-11, an ICE-interacting protease, is essential for the activation of ICE. *Cell* 92:501–9.

83. Schauvliege, R., et al. 2002. Caspase-11 gene expression in response to lipopolysaccharide and interferon-gamma requires nuclear factor-kappa B and signal transducer and activator of transcription (STAT) 1. *J Biol Chem* 277:41624–30.

84. Lin, X. Y., M. S. Choi, and A. G. Porter. 2000. Expression analysis of the human caspase-1 subfamily reveals specific regulation of the CASP5 gene by lipopolysaccharide and interferon-gamma. *J Biol Chem* 275:39920–26.

85. Hitomi, J., et al. 2004. Involvement of caspase-4 in endoplasmic reticulum stress-induced apoptosis and Abeta-induced cell death. *J Cell Biol* 165:347–56.

86. Poyet, J. L., et al. 2001. Identification of Ipaf, a human caspase-1-activating protein related to Apaf-1. *J Biol Chem* 276:28309–13.

87. Tschopp, J., F. Martinon, and K. Burns. 2003. NALPs: A novel protein family involved in inflammation. *Nat Rev Mol Cell Biol* 4:95–104.

88. Ferrari, D., et al. 2006. The P2X7 receptor: A key player in IL-1 processing and release. *J Immunol* 176:3877–83.

89. Martinon, F., et al. 2006. Gout-associated uric acid crystals activate the NALP3 inflammasome. *Nature* 440:237–41.

90. Ozoren, N., et al. 2006. Distinct roles of TLR2 and the adaptor ASC in IL-1beta/IL-18 secretion in response to *Listeria monocytogenes*. *J Immunol* 176:4337–42.

91. Yamamoto, M., et al. 2004. ASC is essential for LPS-induced activation of procaspase-1 independently of TLR-associated signal adaptor molecules. *Genes Cells* 9:1055–67.

92. Pathan, N., et al. 2001. TUCAN, an antiapoptotic caspase-associated recruitment domain family protein overexpressed in cancer. *J Biol Chem* 276:32220–29.

93. Bouchier-Hayes, L., et al. 2001. CARDINAL, a novel caspase recruitment domain protein, is an inhibitor of multiple NF-kappa B activation pathways. *J Biol Chem* 276:44069–77.

94. Razmara, M., et al. 2002. CARD-8 protein, a new CARD family member that regulates caspase-1 activation and apoptosis. *J Biol Chem* 277:13952–58.

95. Martinon, F., K. Burns, and J. Tschopp. 2002. The inflammasome: A molecular platform triggering activation of inflammatory caspases and processing of proIL-beta. *Mol Cell* 10:417–26.

96. Sutterwala, F. S., et al. 2006. Critical role for NALP3/CIAS1/cryopyrin in innate and adaptive immunity through its regulation of caspase-1. *Immunity* 24:317–27.

97. Agostini, L., et al. 2004. NALP3 forms an IL-1beta-processing inflammasome with increased activity in Muckle-Wells autoinflammatory disorder. *Immunity* 20:319–25.

98. Boyden, E. D., and W. F. Dietrich. 2006. Nalp1b controls mouse macrophage susceptibility to anthrax lethal toxin. *Nat Genet* 38:240–44.

99. Franchi, L., et al. 2006. Cytosolic flagellin requires Ipaf for activation of caspase-1 and interleukin 1beta in salmonella-infected macrophages. *Nat Immunol* 7:576–82.
100. Miao, E. A., et al. 2006. Cytoplasmic flagellin activates caspase-1 and secretion of interleukin 1beta via Ipaf. *Nat Immunol* 7:569–75.
101. Zamboni, D. S., et al. 2006. The Bircle cytosolic pattern-recognition receptor contributes to the detection and control of *Legionella pneumophila* infection. *Nat Immunol* 7:318–25.

2 Role of Caspases in Inflammation-Driven Diseases

Kristof Kersse, Tom Vanden Berghe, Saskia Lippens, Wim Declercq, and Peter Vandenabeele

CONTENTS

2.1   INTRODUCTION

Caspases constitute an evolutionarily conserved family of cysteinyl-dependent proteases that fulfill essential roles in apoptosis, inflammation, cell survival, proliferation, and differentiation, and act by aspartate-specific cleavage of a wide number of cellular substrates.[1,2] Ten murine (caspase-1, -2, -3, -6, -7, -8, -9, -11, -12, -14) and eleven human (caspase-1, -2, -3, -4, -5, -6, -7, -8, -9, -10, -14) catalytically active caspases have been reported so far.[3] They are synthesized as inactive proenzymes that include an N-terminal prodomain of variable length, followed by two domains with conserved sequences: a large subunit (~20 kDa, p20) and a small C-terminal sub-unit (~10 kDa, p10). Caspases can be subdivided according to the length of the pro-domain into two main categories. Large-prodomain caspases contain an N-terminal homotypic protein-protein interaction motif belonging to the death domain (DD) superfamily, in particular the caspase recruitment and activation domain (CARD; caspase-1, -2, -4, -5, -9, -11, and -12) and the death effector domain (DED, caspase-8 and -10). These domains are used to recruit inactive monomers to multiprotein com-plexes, each of which is specific for a particular caspase. In these complexes the proenzyme undergoes conformational changes required for its activation.[1,4] The short-prodomain caspases (caspase-3, -6, -7, and -14) contain a prodomain of only a few amino acids and exist as preformed dimers. They acquire enzymatic activity upon proteolytic maturation by large-prodomain caspases or other proteases. Caspases can also be classified according to their functions, which correlate with their phylogeny: the apoptotic initiators, the apoptotic executioners, and the inflammatory caspases. The latter group encompasses caspase-1, -4, -5, -11, and -12; caspase-4 and -5 are most likely the human paralogs of murine caspase-11.[3]

Caspase-1, the prototypic member of the inflammatory caspases, is responsi-ble for the maturation of pro-interleukin (IL)-1β, pro-IL-18, and pro-IL-33, three related cytokines playing critical roles during inflammation.[5-8] Indeed, the require-ment for caspase-1 in maturation of pro-IL-1β and pro-IL-18 was confirmed in caspase-1-deficient mice.[9,10] The enzymatic activation of caspase-1, a member of the CARD-containing large-prodomain caspases, occurs in large, multimeric protein platforms (~700 kDa) commonly referred to as inflammasomes.[11] In general, the assembly of these complexes involves the orchestrated recruitment of three function-ally distinct building blocks: a sensor, an adaptor, and an effector.[12,13] Inflammasome formation is initiated when a sensor-platform protein belonging to the NOD-like receptor (NLR) family detects the presence of an instigating factor through its leucine-rich repeats (LRRs), an evolutionarily conserved ligand-sensing domain. Distinct NLR-family members sense distinct pathogen-associated molecular pat-terns (PAMPs) and danger signals known as damage-associated molecular patterns (DAMPs). NALP1b detects *Bacillus anthracis* toxin; NALP3 senses gout-associated uric acid crystals, bacterial RNA, and pathogens, such as *Staphylococcus aureus* and *Listeria monocytogenes*; Ipaf specifically detects the presence of flagellin of intracellular bacteria, such as *Salmonella typhimurium* and *Legionella pneumo-philia*.[14-21] Direct or indirect binding of a ligand to the LRR induces conformational changes that expose the NACHT domain and thereby trigger oligomerization of the NLR proteins and consequent inflammasome assembly.[22] NLR proteins also contain at their N-termini a homotypic interaction motif belonging to the DD superfamily, either a CARD or a PYRIN. These motifs mediate the recruitment of procaspase-1 in one of two ways: (1) CARD-containing NLR members directly recruit caspase-1 through a CARD-CARD interaction, or (2) PYRIN-containing NLR members recruit caspase-1 via the adaptor protein ASC/PYCARD, which contains both a CARD and a PYRIN domain (Figure 2.1).[23,24] In addition, NALP3 inflammasomes encompass a second adaptor, CARDINAL, which recruits procaspase-1 through its CARD domain.[12] In this way, multiple procaspase-1 molecules are brought close

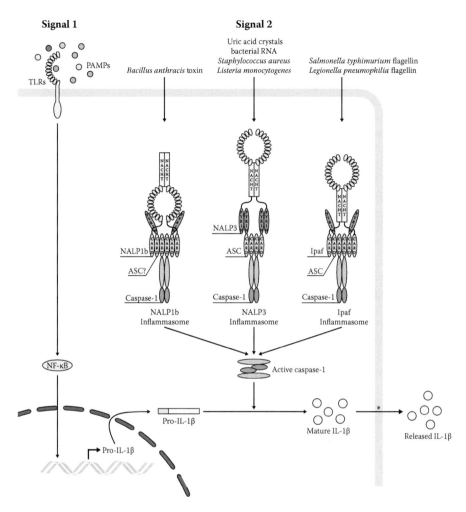

Signal 1

Signal 2

Uric acid crystals
bacterial RNA
Staphylococcus aureus
Listeria monocytogenes

Salmonella typhimurium flagellin
Legionella pneumophilia flagellin

Bacillus anthracis toxin

PAMPs

TLRs

NALP3

NALP1b

ASC?

ASC

Ipaf

ASC

Caspase-1

Caspase-1

Caspase-1

NALP1b
Inflammasome

NALP3
Inflammasome

Ipaf
Inflammasome

NF-κB

Active caspase-1

Pro-IL-1β

Mature IL-1β

Released IL-1β

Pro-IL-1β

FIGURE 2.1 Ligands and molecular composition of the caspase-1-activating inflammasomes. Recognition of PAMPs by TLRs (signal 1) induces synthesis of the IL-1β precursor through NF-κB activation. Pathogens and danger signals (signal 2) are detected by distinct cytosolic NLR family members: NALP1b, NALP3, and Ipaf. Ligand sensing results in oligomerization of the NLR proteins, which enables the recruitment and activation of caspase-1 through the common inflammasome adaptor ASC/PYCARD. Active caspase-1 proteolytically processes the IL-1β precursor to its bioactive form, which is then secreted from the cell. An asterisk indicates that the mechanism of IL-1β secretion has not been clarified.

together, resulting in conformational changes, autoproteolysis, and release of the caspase-1 homodimer in the cytosol, where it mediates the activation and release of substrates such as pro-IL1β and pro-IL-18. However, in contrast to pro-IL-18, which is present in hematopoietic cells under homeostatic conditions, pro-IL-1β has first to be transcriptionally induced.[25] This second signal, which is independent of the inflammasome-mediated caspase-1-activating signal, involves engagement of Toll-like receptor (TLR) signaling and pro-inflammatory cytokine receptors (Figure 2.1).[26]

IL-1β participates in the generation of systemic and local responses to infection, injury, and immunological challenges by generating fever, activating lymphocytes, and promoting infiltration of leukocytes into the sites of injury or infection.[27] IL-1β signaling is mediated by the IL-receptor complex, which contains the type I IL-1 receptor (IL-1RI) and the IL-1R accessory protein (IL-1RAcP).[28] After IL-1β binds to the receptor complex, multiple signaling cascades are initiated. These cascades lead to activation of nuclear factor κB (NF-κB) and mitogen-activated protein kinase (MAPK) cascades, which in turn lead to inflammatory cytokine expression, Ca^{2+} fluxes, cell contraction through reorganization of the actin cytoskeleton, and synthesis of AP-1-dependent proteins by the expression of the early response genes *c-fos* and *c-jun*.[29–35] The latter are drivers of extracellular matrix degradation. Although mature IL-1β has many beneficial effects, excessive production is harmful and needs to be appropriately modulated. A first level of control is regulation of pro-IL-1β transcription. Second, as described above, this proform requires cleavage by caspase-1 for activity. A group of proteins safeguarding IL-1β levels by preventing caspase-1 activation are the caspase-1-related CARD-only proteins, COP, INCA, and ICE-BERG, which arose from recent caspase-1 gene duplications in primates.[36] These CARD-only proteins interact with procaspase-1 and thereby prevent its recruitment and activation in inflammasomes.[37–40] Finally, IL-1β signaling itself is controlled by the type II IL-1R (IL-1RII) decoy receptor, the IL-1 receptor antagonist (IL-1Ra), and soluble receptor isoforms of IL-1RII and IL-1RAcP (Figure 2.2). In contrast to IL-1RI and IL-1RAcP, IL-1RII lacks the intracellular signaling domain, and so it cannot initiate a signaling cascade. Because IL-1RII can still bind IL-1β, it acts as a molecular trap.[41,42] Soluble IL-1RII, generated after stimulus-induced cleavage at the plasma membrane, prevents IL-1 signaling by sequestering circulating IL-1β.[43] This effect is enhanced by the soluble IL-1RAcP variant, which is generated by

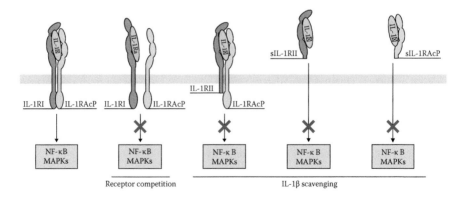

FIGURE 2.2 Preventing IL-1β signaling by the functional IL-1R complex. IL-1β activates NF-κB and MAPKs by binding a heterodimer of IL-1RI and IL-RAcP (left). Two major mechanisms downregulate IL-1β signaling: competition between IL-1β and IL-1Ra for binding to the functional IL-1RI complex, and scavenging of IL-1β by binding to the dysfunctional IL-1RII complex, soluble IL-1RII, or soluble IL-1RAcP.

alternative splicing.[44] In addition, an IL-1 receptor antagonist functions as an inhibitor by competing for the occupancy of the IL-1R complex.[45,46]

Like IL-1β, IL-18 is a member of the large IL-1 family. IL-18 is chiefly known as a factor that induces interferon-γ (IFN-γ) in activated T cells and NK cells, thereby inducing Th1 responses.[6,47–50] However, IL-18 is also involved in the induction of secondary pro-inflammatory cytokines, chemokines, cell adhesion molecules, and nitric oxide synthesis.[51,52] The IL-18 receptor (IL-18R) complex encompasses, like the IL-1R complex, a heterodimer containing an α-chain responsible for extracellular binding of IL-18, and a nonbinding signal-transducing β-chain.[53–55] Binding of IL-18 to the IL-18R complex induces signaling pathways shared with IL-1R, such as activation of NF-κB, AP-1, MAPK, and STAT3.[56–60] In contrast to pro-IL-1β, pro-IL-18 is constitutively present in the cell, which means that it is not regulated at this level.[25] However, it was recently shown that the release of both pro- and mature IL-18 can be controlled by a p38 MAPK-dependent mechanism, at least in keratinocytes.[61,62] Although the second point of regulation, caspase-1 activity, is the same as for IL-1β, the regulation of IL-18 signaling per se differs slightly. More precisely, only a soluble IL-18-binding protein (sIL-18BP) has been discovered.[63,64] Both sIL-18BP and the soluble form of IL-1RII act as decoy receptors to block the binding of their respective ligands to the signaling receptor chains.[63,64] However, IL-18BP has limited structural and amino acid sequence homology with IL-18Rs.[65,66]

In conclusion, caspase-1 exercises an important function in mounting inflammatory responses against harmful stimuli. Caspase-1 and its substrates IL-1β and IL-18 are important mediators in several inflammation-driven disorders. In the next paragraphs we will overview disorders in which the crucial contribution of caspase-1 is supported by substantial evidence.

2.2 ARTHRITIC DISORDERS

2.2.1 OSTEOARTHRITIS

Osteoarthritis (OA) is one of the most prevalent musculoskeletal diseases and affects mostly the elderly.[67] This chronic, degenerative disease with occasional episodes of inflammation is characterized by changes in the structure and function of the whole joint, with a progressive degradation of the hyaline articular cartilage, remodeling of bone, sclerosis, and osteophytosis.[68] This results in constant pain and disability and reduces the quality of life. The disease process is driven by mechanical factors and inflammatory mediators. In homeostatic conditions the inflammatory mediators, together with ambient growth factors, ensure the integrity of articular cartilage by regulating the balance between anabolic and catabolic processes. However, if macrophages, synoviocytes, and chondrocytes of the articular joint tissue produce and release excessive amounts of potent inflammatory cytokines, such as IL-1β, IL-18, and TNF, the result is destruction of the cartilage matrix. This occurs by enhancement of the expression and activation of proteolytic enzymes, mainly matrix metalloproteinases (MMPs) and collagenases, prostaglandin E_2, and inducible NO synthase (iNOS).[69–73] The roles of IL-1β and IL-18 in disease progression directly implicate caspase-1 as the protease responsible for the maturation and release of these cytokines, as both a contributor to the disease and a therapeutic target.[5,6,8,74]

The roles of caspase-1 and the above-mentioned cytokines in OA have been established in many *in vitro* and *in vivo* studies in men and mice. Analysis of human OA tissues demonstrated that both synovial membrane cells and cartilage chondrocytes express caspase-1, and that the number of active caspase-1-positive cells is larger in OA than in normal tissues.[75] Consequently, elevated levels of IL-1β and to a lesser extent IL-18, produced by the chondrocytes of the cartilage, are observed in the synovial fluid of OA patients.[75–77] The crucial role of caspase-1 in OA was further confirmed by the abrogation of IL-1β and IL-18 production in cartilage explants treated *ex vivo* with the caspase-1 inhibitor YVAD-cmk.[75] These observations led to investigation of the effect of the caspase-1-specific, YVAD-peptidomimetic pralnacasan (VX-740, developed by Vertex, Inc.) on joint damage in two mouse models of knee osteoarthritis. In one model, knee joint instability and subsequent OA were induced by injection of highly purified bacterial collagenase from *Clostridium histolyticum* in the joints of female Balb/c mice. The second OA model uses male STR/1N mice, which develop OA spontaneously. Administration of pralnacasan significantly reduced histopathological joint damage in both models.[78] The crucial role of IL-1β in the progression of OA was further substantiated by administering human IL-1Ra, a competitive antagonist of IL-1, either by intra-articular injection of the recombinant protein or by gene transfer in OA animal models and patients.[79–84] In all these studies, disease outcome was ameliorated, as exemplified by the significant improvements observed in patients with symptomatic knee OA who were treated with the commercial variant of recombinant IL-1Ra, anakinra (Kineret™, Amgen).[80]

2.2.2 Rheumatoid Arthritis

Rheumatoid arthritis (RA) is a chronic, systemic, inflammatory autoimmune disease that primarily affects the synovial tissues, cartilage, and bone.[85,86] It affects approximately 1% of adults worldwide and is associated with significant morbidity and mortality when the disease is unchecked.[87] In contrast to OA, RA is not a condition of wear and tear associated with injury or aging; it can develop at any age. In RA, inflammation causes damage to the joints.[86] The smaller joints (hands and feet) are usually affected symmetrically, but the larger joints (knees and hips) are affected asymmetrically.[68,88] Each phase of the pathogenesis of RA involves a number of pro- and anti-inflammatory cytokines that contribute by promoting autoimmunity, maintaining inflammatory synovitis, and driving the destruction of adjacent joint tissue.[89] Numerous reports illustrate the importance of TNF and IL-1 family members, including caspase-1-activated IL-1β and IL-18, in the initiation and progression of the disease.[90]

Synovial fluids of RA patients contain elevated levels of IL-1β produced by macrophages, activated T cells, fibroblasts, and chondrocytes of the synovial tissue.[91,92] These elevated levels are partly due to enhanced NALP3 expression in synovial tissue macrophages.[93] NALP3 is a sensor-platform protein of the caspase-1-activating inflammasomes.[12] IL-1β has been observed in the inflamed synovium of two RA animal models, type II collagen-induced arthritis (CIA) and antigen-induced arthritis. Furthermore, when IL-1β-deficient mice were injected with streptococcal cell walls to induce arthritis, they became resistant to joint destruction but not to inflammation.[94–96] The functional effects of IL-1β in RA are similar to those in OA pathogenesis, and include induction of expression of synovial fibroblast cytokines, chemokines, MMP, PGE_2, and iNOS, which in turn result in leukocyte migration and tissue destruction.[89,97] The involvement of IL-1β in RA was also demonstrated when arthritis was induced in IL-1Rα knockout DBA/1 mice by collagen injection. Because these mice lack one of the endogenous mechanisms regulating IL-1β signaling, they develop a more severe form of arthritis.[98] Further studies revealed that the absence of IL-1Ra leads to arthritis by enhancing IL-1β-induced T-cell-dependent antibody secretion.[99]

On the other hand, IL-18 mRNA and protein are significantly more abundant in the lining and sublining of synovial tissues of patients with RA than in those with OA.[100–103] Levels of IL-18 in synovial fluid as well as in serum correlate with disease activity and response to therapy.[101–104] Moreover, compared to heterozygous and wild-type mice, IL-18-deficient DBA/1 mice display a marked decrease in incidence and severity of collagen-induced arthritis, with histological evidence for amelioration of joint inflammation and destruction.[105] In addition, when these mice were treated with recombinant IL-18, they developed the disease just like wild-type mice. Similarly, the acute arthritis that developed after administration of streptococcal cell wall was significantly ameliorated by treatment with a neutralizing anti-IL-18 antibody.[106] A marked reduction of joint swelling and serum levels of TNF and IL-1β was observed following anti-IL-18 administration, pointing to an upstream role for IL-18 in the production of TNF and IL-1β.[106] In turn, TNF and IL-1β induce the release of IL-18 in RA joints, thereby establishing a positive feedback loop that not only triggers but also maintains the inflammatory joint disorder.[100,107,108] Secreted IL-18 mainly targets T cells, macrophages, synovial fibroblasts, chondrocytes, and NK cells to induce the release of cytokines, chemokines, and adhesion molecules in a sustained Th1 response, with consequent chronic inflammation, and cytotoxicity.[105,109]

Current therapeutic strategies aim at blocking the biological activities of IL-1β and IL-18, either by targeting the activation of these cytokines, thus inhibiting the activity of their common activator caspase-1, or by preventing signaling through their respective receptors. Of the multiple caspase-1-specific inhibitors developed to date, pralnacasan and VX-765 were the most promising in animal models of RA.[110,111] Administration of pralnacasan in the CIA model significantly delayed the onset of forepaw inflammation and reduced disease severity by 50%–70%. Histological analysis of wrist joints showed that pralnacasan treatment reduced the incidence of cartilage damage and bone erosion by approximately 60% and 80%, respectively.[112] Consequently, pralnacasan entered a 12-week phase IIa clinical trial involving 285 RA patients.[110] A dose-dependent trend toward improvement was observed in signs and symptoms measured by ACR20 response rates (the American College of Rheumatology [ACR] responder index, defined as a 20% improvement in several measures expressed as a composite score). This trend, however, did not reach statistical significance.[110] Further clinical studies were aborted due to the development of liver abnormalities in long-term animal toxicology studies. The other Vertex compound, VX-765, has similar beneficial effects when used in the CIA model, namely, a statistically significant reduction in forepaw inflammation scores that correlates with the degree of joint structural damage as measured by histological analysis.[111] So far, no results of clinical studies on the use of VX-765 in RA patients have been communicated.

The second therapeutic strategy is based on interfering with IL-1β and IL-18 signaling by preventing the binding of these cytokines to their receptors. For IL-1β this is done by administering recombinant IL-1Ra, which binds to IL-1R without inducing signaling. It was demonstrated that IL-1Rα can be used to reduce cartilage and bone destruction in immune complex–induced arthritis in mice, antigen-induced arthritis in rabbits, and CIA in mice.[95,113–115] Furthermore, anakinra has also been used in several clinical trials in RA patients.[116–121] These studies demonstrated that anakinra treatment confers protection against cartilage degradation and bone erosion.[117,120] Combination therapy using anakinra together with methotrexate in patients with severely active RA led to a significantly higher rate of ACR responses (ACR20 to ACR70) compared to methotrexate alone.[116,118,119,121] However, though these treatments are promising, they are less effective than anti-TNF agents such as infliximab, etanercept, and adalimumab.[122,123] IL-18 activity can be neutralized by administration of recombinant IL-18BP, a naturally occurring protein that binds IL-18 with high affinity and competes with IL-18R for binding to IL-18. IL-18BP treatment of mice with established CIA reduced the activity of clinical disease and the histological scores of joint damage by 50%.[124]

2.2.3 GOUT AND PSEUDOGOUT

Arthritis caused by gout has been known since ancient times and is regarded as one of the most painful conditions one can experience. The disease is more common in men older than 40, and the increase in its worldwide prevalence over the last several years has been attributed to obesity and aging of the population.[125,126] Pathophysiologically, gout is considered an autoinflammatory disorder associated with hyperuricemia, i.e., elevated levels of serum uric acid, and deposition of monosodium urate (MSU) crystals in joints and periarticular tissues.[127] Uric acid is a metabolic by-product of purine catabolism. In most mammals purines are converted by uricase to the more soluble and better secreted allantoin. However, in humans and apes, the genes encoding uricase have acquired mutations that rendered them dysfunctional.[128] High purine diet, alcohol use, diuretic therapy, obesity, and a reduced renal clearance are the most important risk factors for development of gout.[129,130] On the other hand, pseudogout, a second crystal-induced arthropathy, is associated with deposition of calcium pyrophosphate dehydrate (CPPD) crystals in cartilage and synovium when the cartilage matrix becomes supersaturated with pyrophosphate (PPi) released from dysregulated chondrocytes.[131–134]

It has been suggested from a long time ago that gout and pseudogout are linked to IL-1β,[135] but their connection to caspase-1 and to the inflammasome was revealed only recently.[20] When peritoneal macrophages deficient in NALP3, ASC, or caspase-1 were stimulated with MSU and CPPD crystals, maturation of IL-1β was abolished.[20] Furthermore, an impaired neutrophil influx was observed when mice lacking ASC, caspase-1, or IL-1R were intraperitoneally injected with pathogenic crystals, a model mimicking the infiltration of neutrophils in the intra- and periarticular spaces.[20] These data, together with the substantial levels of IL-1β and IL-18 found in synovial fluids of patients suffering from gout or pseudogout, reinforce the notion of the pivotal role of these cytokines in these pathologies.[135,136] Consequently, a pilot open-label study was initiated, in which patients with acute gout that were refractory to conventional therapies were treated with anakinra. After 3 days of treatment inflammatory symptoms were relieved in all the patients and the pain was reduced by 79% according to their subjective assessment.[137]

2.3 HEREDITARY PERIODIC FEVER SYNDROMES

2.3.1 CRYOPYRIN-ASSOCIATED PERIODIC SYNDROMES

Cryopyrin-associated periodic syndromes (CAPS) encompass three dominantly inherited syndromes that, in order of severity from mild to severe, include familial cold autoinflammatory syndrome (FCAS), Muckle-Wells syndrome (MWS), and chronic infantile neurologic cutaneous articular syndrome (CINCA), which is also known as neonatal-onset multisystem inflammatory disease (NOMID).[138–140] All these syndromes are all linked to mutations in the gene encoding cryopyrin (also known as NALP3).[141–146] Common symptoms of these rare diseases include periods of fever and inflammation, conjunctivitis, arthralgia, and recurrent urticaria-like skin rash, usually occurring during infancy. In addition, each syndrome is characterized by its unique, distinguishing features. These include precipitation of attacks following exposure to cold in FCAS, sensorineural hearing loss, and an intense acute-phase response in MWS, and more severe neurological symptoms ranging from chronic aseptic meningitis to mental retardation to facial dysmorphia and optic nerve irritation in CINCA.[138–140,147,148]

All the known disease-associated mutations in CAPS are in the NACHT domain of cryopyrin, which is responsible for oligomerization and ATP binding; these mutations confer a gain of function by enhancing the tendency for inflammasome assembly.[12,146] This sensitized inflammasome formation is explained by either of two mechanisms: (1) certain mutations increase the potential of cryopyrin to oligomerize and interact with adaptor proteins, or (2) other mutations facilitate ATP binding and inhibit the intrinsic ATPase activity of cryopyrin.[12,149] Although the contribution of ATP hydrolysis to inflammasome assembly or disassembly has not been characterized, studies in plant R proteins, encompassing NLR homologs, indicated that the ATP-bound state rather than the ADP-bound state confers the active form.[150] Disease-associated mutations with increased oligomerization potential probably induce conformational changes that decrease the binding of the autoinhibitory LRR domain with the NACHT domain. LRR-deficient cryopyrin can, in contrast to wild-type cryopyrin, readily interact with the adaptor protein CARDINAL in the absence of any stimulus.[12] Binding of ATP to cryopyrin constitutes another regulatory step in the tight control of inflammasome assembly. Experimental mutation of the ATP-recognizing Walker A motif in wild-type and disease-associated cryopyrin variants completely abolishes IL-1β secretion, indicating that the disease-associated mutations do not circumvent the requirement for nucleotide binding.[149] However, wild-type cryopyrin requires higher ATP concentrations in order to release as much IL-1β as disease-associated mutants.[12,151] These data indicate that the disease-associated mutations either increase the affinity for ATP, promote ADP/ATP exchange, or inactivate the ATPase activity.[149,151] Ultimately, the disease-associated mutations give rise to a high caspase-1 activity and concomitant release of IL-1β and IL-18, which explains the enhanced inflammatory responses observed in CAPS patients. Indeed, monocytes and macrophages from CAPS patients secrete, spontaneously or upon stimulation, large amounts of IL-1β and IL-18.[12,141,152]

These observations supported the initiation of empiric therapeutic trials using anakinra to inhibit IL-1β activity. So far, more than ten studies reported that anakinra treatment was beneficial for all CAPS patients.[148,151–161] One comparative study investigated the effect of anakinra on the acute inflammatory mechanisms occurring after cold challenge (45 minutes at 4°C) in patients with FCAS, the mildest form of CAPS.[159] Subcutaneous administration of 100 mg anakinra 24 hours and 1 hour before the cold challenge prevented rash of the limbs and trunk, fever, arthralgia, and the increase in white blood cell counts and serum IL-6 levels.[159] Renal amyloidosis, a feature rarely associated with FCAS, can also be successfully treated with recombinant IL-1Ra.[152] Likewise, sustained improvements were obtained in three studies in which anakinra was used to treat MWS patients who were unresponsive to different drugs, such as colchicine, corticosteroids, azathioprine, and infliximab (a chimeric monoclonal anti-TNF antibody).[148,158,160] Within 4 hours of anakinra administration, all clinical and serological features associated with active inflammation ceased. The mean plasma serum concentrations of amyloid A protein (SAA) and C-reactive protein (CRP) even decreased up to 100-fold and reached the normal range. Moreover, numerous reports indicate that even the most severe form of CAPS, CINCA, can be successfully treated with anakinra.[151,153–157,160,161] In a study involving eighteen CINCA patients with active disease, a rapid response to anakinra therapy was observed.[156] Rash and conjunctivitis disappeared within 3 days, serum levels of SAA and CRP declined, and the erythrocyte sedimentation rate fell significantly in all patients. Symptoms related to the central nervous system, such as headache, elevated white blood cell counts in cerebrospinal fluids, and cochlear and leptomeningeal lesions were ameliorated significantly after 3 months of therapy. IL-6, TNF, and E-selectin levels in serum and cerebrospinal fluids also decreased. Additionally, spontaneous and stimulated secretion of IL-1β by periferal blood monocytes (PBMCs) obtained from anakinra-treated patients declined gradually. Withdrawal of anakinra uniformly resulted in relapse of all symptoms within days, but resumption of treatment again led to rapid improvement, signifying the necessity for a sustained anakinra therapy.[156] Because anakinra can be administered only by injection, there is a need for a drug that can be administered orally. Therefore, VX-765, an orally bioavailable caspase-1-specific inhibitor, could represent an attractive new therapeutic approach for the treatment of these disorders because it can block IL-1β secretion in LPS-stimulated PBMCs from FCAS patients.[162]

2.3.2 Familial Mediterranean Fever

Familial Mediterranean fever (FMF), the most prevalent of the hereditary periodic fever syndromes, is an autosomal recessive disease characterized by recurrent episodes of fever, arthralgia, rash, and abdominal and chest pain.[163] If left untreated, patients frequently develop organ-threatening amyloidosis resulting in renal failure.[163] The disease is more commonly found in people originating from a particular area around the Mediterranean Sea, including non-Ashkenazi Jews, Armenians, Arabs, Turks, Greeks, and Italians.[164] FMF is caused by mutations in the *MEFV* gene, which encodes pyrin. The pyrin protein is composed of an N-terminal PYD, two B-boxes (BBs), a coiled-coil (CC) region, and a C-terminal B30.2 domain (also known as the PRYSPRY domain). The mutations, which are predominantly missense, are found throughout the gene, but mutations associated with the most severe phenotypes are clustered in exon 10, which codes for the B30.2 domain.[165]

The function of wild-type pyrin and of the FMF-associated mutants in the regulation of inflammatory responses remains controversial. One view emphasizes an IL-1β-inhibiting, anti-inflammatory role for pyrin, whereas the other view imputes an IL-1β-inducing, pro-inflammatory role. According to the two hypotheses, the excessive IL-1β maturation observed in FMF patients is caused, respectively, by loss or gain of function of pyrin. Evidence supporting the inhibitory effect of pyrin on caspase-1-mediated IL-1β production came from multiple overexpression, knockdown, and *in vivo* and *ex vivo* mouse studies. More specifically, pyrin interacts at an endogenous level with ASC through their respective PYD domains, and with active caspase-1, NALP family members, and pro-IL-1β through its B30.2 domain.[166–168] In this way, once the inflammasome is activated, pyrin sequesters all known inflammasome components, thereby tightly controlling the release of IL-1β.[168] Accordingly, overexpression of pyrin in the murine RAW264.7 and human THP.1 cell lines suppresses the release of IL-1β in response to stimulation of the inflammasomes with crude LPS, PGN, or MSU.[166,168] Consistent with these observations, peritoneal macrophages from mice expressing a truncated pyrin had higher IL-1β production in response to LPS, PGN, and MSU, and RNAi-mediated pyrin knock-down in THP.1 cells had similar results.[166–168] In contrast, Yu et al.[177] suggested that pyrin promotes inflammation. Using a HEK293t cell-based reconstitution system, in which ASC and procaspase-1 are stably overexpressed, they demonstrated that pyrin, in analogy to cryopyrin, can assemble an ASC-containing inflammasome, which leads to caspase-1 activation and IL-1β maturation (Figure 2.3).[169]

The second point of debate is the effect of the FMF-associated pyrin mutants on the modulation of caspase-1 activity and IL-1β production. Taking the IL-1β-regulating propensity of wild-type pyrin as a reference, one study demonstrated attenuation of the IL-1β-inhibiting capacity of the FMF-associated pyrin mutants, which explains the enhanced IL-1β levels in FMF patients.[167] A second study showed that the ability of pyrin to inhibit IL-1β is not affected by the FMF-associated M694V mutation in pyrin.[168] Finally, a third report stated that the caspase-1-activating and IL-1β-producing capacity of pyrin is not altered in the mutants.[169] These results are all the more puzzling because they were obtained in similar HEK293t reconstitution experiments.[167–169] The lack of a convincing functional difference between wild-type and FMF-associated pyrin variants in these experiments may be explained by the absence of a pathogenic stimulus, because these mutations might have been acquired during evolution to confer enhanced responsiveness to certain PAMPs. The B30.2 domain, harboring the most prevalent FMF-associated mutations, could represent a regulatory domain mediating responses to certain pathogens.[170]

Model I: Pyrin inhibits inflammasome assembly (loss-of-function mutations in PAPA syndrome)

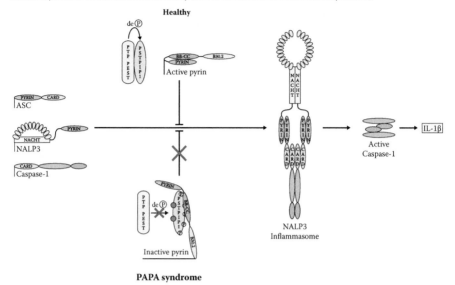

Model II: Pyrin assembles a distinct inflammasome (gain-of-function mutations in PAPA syndrome)

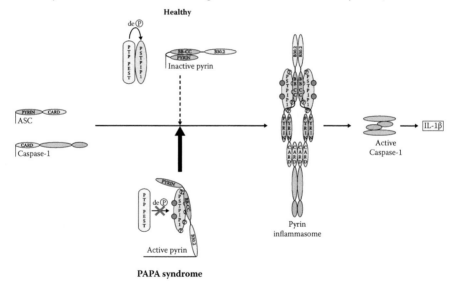

FIGURE 2.3 Proposed mechanisms for pyrin action in the PAPA syndrome. Model I: Pyrin functions as an inhibitor of the assembly of inflammasome complexes, implying a loss of pyrin function for the PAPA-associated PSTPIP1 mutations.[166–168] In healthy individuals, pyrin has a low affinity for PSTPIP1 due to the low degree of PSTPIP1 phosphorylation, which is dependent on the interaction of PSTPIP1 with the phosphatase PTP-PEST. In this situation, pyrin can prevent the assembly of functional caspase-1-activating inflammasome complexes and the concomitant release of bioactive IL-1β. However, PAPA syndrome patients harbor mutations (depicted by circles) in the PSTPIP1 that disrupt its interaction with PTP-PEST.

Two case reports show that the IL-1-receptor antagonist, anakinra, was effective in the treatment of two FMF patients who were unresponsive to classical colchicine treatment; colchicine unresponsiveness occurs in 5%–10% of the patients.[167,171] Monitoring of acute-phase reactants demonstrated that both SAA and CRP levels returned to base levels in both patients.[167,171] Episodes of fever accompanied by headache, abdominal pain, and chest pain were also scarce during treatment.[171] However, discontinuation of treatment resulted in relapse of the disease.[167,171] Viewed together, these results confirm the importance of caspase-1 and IL-1β in FMF pathology.

2.3.3 PAPA Syndrome

The syndrome of pyrogenic aseptic arthritis, pyoderma gangrenosum, and cystic acne (PAPA) is a rare autosomal dominant autoinflammatory disorder of early onset. It is characterized by recurrent sterile neutrophil-rich effusions in the joints, sterile abscesses, and expanding purulent ulcerating lesions.[172,173] The disease is caused by missense mutations in the gene encoding the proline-serine-threonine phosphatase-interacting protein 1 (PSTPIP1), an actin-binding cytoskeleton-organizing protein.[174,175] The PAPA-associated mutations, A230T and E250Q, increase the degree of interaction between PSTPIP1 and pyrin. This interaction is positively correlated with the tyrosine phosphorylation status of PSTPIP1, which is counteracted by the PEST-type protein tyrosine phosphatases (PTPs) (Figure 2.3).[176,177] Actually, both mutations, located at the PTP-PEST binding interface, severely reduce PTP-PEST recruitment, and thus abrogate PSTPIP1 dephosphorylation.[175] In accordance with the opposing functions attributed to pyrin, i.e., inhibiting or stimulating IL-1β release (see above), there are two ways to explain the increased IL-1β production by peripheral blood leukocytes of PAPA patients harboring the A230T mutation, and by cell lines transfected with PAPA-associated PSTPIP1 mutants.[176] On the one hand, increased PSTPIP1-pyrin binding would alleviate its inhibiting, inflammasome sequestering potential, which would enable proper inflammasome assembly, caspase-1 activation, and IL-1β maturation (Figure 2.3).[168,176] In this regard, the PAPA-associated mutations in PSTPIP1 impute a loss of pyrin function. On the other hand, Yu et al. reported that pyrin is a cytosolic receptor for PSTPIP1, which liberates pyrin from its autoinhibited state and allows the assembly of pyrin inflammasomes, another caspase-1-activating platform (Figure 2.3).[177,178] In this respect, the PAPA-associated mutations in PSTPIP1 represent a gain of function for pyrin. Although further research is required to clarify these ambiguities, the contribution of IL-1β to

FIGURE 2.3 (Continued) This leads to PSTPIP1 hyperphosphorylation, which enables the recruitment of pyrin and relieves the inhibition of inflammasome assembly. This ultimately results in caspase-1 activation and excessive IL-1β maturation. Model II: Pyrin assembles a distinct inflammasome, implying a gain of pyrin function for the PAPA-associated PSTPIP1 mutations.[169,177] In healthy individuals, pyrin is kept in an autoinactivated state, unable to assemble inflammasome complexes. PAPA-associated mutations enable the interaction between PSTPIP1 and pyrin and relieve the autorepressed status of pyrin; PSTPIP1 thus acts as a ligand. Active pyrin can then assemble inflammasome complexes by its ability to interact with ASC/PYCARD, which recruits procaspase-1. Ultimately, this leads to caspase-1 activation and enhanced secretion of bioactive IL-1β.

PAPA pathology was substantiated in a case study in which anakinra proved to be an effective treatment for flares in PAPA syndrome.[179]

2.4 INFLAMMATORY BOWEL DISEASES

Inflammatory bowel disease (IBD) is a family of chronic disorders that cause inflammation of the digestive tract due to an inappropriate response of the immune system.[180-182] The main forms of idiopathic IBD are Crohn's disease (CD) and ulcerative colitis (UC). With a prevalence rate of 0.4%, IBD is becoming a significant problem in the Western world.[183] UC is characterized by a diffuse inflammation that affects the mucosal lining of the rectum and extends proximally to affect a variable length of the colon. Histopathologically, UC is associated with development of superficial crypt abscesses. CD, on the other hand, is a chronic transmural granulomatous inflammatory condition potentially involving any part of the alimentary tract from mouth to anus, but with a propensity for the distal small bowel, the proximal large bowel, and the peri-anal region. Inflammation in CD is often discontinuous along the length of the gut. Although the etiologies of UC and CD are not known, it is thought that IBD results from the inappropriate and ongoing activation of the mucosal immune system by the normal luminal flora. This aberrant response is most likely facilitated by defects both in the barrier function of the intestinal epithelium and in the mucosal immune system. A strong genetic influence, particularly in CD, has been shown by familial clustering, and so genome-wide searches for IBD susceptibility loci have been performed.[184] This led to the identification of several genes implicated in IBD, including *nod2*, *il23r*, *atg16l1*, *nell1*, *dlg5*, *octn1*, and *octn2*.[185-191]

The first identified and most thoroughly studied susceptibility gene, *NOD2*, encodes a member of the NLR intracellular pattern recognition receptor family, and fulfills an important role in host defense responses.[192] NOD2 is expressed in the cytosol of antigen-presenting cells, and in the cytosol and at the plasma membrane of epithelial cells at the base of small intestinal crypts, known as Paneth cells.[193,194] The protein is composed of two N-terminal CARD domains, a central NACHT domain, and a C-terminal LRR domain that senses muramyl dipeptide (MurNAc-L-Ala-D-isoGln, MDP), the largest peptidoglycan motif common to Gram-negative and Gram-positive bacteria.[195-198] Several *in vitro* studies have implicated NOD2 as a sensor of intracellular pathogens, such as *Listeria monocytogenes*, *Streptococcus pneumoniae*, and *Mycobacterium tuberculosis*.[199-201] Analysis in NOD2-deficient mice further confirmed the role of NOD2 in the clearance of orally administered *Listeria monocytogenes*, which are not cleared when delivered intravenously or into the peritoneum.[200] Functionally, ligand binding activates NOD2 and induces recruitment of the serine-threonine kinase, RIP2, through a homotypic CARD-CARD interaction.[198] RIP2 mediates the activation of NF-κB and the MAP kinases p38, ERK, and JNK.[198,200,202-204] Furthermore, optimal p38 and JNK activation in response to pathogen infection requires recruitment of the CARD-containing adaptor, CARD9, through RIP2.[202,205] The NF-κB and MAP kinase pathways cooperatively upregulate the transcription of pro-inflammatory cytokines and chemokines that contribute to CD through innate and adaptive immune responses.

Whether the three most common CD-associated NOD2 mutations (R702W, G908R, and L1007insC) confer gain-of-function or loss-of-function mutations remains a matter of debate, as different experimental setups and readout systems provided evidence in support of both views.[206] In the gain-of-function concept, it is simple to assume excessive inflammatory signaling in CD patients. However, PBMCs from CD patients exhibit a marked decrease in NF-κB activation when stimulated with MDP.[207] The L1007insC mutation causes plasma membrane recruitment, which appears to be crucial for MDP-induced NOD2-mediated NF-κB signaling, to become dysfunctional.[193,194] Moreover, bacterial products induce cytoprotective and reparative factors in the intestine, and mice depleted of commensal bacteria develop severe inflammation following epithelial injury by dextran sodium sulfate (DSS) administration; these findings led to the concept that basal NOD2-dependent signaling is required for epithelial homeostasis and for preventing bacterial invasion and subsequent dysregulated inflammation in the intestine.[208–210] Indeed, NOD2 deletion results in an increased susceptibility to bacterial invasion, and reduced production of antibacterial peptides in the intestine.[200] Other studies suggested that NOD2 suppresses TLR2-induced NF-κB activation in wild-type conditions.[211] In CD patients, TLR2 signaling is unleashed, leading to an inflammatory Th1 response. These observations indicate that steady-state NOD2-dependent signaling is crucial for homeostasis of the intestinal epithelium.[209] A possible mechanism explaining the heightened inflammation observed in CD patients is based on the enhanced production of IL-1β observed in DSS-treated knock-in mice harboring the murine homolog of the most common susceptibility allele, NOD2^{L1007insC}.[212] Given the importance of caspase-1 activity in the maturation and release of IL-1β, there must exist a mechanism by which NOD2 controls caspase-1 activation. In analogy to other NLR family members, NOD2 could assemble a distinct NOD2 inflammasome by directly recruiting caspase-1 through mutual CARD-CARD domain interaction.[213] Alternatively, other caspase-1-activating inflammasomes could become activated by one of two mechanisms: (1) NOD2 activates the inflammasomes by its ability to interact with them, or (2) alleviation of the protective steady-state NF-κB activation, mediated by NOD2 signaling, enables the assembly of functional inflammasomes.[211,214] The latter hypothesis is in accord with the excessive TLR2 signaling observed when NOD2 is deleted.[215] Intriguingly, monocytes from patients harboring the same L1007insC mutation, as well as macrophages from NOD2-deficient mice, display a marked defect in IL-1β production in response to MDP.[200,207,216,217]

While these discrepancies raise questions about the contribution of NOD2 to CD, strong evidence shows that caspase-1 and its second pro-inflammatory substrate, IL-18, play a substantial role.[218,219] In particular, the capacity of IL-18 to drive Th1 responses, primarily by inducing IFN-γ production in T and NK cells, is of importance in CD, where Th1 polarization appears to be the key pathogenic mechanism.[220–222] Consequently, increased IL-18 production, coinciding with detection of active caspase-1, has been observed in mucosal macrophages and epithelial cells of chronic CD lesions.[220–224] Moreover, caspase-1-deficient mice exhibited >50% decrease in the clinical scores of weight loss, diarrhea, rectal bleeding, and colon length during acute DSS-induced colitis.[219] Daily treatment with IL-1Ra, which blocks only IL-1 signaling, produced only a modest reduction in disease severity.[225]

By contrast, intraperitoneal administration of the caspase-1 inhibitor pralnacasan, which prevents production of IL-1β and IL-18, resulted in significant amelioration of the disease score as measured by loss of body weight, stool consistency, and colon length.[226,227] In conclusion, inhibition of caspase-1 is an important strategy for treatment of inflammatory bowel diseases, because IL-18 and, to a lesser extent, IL-1β are important drivers of the disease.

2.5 SEPSIS

Sepsis is regarded as a systemic inflammatory response syndrome (SIRS) with a proven or suspected infectious etiology. It can be manifested in several ways, including increased or decreased body temperature, increased heart rate, elevated white blood cell counts, rapid respiratory rate, altered mental status, shock, and multiple organ dysfunction.[228,229] Sepsis is defined as severe when there are signs of organ dysfunction or organ hypoperfusion, and as septic shock if hypotension persists regardless of adequate fluid resuscitation.[230] Despite extensive research and clinical trials, sepsis still represents a major health problem affecting about 37% of intensive care unit patients in Europe.[231] As an inflammation-driven disorder, sepsis is orchestrated by pro-inflammatory cytokines, such as TNF, IL-1β, IL-6, IL-8, and IL-18. Accordingly, mice deficient in caspase-1, ASC, and NALP3 are protected in models of sepsis induced by endotoxin or live *Escherichia coli*.[10,19,23,232] However, IL-1β-deficient mice, as well as IL-1β/IL-18 double-deficient animals, are not protected against endotoxin or live bacterial challenge, indicating that the protective effect of caspase-1 blockade is independent of its role as the main activator of IL-1β and IL-18.[232,233] The involvement of caspase-1 in progression of sepsis could be attributed to its recently identified capacity to mediate apoptosis, directly or indirectly, in splenic B lymphocytes: caspase-1-deficient animals exhibit a remarkably low level of B lymphocyte apoptosis in the spleen, in contrast to wild-type, IL-1β knockout, and IL-1β/IL-18 double-knockout mice.[232] Accordingly, inhibition of apoptosis by different means, including caspase inhibition, overexpression of the anti-apoptotic Bcl-2 protein, or interference with the FasL-induced pro-apoptotic pathway, already proved its applicability in animal models of sepsis.[234] In contrast, clinical trials using anakinra were not successful.[235,236] However, the protection observed in these sepsis models could also be due to other caspase-1 functions that have not been identified yet.

2.6 INFLAMMATORY SKIN DISEASES

Our bodies are continuously exposed to environmental hazards, and the skin serves as the interface between an organism and the outside world to protect against chemical, physical, and microbial insults. These insults result in the generation of PAMPs and DAMPs that can activate innate immune responses, our first line of defense.[237] As keratinocytes are the main cell type in the epidermis, they are pivotal in the initiation, maintenance, and regulation of immune responses in the skin.[238] As described above, mutations in the NALP3 gene are associated with severe inflammatory disorders. Interestingly, patients affected by these diseases have inflammatory skin lesions, suggesting a central role for inflammasome signaling in skin inflammation.

Psoriasis affects approximately 2% of Western populations and is characterized by epidermal inflammation and hyperplasia, and abnormal keratinocyte differentiation. The etiology of psoriasis is not understood, but the disease is characterized by elevated levels of caspase-1, IL-1β, and IL-18, indicating that caspase-1 can be a valuable therapeutic target in the treatment of inflammatory skin diseases.[61,239,240] Furthermore, the IL-1-induced transcriptome in keratinocytes resembles that of psoriatic lesions, and epidermis-specific caspase-1 transgenic mice develop dermatitis with psoriasis-like changes.[241] Interestingly, vitamin D analogs, well known for their anti-inflammatory and antiproliferative activities, are often used to treat psoriasis; they have been shown to increase the ratio of icIL-1Ra to IL-1, and to directly suppress IL-18 production in keratinocytes. This suggests that vitamin D analogs act, at least in part, by suppressing IL-1 and IL-18 activities in keratinocytes.[242] BALB/c mice deficient in IL-1Ra developed cutaneous inflammation with many features resembling human psoriasis.[243] The epidermis became thickened and the stratum corneum showed parakeratotsis. Clinical pilot studies on the use of anakinra to treat psoriasis are ongoing. Pralnacasan and VX-765 are the only caspase-1 inhibitors used so far in clinical trials, but pralnacasan was discontinued in 2005 due to toxicity. The results of a phase II clinical trial on the therapeutic efficacy of VX-765 in psoriasis have yet to be disclosed.

Caspase-1 and other inflammasome constituents, such as NALP3 and ASC, are reportedly also required for the eczema-like **contact allergen sensitivity** in mouse models.[244,245] It was recently shown that IL-1β but not IL-18 can rescue the defective contact hypersensitivity response observed in caspase-1–/– mice, which have no functional IL-1β or IL-18.[246] Wannamaker and colleagues found that orally administered VX-765 alleviates oxazolone-induced contact sensitization as effectively as prednisolone, a corticosteroid commonly used to treat inflammatory diseases.[111] It may be worthwhile to test caspase-1 inhibitors in other common inflammatory skin diseases, such as atopic dermatitis, acne vulgaris, and vitiligo-associated multiple autoimmune disease. The latter disease is characterized by depigmented skin lesions resulting from a reduction in the number and function of melanocytes, and it has been shown that mutations in the DNA sequence in the coding and promoter regions of the NALP1 gene correlated with an increased risk of developing the disease.[247] As mentioned above, NALP1 is one of the sensor-platform proteins promoting inflammasome formation and activation.

2.7 NEUROLOGICAL DISEASES

Inflammatory mediators, particularly the caspase-1 substrates IL-1β and IL-18, are pivotal mediators of the neuronal injury that occurs in a wide range of acute neurodegenerative disorders, such as stroke, trauma, and epilepsy, as well as in chronic forms such as Alzheimer's disease (AD), Parkinson's disease (PD), multiple sclerosis (MS), and amyotrophic lateral sclerosis (ALS).[248–251] In addition, the CAPS diseases discussed above, including FCAs, MWS, and CINCA, are also often associated with neuronal injury. Although some clinical studies reported higher levels of IL-1β in postmortem examination of brains and in patient cerebrospinal fluids, most evidence supporting a direct role of IL-1β in disease pathology has been obtained in experimental models of neuronal injury in which the expression, release, or effects of IL-1β were inhibited.[252,253]

2.7.1 ACUTE NEURODEGENERATIVE DISORDERS

Stroke is the major cause of ischemic brain injury. Various animal models of cerebral ischemia, including focal, global, transient, or permanent ischemia, have been used to gain insight into the contribution of IL-1β and IL-18 to stroke. Transgenic mice expressing a dominant-negative caspase-1 mutant and mice deficient in caspase-1 show significantly reduced ischemic brain injury after transient and permanent focal ischemia induced by middle cerebral artery occlusion.[254–256] Similarly, intracerebroventricular (icv) injection of the specific caspase-1 peptide inhibitor, Ac-YVAD-cmk, or recombinant IL-1Ra significantly reduced lesion volume after, respectively, permanent and transient focal cerebral ischemia in the rat.[257,258] Furthermore, neuroprotection is maintained when, 3 hours after occlusion, high doses of IL-1Ra are administered peripherally, whereupon IL-1Ra is transported across the blood-brain barrier by an active mechanism.[258–262] Accordingly, a randomized phase IIa trial was initiated in which patients suffering an acute stroke were treated with IL-1Ra within 6 hours. This trial demonstrated an improvement in the biological endpoints, such as counts of total white blood cells and neutrophils, and concentrations of IL-6 and CRP.[263] Although promising, IL-1Ra is a large protein with limited brain-penetrating capacity. However, a recent report clearly showed that icv injection of VRT-018858, the active metabolite of the selective nonpeptide caspase-1 inhibitor, pralnacasan, markedly reduced ischemic brain injury in rats even when administered up to 3 hours after transient focal cerebral ischemia.[264] VRT-018858 has >100-fold selectivity for caspase-1 and caspase-11 than for other caspases.[112] The use of pralnacasan for chronic disorders may have been excluded by the finding of liver abnormalities in long-term animal toxicology studies, but it may find an alternative application in the treatment of acute disorders.

Traumatic brain injury (TBI) is an important cause of death and morbidity, especially in children under the age of 6, whose brains appear to be particularly vulnerable.[265,266] The main animal models used to simulate the key features of TBI are the weight-drop model, the aseptic cryogenic model, and the closed head injury model.[267] These models, respectively, involve dropping a 10 g weight from a height of 8 cm on the right parietal bone, applying a metal rod precooled with liquid nitrogen for 15 seconds to the right parietal cortex, and dropping a 75 g weight from a height of 18 cm on a cone placed 1 mm beside the midline, in the midcoronal plane.[268–270] In the weight-drop TBI model, as in stroke, neuroprotection was achieved by overexpressing the dominant negative inhibitor of caspase-1 or by icv injection of Ac-YVAD-cmk.[271] Transgenic overexpression or icv administration of recombinant IL-Ra in the closed head injury and the aseptic cryogenic models, respectively, significantly improved neurological recovery, reduced lesion volume, and improved behavioral deficits, demonstrating the importance of IL-1β signaling in TBI.[272,273] Likewise, IL-18 is an important contributor to TBI, as mice subjected to closed head injury and injected with IL-18BP showed significantly improved neurological recovery after 7 days.[274] In accordance, elevated IL-1β and IL-18 levels are found in cerebral fluids of TBI patients.[274,275] However, all previous studies were conducted in adult mice, while research focusing on TBI in the developing brain, reflecting the largest target group of young children, has been rather scarce. Recently, Sifringer et al. showed that 7-day-old mice deficient in IL-18 or intraperitoneally injected with recombinant IL-18BP attenuated TBI when subjected to the weight-drop model. This provided the first evidence that common inflammatory cytokines mediate TBI in the developing and adult brains.[276]

A third acute neurodegenerative disorder, **epilepsy**, affects about 50 million people worldwide. It is characterized by recurrent episodes of seizures due to sudden, disordered, excessive neuronal discharges, wherein glutamate excitotoxicity plays a key role.[277] The most common animal model for human temporal lobe epilepsy is injecting kainic acid, an excitotoxic analog of glutamic acid, in the hippocampus of rodents.[278,279] Treatment of kainic acid–injected animals with the caspase-specific nonpeptide inhibitor pralnacasan or VX-765 represents an effective anticonvulsive strategy, as it results in a twofold delay in the onset of seizures and 50% reduction in their duration.[280] Moreover, susceptibility to seizure was further attenuated by deleting the caspase-1 gene.[280] In accordance, IL-1Ra mediates powerful anticonvulsant effects, and mice overexpressing IL-1Ra in astrocytes are much less susceptible to seizures.[281–283]

2.7.2 CHRONIC NEURODEGENERATIVE DISORDERS

Alzheimer's disease is characterized by a progressive loss of cognitive function and affects a growing number of the aging population. It is associated with the accumulation of the amyloid-β (Aβ) peptide in amyloid plaques in the extracellular brain parenchyma.[284] Because the expression of several inflammatory mediators, including caspase-1 and its substrates IL-1β and IL-18, is increased in postmortem brains of people with AD, it was believed that neuroinflammation was one of the principal pathology drivers.[285–289] However, recent evidence also points to a potentially beneficial and protective role for certain inflammatory mediators in AD, such as IL-1β.[287] APP/PS1 transgenic mice partially mimic AD pathology by overexpressing the amyloid precursor protein (APP) and presenilin 1 (PS1), and specific hippocampal overexpression of IL-1β in these mice was found to ameliorate the pathology by reducing Aβ plaque deposition and lowering soluble Aβ levels.[290] Such results may explain the failure of many anti-inflammatory therapies in AD.

Parkinson's disease is a multigenic neurodegenerative movement disorder affecting 1% of the population over the age of 65 and 5% of 85-year-olds.[291–293] Although the clinical presentation of the pathology is varied, the major symptoms are due to the loss of dopamine-producing neurons in the *substantia nigra*.[294,295] As in AD, PD is associated with persistent neuroinflammation, which in this case is associated with a robust activation of microglia in patients and animal models.[296] Indeed, elevated levels of IL-1β have been observed in the brain and cerebrospinal fluids of patients with PD.[297,298] Moreover, sustained expression of IL-1β in the *substantia nigra* of adult rats mediates neurotoxic effects by inducing a progressive dopaminergic cell death that resembles PD.[299] Similarly, enhanced caspase-1 activity was observed in the *substantia nigra* of PD patients.[300] Interestingly, caspase-1 can directly cleave and inactivate Parkin, an E3 ubiquitin protein ligase that plays a vital role in the survival of dopaminergic neurons, as gene mutations are associated with the loss of these neurons.[301–303] Consequently, caspase-1-mediated Parkin cleavage leads to the accumulation of toxic Parkin cleavage products that trigger dopaminergic cell death.[301] Targeting caspase-1 could be therapeutic because it would block at least two mechanisms that promote dopaminergic cell death. A first indirect proof of principle was provided by a study using the 1-methyl-4-phenyl-1,2,3,6-tetrahydropyridine (MPTP) mouse model of PD. Treatment of these mice with minocycline, a synthetic tetracycline antibiotic that, among other effects, markedly reduces caspase-1 expression and inhibits caspases, prevented excessive loss of dopaminergic neurons.[304]

Multiple sclerosis is an immune-mediated demyelinating disease of the central nervous system (CNS) of unknown etiology. The pathological hallmark of the disease is the presence in the CNS of autoreactive T cells that specifically target myelin antigens, thereby mediating destruction of the myelin sheath, which leads to damage of the underlying axon and accumulation of white matter myelin plaques.[305,306] Experimental autoimmune encephalomyelitis (EAE) is an animal model for MS, as it has similar pathological features, including myelin autoreactive T cells and degradation of the myelin sheet in the CNS.[307] Several pro-inflammatory cytokines, such as TNF, IL-1β, IL-18, and IFN-γ, are crucial for the development of MS, as they participate in the activation of the autoreactive T cells.[308] Indeed, elevated levels of these cytokines were observed in demyelinating plaques, cerebrospinal fluids, and serum of MS patients and in the EAE model.[309–315] Caspase-1 mRNA transcription as well as caspase-1 protein expression in resident oligodendrocytes of MS lesions are also significantly increased in MS patients, which suggests that these cells are the main sources for IL-1β and IL-18.[316] The importance of IL-1β for disease pathology was underscored by the results of experiments in which mice treated with recombinant IL-Ra and mice deficient in type I IL-1 receptor were used in the EAE model. Both strategies resulted in attenuation of EAE development.[317–321] Similarly, neutralizing antibodies to IL-18 prevented the onset of EAE.[322] Blocking caspase-1 activity either by deletion of its gene or by pharmacological inhibition with Ac-YVAD-cmk reduced the incidence and severity of EAE.[315] In an attempt to circumvent the feeble ability of recombinant IL-Ra protein to cross the blood-brain barrier, Furlan et al. designed an IL-1Ra gene therapy approach. Intracisternal administration of an HSV-1-derived vector carrying the IL-1Ra gene effectively delayed EAE onset by almost a week and substantially decreased disease severity.[323]

Another chronic neurodegenerative disorder in which caspase-1 is believed to play a key pathophysiological role in **amyotrophic lateral sclerosis** (ALS), also known as Lou Gehrig's disease. It is the most common adult-onset motor neuron disease, and it is characterized by progressive degeneration of motor neurons and their axons in the corticospinal tracts, which leads to spasticity, muscle atrophy, and paralysis.[324] The transgenic mouse expressing a human ALS-associated mutant of Cu/Zn superoxide dismutase (mSOD1) is used as a model because it develops a motor neuron disease closely resembling human ALS.[325–327] As in other neurodegenerative disorders, neuronal cell death in ALS is also believed to be mediated by inflammatory components, such as TNF and IL-1β, which are upregulated in the spinal cord of ALS patients and mSOD1 transgenic mice.[328,329] However, the main contribution of caspase-1 to the disease was shown to be due to its ability to induce apoptosis in motor neurons.[328,330,331]

Active caspase-1 and caspase-3 are detected in spinal cord samples of ALS patients.[328,332,333] This is also reflected in mSOD1 mice, in which caspase-1 activity precedes caspase-3 activity, indicating that caspase-1 acts slowly as a chronic initiator while caspase-3 functions as the effector in ALS-associated neuronal cell death.[328,331] Furthermore, intracerebroventricular administration of zVAD-fmk, a broad-spectrum caspase inhibitor, is neuroprotective and extends survival of ALS mSOD1 mice.[328] The importance of caspase-1 in the pathogenesis of ALS, regardless of its pro-apoptotic or IL-1β-activating capacity, is underscored by the observation that a double transgenic mouse expressing both a dominant-negative caspase-1 gene and mSOD1 demonstrated extended survival and more than 50% slower disease progression.[334]

2.8   CONCLUDING REMARKS

Increasing knowledge of the molecular mechanisms involved in inflammatory and neurodegenerative pathologies reveals caspase-1 and its substrates, IL-1β and IL-18, as potential targets for treatment of inflammatory diseases. Enhanced IL-1β and IL-18 levels in body fluids are hallmark features in disease models and patients. Indeed, application of caspase-1-specific inhibitors or blockage of IL-1β and IL-18 signaling in animal models and clinical trials yielded promising results (see Table 2.1 for overview). Targeting IL-1β signaling by anakinra treatment was particularly successful in several studies of CAPS, FMF, and PAPA patients. These patients responded promptly to treatment and demonstrated a near complete alleviation of their symptoms, indicating a crucial role for IL-1β. A similar strategy also revealed the involvement of IL-1β in gout, a disease with an increasing worldwide prevalence. However, therapy with the recombinant form of IL-Ra is very expensive, needs to be injected repeatedly, and can cause allergic reactions. Chemical compounds are preferred because they can be administered orally and are much cheaper to produce. Two compounds meeting these properties, pralnacasan and VX-765, are potent caspase-1-specific inhibitors that entered clinical trials for rheumatoid arthritis and psoriasis. Although the results for pralnacasan obtained in a phase IIa clinical trial in RA patients were encouraging, the compound was excluded from further patient studies due to the development of liver abnormalities in long-term animal toxicology studies. Indeed, minimization of short-term and long-term toxicity is an important requirement in the design of effective therapeutic drugs. However, despite the observed toxicity, recent results from pralnacasan and VX-765 therapeutic studies in short-term animal models for stroke, epilepsy, Crohn's disease, and contact allergen sensitivity provided additional evidence for their effectiveness. In addition, strategies that target pro-inflammatory molecules should be considered with great care, because these molecules are of key importance in host defenses against infections, and inhibiting them may immunocompromise the patients. For example, interfering with TNF signaling is associated with serious side effects: infections (e.g., tuberculosis), sepsis, and even death. Future strategies could involve interference with specific disease-associated inflammasome complexes, because mounting evidence implicates distinct inflammasomes in different pathologies, e.g., the involvement of NALP3 inflammasomes in CAPS. After elucidating the composition of these caspase-1-activating platforms, compounds could be developed to specifically target these inflammasomes. In this

TABLE 2.1
Effect of Caspase-1 Inhibitors and the IL-1RI Antagonist in Animal Models and Clinical Trials of Inflammation-Driven Disorders

Compound/Company	Disease	Outcome in Animal Models	Outcome in Clinical or Preclinical Trials
Pralnacasan (VX-740)/ Vertex, Inc.	Osteoarthritis	Reduced joint damage[78]	N/A
	Rheumatoid arthritis	Reduced inflammation and joint destruction[112]	Trend toward improvement in a phase IIa trial[110]
	Crohn's disease	Amelioration of body weight, stool consistency, and colon length[226, 227]	N/A
	Psoriasis	N/A	No results reported from a phase II trial
	Stroke	Reduced brain damage[264]	N/A
	Epilepsy	Delayed seizure onset and reduced duration[280]	N/A
VX-765/Vertex, Inc.	Rheumatoid arthritis	Delayed onset of forepaw inflammation and reduced disease severity[111]	N/A
	Familial cold autoinflammatory syndrome	Blocks IL-1β secretion from PBMC[162]	N/A
	Psoriasis	N/A	No results reported from a phase II trial
	Contact allergen sensitivity	Alleviates oxazolone-induced contact sensitization[111]	N/A
	Epilepsy	Delayed seizure onset and reduced duration[280]	N/A
Anakinra (Kiniret ™)/Amgen	Osteoarthritis	Decreased knee joint instablity[79,81–84]	Alleviation of pain[80]
	Rheumatoid arthritis	Reduced destruction of cartilage and bone[113–115]	Reduction of cartilage degradation and bone erosion[116–121]
	Gout	N/A	Reduction of pain and inflammatory symptoms[137]

(Continued)

TABLE 2.1 (CONTINUED)
Effect of Caspase-1 Inhibitors and the IL-1RI Antagonist in Animal Models and Clinical Trials of Inflammation-Driven Disorders

Compound/Company	Disease	Outcome in Animal Models	Outcome in Clinical or Preclinical Trials
	Familial cold autoinflammatory syndrome	N/A	Reduced fever, athralgia, and rash of limbs and trunk[152,159,160]
	Muckle-Wells syndrome	N/A	Complete disappearance of skin rash, irritability, periodic fever, conjunctivitis, and abdominal pain[148,151,158,160,161]
	Chronic infantile neurologic cutaneous articular syndrome	N/A	Disappearance of rash, conjunctivitis, and central nervous system symptoms such as headache[151,153–157,160]
	Familial Mediterranean fever	N/A	Alleviation of fever, headache, and abdominal and chest pain[167,171]
	Pyrogenic aseptic arthritis, pyoderma gangrenosum, and cystic acne	N/A	Fewer episodes of flares, pain, and joint swelling[179]
	Crohn's disease	N/A	Treatment aggravated the disease outcome[225]
	Sepsis	N/A	No significant effect[235,236]
	Stroke	Reduced lesion volume[258]	N/A
	Traumatic brain injury	Amelioration of behavioral defects and reduction of lesion volume[272,273]	N/A
	Epilepsy	Reduced susceptibility to seizures[282]	N/A
	Multiple sclerosis	Delayed disease onset[317,318,321,323]	N/A

way, specific treatment of chronic disorders would become possible, while the risk for general immunosuppression would be minimized.

ABBREVIATIONS

Aβ, amyloid-β; AD, Alzheimer's disease; ALS, amyotrophic lateral sclerosis; APP, amyloid precursor protein; ASC, apoptotic speck-like protein containing a CARD; CAPS, cryopyrin-associated periodic syndromes; CARD, caspase recruitment and activation domain; CD, Crohn's disease; CIA, collagen-induced arthritis; CINCA, chronic infantile neurologic cutaneous articular syndrome; CNS, central nervous system; COP, CARD-only protein; CPPD, calcium pyrophosphate dehydrate; CRP, C-reactive protein; DAMP, danger-associated molecular pattern; DD, death domain; DED, dead effector domain; DSS, dextran sodium sulfate; EAE, experimental auto-immune encephalomyelitis; FCAS, familial cold autoinflammatory syndrome; FMF, familial Mediterranean fever; IBD, inflammatory bowel disease; icv, intracerebro-ventricular; IFN-γ, interferon-γ; IL, interleukin; IL-18R, IL-18 receptor; IL-1Ra, IL-1 receptor antagonist; IL-1RAcP, IL-1R accessory protein; IL-1RI, type I IL-1 receptor; IL-1RII, type II IL-1R; INCA, inhibitory CARD; iNOS, inducible NO synthase; LPS, lipopolysaccharide; LRR, leucine-rich repeat; MAPK, mitogen-activated protein kinase; MDP, muramyl dipeptide; MMP: matrix metalloproteinases, MPTP, 1-methyl-4-phenyl-1,2,3,6-tetrahydropyridine; MS, multiple sclerosis; MSU, monosodium urate; MWS, Muckle-Wells syndrome; NF-κB, nuclear factor κB; NLR, NOD-like receptor; NOD, nucleotide-binding oligomerization domain; NOMID, neonatal-onset multisystem inflammatory disease; OA, osteoarthritis; PAMP, pathogen-associated molecular pattern; PD, Parkinson's disease; PPi, pyrophosphate; PS1, presenilin 1; RA, rheumatoid arthritis; SAA, serum amyloid A protein; sIL-18BP, IL-18-binding protein; SOD1, Cu/Zn superoxide dismutase; TBI, traumatic brain injury; TLR, Toll-like receptor; TNF, tumor necrosis factor; UC, ulcerative colitis.

ACKNOWLEDGMENTS

We thank A. Bredan for editing the manuscript and P. Rottiers of Actogenix for valuable discussion. This research has been supported by Flanders Institute for Biotechnology (VIB) and several grants. European grants: EC Marie Curie Training and Mobility Program, FP6; ApopTrain, MRTN-CT-035624; EC RTD Integrated Project, FP6; Epistem, LSHB-CT-2005-019067. Belgian grants: Interuniversity Attraction Pools, IAP 6/18. Flemish grants: Fonds Wetenschappelijke Onderzoek Vlaanderen, 3G.0218.06. Ghent University grants: BOF-GOA–12.0505.02.

REFERENCES

1. Festjens, N., et al. 2006. Caspase-containing complexes in the regulation of cell death and inflammation. *Biol Chem* 387:1005–16.
2. Lamkanfi, M., et al. 2007. Caspases in cell survival, proliferation and differentiation. *Cell Death Differ* 14: 44–55.
3. Lamkanfi, M., et al. 2002. Alice in caspase land. A phylogenetic analysis of caspases from worm to man. *Cell Death Differ* 9:358–61.

4. Boatright, K. M., et al. 2003. A unified model for apical caspase activation. *Mol Cell* 11:529–41.
5. Cerretti, D. P., et al. 1992. Molecular cloning of the interleukin-1 beta converting enzyme. *Science* 256:97–100.
6. Ghayur, T., et al. 1997. Caspase-1 processes IFN-gamma-inducing factor and regulates LPS-induced IFN-gamma production. *Nature* 386:619–23.
7. Schmitz, J., et al. 2005. IL-33, an interleukin-1-like cytokine that signals via the IL-1 receptor-related protein ST2 and induces T helper type 2-associated cytokines. *Immunity* 23:479–90.
8. Thornberry, N. A., et al. 1992. A novel heterodimeric cysteine protease is required for interleukin-1 beta processing in monocytes. *Nature* 356:763–74.
9. Kuida, K., et al. 1995. Altered cytokine export and apoptosis in mice deficient in interleukin-1 beta converting enzyme. *Science* 267: 2000–3.
10. Li, P., et al. 1995. Mice deficient in IL-1 beta-converting enzyme are defective in production of mature IL-1 beta and resistant to endotoxic shock. *Cell* 80:401–11.
11. Martinon, F., and Tschopp, J. 2007. Inflammatory caspases and inflammasomes: Master switches of inflammation. *Cell Death Differ* 14: 10–22.
12. Agostini, L., et al. 2004. NALP3 forms an IL-1beta-processing inflammasome with increased activity in Muckle-Wells autoinflammatory disorder. *Immunity* 20:319–25.
13. Martinon, F., Burns, K., and Tschopp, J. 2002. The inflammasome: A molecular platform triggering activation of inflammatory caspases and processing of proIL-beta. *Mol Cell* 10:417–26.
14. Amer, A., et al. 2006. Regulation of Legionella phagosome maturation and infection through flagellin and host Ipaf. *J Biol Chem* 281: 35217–23.
15. Boyden, E. D., and Dietrich, W. F. 2006. Nalp1b controls mouse macrophage susceptibility to anthrax lethal toxin. *Nat Genet* 38: 240–4.
16. Franchi, L., et al. 2006. Cytosolic flagellin requires Ipaf for activation of caspase-1 and interleukin 1beta in salmonella-infected macrophages. *Nat Immunol* 7: 576–82.
17. Kanneganti, T. D., et al. 2006. Critical role for cryopyrin/Nalp3 in activation of caspase-1 in response to viral infection and double-stranded RNA. *J Biol Chem* 281: 36560–8.
18. Kanneganti, T. D., et al. 2006. Bacterial RNA and small antiviral compounds activate caspase-1 through cryopyrin/Nalp3. *Nature* 440:233–6.
19. Mariathasan, S., et al. 2006. Cryopyrin activates the inflammasome in response to toxins and ATP. *Nature* 440: 228–32.
20. Martinon, F., et al. 2006. Gout-associated uric acid crystals activate the NALP3 inflammasome. *Nature* 440:237–41.
21. Miao, E. A., et al. 2006. Cytoplasmic flagellin activates caspase-1 and secretion of interleukin 1beta via Ipaf. *Nat Immunol* 7:569–75.
22. Martinon, F., and Tschopp, J. 2005. NLRs join TLRs as innate sensors of pathogens. *Trends Immunol* 26:447–54.
23. Mariathasan, S., et al. 2004. Differential activation of the inflammasome by caspase-1 adaptors ASC and Ipaf. *Nature* 430:213–8.
24. Ting, J. P., and Davis, B. K. 2005. CATERPILLER: A novel gene family important in immunity, cell death, and diseases. *Annu Rev Immunol* 23:387–414.
25. Puren, A. J., Fantuzzi, G., and Dinarello, C. A. 1999. Gene expression, synthesis, and secretion of interleukin 18 and interleukin 1beta are differentially regulated in human blood mononuclear cells and mouse spleen cells. *Proc Natl Acad Sci USA* 96: 2256–61.
26. Kanneganti, T. D., et al. 2007. Pannexin-1-mediated recognition of bacterial molecules activates the cryopyrin inflammasome independent of Toll-like receptor signaling. *Immunity* 26:433–43.

27. Dinarello, C. A. 1996. Biologic basis for interleukin-1 in disease. *Blood* 87: 2095–147.

28. Dinarello, C. A. 1998. Interleukin-1, interleukin-1 receptors and interleukin-1 receptor antagonist. *Int Rev Immunol* 16:457–99.

29. Bergman, M. R., et al. 2003. A functional activating protein 1 (AP-1) site regulates matrix metalloproteinase 2 (MMP-2) transcription by cardiac cells through interactions with JunB-Fra1 and JunB-FosB heterodimers. *Biochem J* 369:485–96.

30. Luo, L., Cruz, T., and McCulloch, C. 1997. Interleukin 1-induced calcium signalling in chondrocytes requires focal adhesions. *Biochem J* 324 (Pt 2):653–8.

31. Matthews, J. S., and O'Neill, L. A. 1999. Distinct roles for p42/p44 and p38 mitogen-activated protein kinases in the induction of IL-2 by IL-1. *Cytokine* 11: 643–55.

32. Palsson, E. M., et al. 2000. Divergent roles for Ras and Rap in the activation of p38 mitogen-activated protein kinase by interleukin-1. *J Biol Chem* 275:7818–25.

33. Shirakawa, F., et al. 1989. Interleukin 1 and cyclic AMP induce kappa immunoglobulin light-chain expression via activation of an NF-kappa B-like DNA-binding protein. *Mol Cell Biol* 9:959–64.

34. Shirakawa, F., and Mizel, S. B. 1989. In vitro activation and nuclear translocation of NF-kappa B catalyzed by cyclic AMP-dependent protein kinase and protein kinase C. *Mol Cell Biol* 9:2424–30.

35. Zhu, P., et al. 1998. Regulation of interleukin 1 signalling through integrin binding and actin reorganization: Disparate effects on NF-kappaB and stress kinase pathways. *Biochem J* 330 (Pt 2):975–81.

36. Kersse, K., et al. 2007. A phylogenetic and functional overview of inflammatory caspases and caspase-1-related CARD-only proteins. *Biochem Soc Trans* 35:1508–11.

37. Druilhe, A., et al. 2001. Regulation of IL-1beta generation by pseudo-ICE and ICE-BERG, two dominant negative caspase recruitment domain proteins. *Cell Death Differ* 8:649–57.

38. Humke, E. W., et al. 2000. ICEBERG: A novel inhibitor of interleukin-1beta generation. *Cell* 103:99–111.

39. Lamkanfi, M., et al. 2004. INCA, a novel human caspase recruitment domain protein that inhibits interleukin-1beta generation. *J Biol Chem* 279: 51729–38.

40. Lee, S. H., Stehlik, C., and Reed, J. C. 2001. Cop, a caspase recruitment domain-containing protein and inhibitor of caspase-1 activation processing. *J Biol Chem* 276: 34495–500.

41. Colotta, F., et al. 1993. Interleukin-1 type II receptor: A decoy target for IL-1 that is regulated by IL-4. *Science* 261:472–5.

42. Lang, D., et al. 1998. The type II IL-1 receptor interacts with the IL-1 receptor accessory protein: A novel mechanism of regulation of IL-1 responsiveness. *J Immunol* 161:6871–7.

43. Giri, J. G., et al. 1994. Elevated levels of shed type II IL-1 receptor in sepsis. Potential role for type II receptor in regulation of IL-1 responses. *J Immunol* 153:5862–9.

44. Smith, D. E., et al. 2003. The soluble form of IL-1 receptor accessory protein enhances the ability of soluble type II IL-1 receptor to inhibit IL-1 action. *Immunity* 18:87–96.

45. Carter, D. B., et al. 1990. Purification, cloning, expression and biological characterization of an interleukin-1 receptor antagonist protein. *Nature* 344: 633–8.

46. Hannum, C. H., et al. 1990. Interleukin-1 receptor antagonist activity of a human interleukin-1 inhibitor. *Nature* 343:336–40.

47. Dao, T., Mehal, W. Z., and Crispe, I. N. 1998. IL-18 augments perforin-dependent cytotoxicity of liver NK-T cells. *J Immunol* 161:2217–22.

48. Micallef, M. J., et al. 1996. Interferon-gamma-inducing factor enhances T helper 1 cytokine production by stimulated human T cells: Synergism with interleukin-12 for interferon-gamma production. *Eur J Immunol* 26:1647–51.

49. Okamura, H., et al. 1995. Cloning of a new cytokine that induces IFN-gamma production by T cells. *Nature* 378:88–91.
50. Yoshimoto, T., et al. 1998. IL-12 up-regulates IL-18 receptor expression on T cells, Th1 cells, and B cells: Synergism with IL-18 for IFN-gamma production. *J Immunol* 161:3400–7.
51. Horwood, N. J., et al. 1998. Interleukin 18 inhibits osteoclast formation via T cell production of granulocyte macrophage colony-stimulating factor. *J Clin Invest* 101:595–603.
52. Olee, T., et al. 1999. IL-18 is produced by articular chondrocytes and induces proinflammatory and catabolic responses. *J Immunol* 162:1096–100.
53. Hoshino, K., et al. 1999. Cutting edge: Generation of IL-18 receptor-deficient mice: Evidence for IL-1 receptor-related protein as an essential IL-18 binding receptor. *J Immunol* 162:5041–4.
54. Parnet, P., et al. 1996. IL-1Rrp is a novel receptor-like molecule similar to the type I interleukin-1 receptor and its homologues T1/ST2 and IL-1R AcP. *J Biol Chem* 271:3967–70.
55. Torigoe, K., et al. 1997. Purification and characterization of the human interleukin-18 receptor. *J Biol Chem* 272:25737–42.
56. Barbulescu, K., et al. 1998. IL-12 and IL-18 differentially regulate the transcriptional activity of the human IFN-gamma promoter in primary CD4+ T lymphocytes. *J Immunol* 160:3642–7.
57. Kalina, U., et al. 2000. IL-18 activates STAT3 in the natural killer cell line 92, augments cytotoxic activity, and mediates IFN-gamma production by the stress kinase p38 and by the extracellular regulated kinases p44erk-1 and p42erk-21. *J Immunol* 165:1307–13.
58. Matsumoto, S., et al. 1997. Interleukin-18 activates NF-kappaB in murine T helper type 1 cells. *Biochem Biophys Res Commun* 234:454–7.
59. Nakahira, M., et al. 2002. Synergy of IL-12 and IL-18 for IFN-gamma gene expression: IL-12-induced STAT4 contributes to IFN-gamma promoter activation by up-regulating the binding activity of IL-18-induced activator protein 1. *J Immunol* 168:1146–53.
60. Robinson, D., et al. 1997. IGIF does not drive Th1 development but synergizes with IL-12 for interferon-gamma production and activates IRAK and NFkappaB. *Immunity* 7:571–81.
61. Johansen, C., et al. 2007. The activity of caspase-1 is increased in lesional psoriatic epidermis. *J Invest Dermatol* 127:2857–64.
62. Niyonsaba, F., et al. 2005. The human beta-defensins (-1, -2, -3, -4) and cathelicidin LL-37 induce IL-18 secretion through p38 and ERK MAPK activation in primary human keratinocytes. *J Immunol* 175:1776–84.
63. Aizawa, Y., et al. 1999. Cloning and expression of interleukin-18 binding protein. *FEBS Lett* 445:338–42.
64. Novick, D., et al. 1999. Interleukin-18 binding protein: A novel modulator of the Th1 cytokine response. *Immunity* 10:127–36.
65. Kim, S. H., et al. 2002. Identification of amino acid residues critical for biological activity in human interleukin-18. *J Biol Chem* 277:10998–1003.
66. Kim, S. H., et al. 2000. Structural requirements of six naturally occurring isoforms of the IL-18 binding protein to inhibit IL-18. *Proc Natl Acad Sci USA* 97:1190–5.
67. Lawrence, R. C., et al. 1998. Estimates of the prevalence of arthritis and selected musculoskeletal disorders in the United States. *Arthritis Rheum* 41:778–99.
68. Felson, D. T. 2006. Clinical practice. Osteoarthritis of the knee. *N Engl J Med* 354:841–8.
69. Fernandes, J. C., Martel-Pelletier, J., and Pelletier, J. P. 2002. The role of cytokines in osteoarthritis pathophysiology. *Biorheology* 39:237–46.
70. Goldring, M. B. 1999. The role of cytokines as inflammatory mediators in osteoarthritis: Lessons from animal models. *Connect Tissue Res* 40:1–11.

71. Goldring, M. B. 2000. Osteoarthritis and cartilage: The role of cytokines. *Curr Rheumatol Rep* 2:459–65.
72. Pelletier, J. P., et al. 1995. Synthesis of metalloproteases and interleukin 6 (IL-6) in human osteoarthritic synovial membrane is an IL-1 mediated process. *J Rheumatol* Suppl 43:109–14.
73. Vuolteenaho, K., et al. 2002. Effects of TNFalpha-antagonists on nitric oxide production in human cartilage. *Osteoarthritis Cartilage* 10:327–32.
74. Gu, Y., et al. 1997. Activation of interferon-gamma inducing factor mediated by interleukin-1beta converting enzyme. *Science* 275:206–9.
75. Saha, N., et al. 1999. Interleukin-1beta-converting enzyme/caspase-1 in human osteoarthritic tissues: Localization and role in the maturation of interleukin-1beta and interleukin-18. *Arthritis Rheum* 42:1577–87.
76. Futani, H., et al. 2002. Relation between interleukin-18 and PGE2 in synovial fluid of osteoarthritis: A potential therapeutic target of cartilage degradation. *J Immunother* 25 (Suppl 1):561–4.
77. Nouri, A. M., Panayi, G. S., and Goodman, S. M. 1984. Cytokines and the chronic inflammation of rheumatic disease. I. The presence of interleukin-1 in synovial fluids. *Clin Exp Immunol* 55:295–302.
78. Rudolphi, K., et al. 2003. Pralnacasan, an inhibitor of interleukin-1beta converting enzyme, reduces joint damage in two murine models of osteoarthritis. *Osteoarthritis Cartilage* 11:738–46.
79. Caron, J. P., et al. 1996. Chondroprotective effect of intraarticular injections of interleukin-1 receptor antagonist in experimental osteoarthritis. Suppression of collagenase-1 expression. *Arthritis Rheum* 39:1535–44.
80. Chevalier, X., et al. 2005. Safety study of intraarticular injection of interleukin 1 receptor antagonist in patients with painful knee osteoarthritis: A multicenter study. *J Rheumatol* 32:1317–23.
81. Evans, C. H., et al. 2004. Osteoarthritis gene therapy. *Gene Ther* 11:379–89.
82. Gouze, J. N., et al. 2003. A comparative study of the inhibitory effects of interleukin-1 receptor antagonist following administration as a recombinant protein or by gene transfer. *Arthritis Res Ther* 5:R301–9.
83. Iqbal, I., and Fleischmann, R. 2007. Treatment of osteoarthritis with anakinra. *Curr Rheumatol Rep* 9:31–5.
84. Zhang, X., Mao, Z., and Yu, C. 2004. Suppression of early experimental osteoarthritis by gene transfer of interleukin-1 receptor antagonist and interleukin-10. *J Orthop Res* 22:742–50.
85. Firestein, G. S. 2003. Evolving concepts of rheumatoid arthritis. *Nature* 423:356–61.
86. Koch, A. E. 2007. The pathogenesis of rheumatoid arthritis. *Am J Orthop* 36:5–8.
87. Gabriel, S. E. 2001. The epidemiology of rheumatoid arthritis. *Rheum Dis Clin North Am* 27:269–81.
88. Lane, N. E. 2007. Clinical practice. Osteoarthritis of the hip. *N Engl J Med* 357: 1413–21.
89. McInnes, I. B., and Schett, G. 2007. Cytokines in the pathogenesis of rheumatoid arthritis. *Nat Rev Immunol* 7:429–42.
90. Burger, D., et al. 2006. Is IL-1 a good therapeutic target in the treatment of arthritis? *Best Pract Res Clin Rheumatol* 20:879–96.
91. Kay, J., and Calabrese, L. 2004. The role of interleukin-1 in the pathogenesis of rheumatoid arthritis. *Rheumatology* (Oxford) 43(Suppl 3):iii2–iii9.
92. Westacott, C. I., et al. 1990. Synovial fluid concentration of five different cytokines in rheumatic diseases. *Ann Rheum Dis* 49:676–81.
93. Rosengren, S., et al. 2005. Expression and regulation of cryopyrin and related proteins in rheumatoid arthritis synovium. *Ann Rheum Dis* 64:708–14.

94. Gabay, C., et al. 2001. Increased production of intracellular interleukin-1 receptor antagonist type I in the synovium of mice with collagen-induced arthritis: a possible role in the resolution of arthritis. *Arthritis Rheum* 44:451–62.

95. Van Lent, P. L., et al. 1995. Major role for interleukin 1 but not for tumor necrosis factor in early cartilage damage in immune complex arthritis in mice. *J Rheumatol* 22:2250–8.

96. van den Berg, W. B., et al. 1999. Role of tumour necrosis factor alpha in experimental arthritis: Separate activity of interleukin 1beta in chronicity and cartilage destruction. *Ann Rheum Dis* 58(Suppl 1):140–8.

97. Milner, J. M., and Cawston, T. E. 2005. Matrix metalloproteinase knockout studies and the potential use of matrix metalloproteinase inhibitors in the rheumatic diseases. *Curr Drug Targets Inflamm Allergy* 4:363–75.

98. Ma, Y., et al. 1998. Altered susceptibility to collagen-induced arthritis in transgenic mice with aberrant expression of interleukin-1 receptor antagonist. *Arthritis Rheum* 41:1798–865.

99. Nakae, S., et al. 2001. IL-1 enhances T cell-dependent antibody production through induction of CD40 ligand and OX40 on T cells. *J Immunol* 167:90–7.

100. Gracie, J. A., et al. 1999. A proinflammatory role for IL-18 in rheumatoid arthritis. *J Clin Invest* 104:1393–401.

101. Joosten, L. A., et al. 2003. Association of interleukin-18 expression with enhanced levels of both interleukin-1beta and tumor necrosis factor alpha in knee synovial tissue of patients with rheumatoid arthritis. *Arthritis Rheum* 48:339–47.

102. Rooney, T., et al. 2004. Synovial tissue interleukin-18 expression and the response to treatment in patients with inflammatory arthritis. *Ann Rheum Dis* 63:1393–8.

103. Yamamura, M., et al. 2001. Interferon-gamma-inducing activity of interleukin-18 in the joint with rheumatoid arthritis. *Arthritis Rheum* 44:275–85.

104. Petrovic-Rackov, L., and Pejnovic, N. 2006. Clinical significance of IL-18, IL-15, IL-12 and TNF-alpha measurement in rheumatoid arthritis. *Clin Rheumatol* 25:448–52.

105. Wei, X. Q., et al. 2001. Reduced incidence and severity of collagen-induced arthritis in mice lacking IL-18. *J Immunol* 166:517–21.

106. Joosten, L. A., et al. 2000. An IFN-gamma-independent proinflammatory role of IL-18 in murine streptococcal cell wall arthritis. *J Immunol* 165:6553–8.

107. Pittoni, V., et al. 2002. Anti-tumour necrosis factor (TNF) alpha treatment of rheumatoid arthritis (infliximab) selectively down regulates the production of interleukin (IL) 18 but not of IL12 and IL13. *Ann Rheum Dis* 61:723–5.

108. Tanaka, M., et al. 2001. Mature form of interleukin 18 is expressed in rheumatoid arthritis synovial tissue and contributes to interferon-gamma production by synovial T cells. *J Rheumatol* 28:1779–87.

109. Dai, S. M., et al. 2007. Cellular targets of interleukin-18 in rheumatoid arthritis. *Ann Rheum Dis* 66:1411–8.

110. Pavelka, K., et al. 2002. Clinical effects of pralnacasan (PRAL), an orally-active interleukin-1b converting enzyme (ICE) inhibitor, in a 285 patient PhII trial in rheumatoid arthritis (RA). *Arthritis Rheum* 46:LB02.

111. Wannamaker, W., et al. 2007. (S)-1-((S)-2-{[1-(4-amino-3-chloro-phenyl)-methanoyl]-amino}-3,3-dimethyl-butanoyl)-pyrrolidine-2-carboxylicacid((2R,3S)-2-ethoxy-5-oxo-tetrahydro-furan-3-yl)-amide (VX-765), an orally available selective interleukin (IL)-converting enzyme/caspase-1 inhibitor, exhibits potent anti-inflammatory activities by inhibiting the release of IL-1beta and IL-18. *J Pharmacol Exp Ther* 321:509–16.

112. Ku, G., Ford, P., and Raybuck, S. A. 2002. Selective interleukin-1beta-converting enzyme (ICE/CASPASE-1) inhibition with pralnacasan (HMR 3480/VX-740) reduces inflammation and joint destruction in murine type II collegenase-induced arthritis. *Arthritis Rheum* 44:5241.

48   Design of Caspase Inhibitors as Potential Clinical Agents

113. Arner, E. C., et al. 1995. Interleukin-1 receptor antagonist inhibits proteoglycan break-
     down in antigen induced but not polycation induced arthritis in the rabbit. *J Rheumatol*
     22:1338–46.
114. Joosten, L. A., et al. 1999. IL-1 alpha beta blockade prevents cartilage and bone
     destruction in murine type II collagen-induced arthritis, whereas TNF-alpha block-
     ade only ameliorates joint inflammation. *J Immunol* 163:5049–55.
115. Joosten, L. A., et al. 1996. Anticytokine treatment of established type II collagen-
     induced arthritis in DBA/1 mice. A comparative study using anti-TNF alpha, anti-
     IL-1 alpha/beta, and IL-1Ra. *Arthritis Rheum* 39:797–809.
116. Botsios, C., et al. 2007. [Anakinra, a recombinant human IL-1 receptor antago-
     nist, in clinical practice. Outcome in 60 patients with severe rheumatoid arthritis].
     Reumatismo 59:32–7.
117. Bresnihan, B., et al. 1998. Treatment of rheumatoid arthritis with recombinant human
     interleukin-1 receptor antagonist. *Arthritis Rheum* 41:2196–204.
118. Cohen, S., et al. 2002. Treatment of rheumatoid arthritis with anakinra, a recom-
     binant human interleukin-1 receptor antagonist, in combination with methotrexate:
     Results of a twenty-four-week, multicenter, randomized, double-blind, placebo-con-
     trolled trial. *Arthritis Rheum* 46:614–24.
119. Cohen, S. B., Woolley, J. M., and Chan, W. 2003. Interleukin 1 receptor antago-
     nist anakinra improves functional status in patients with rheumatoid arthritis.
     J Rheumatol 30:225–31.
120. Jiang, Y., et al. 2000. A multicenter, double-blind, dose-ranging, randomized, pla-
     cebo-controlled study of recombinant human interleukin-1 receptor antagonist in
     patients with rheumatoid arthritis: Radiologic progression and correlation of Genant
     and Larsen scores. *Arthritis Rheum* 43:1001–9.
121. Nixon, R., Bansback, N., and Brennan, A. 2007. The efficacy of inhibiting tumour
     necrosis factor alpha and interleukin 1 in patients with rheumatoid arthritis: A meta-
     analysis and adjusted indirect comparisons. *Rheumatology* (Oxford), 46:1146–7.
122. Dinarello, C. A. 2004. Therapeutic strategies to reduce IL-1 activity in treating local
     and systemic inflammation. *Curr Opin Pharmacol* 4:378–85.
123. Toussirot, E., and Wendling, D. 2007. The use of TNF-alpha blocking agents in rheu-
     matoid arthritis: An update. *Expert Opin Pharmacother* 8:2089–107.
124. Plater-Zyberk, C., et al. 2001. Therapeutic effect of neutralizing endogenous IL-18
     activity in the collagen-induced model of arthritis. *J Clin Invest* 108:1825–32.
125. Mikuls, T. R., and Saag, K. G. 2006. New insights into gout epidemiology. *Curr Opin
     Rheumatol* 18:199–203.
126. Wallace, K. L., et al. 2004. Increasing prevalence of gout and hyperuricemia over 10
     years among older adults in a managed care population. *J Rheumatol* 31:1582–7.
127. Dalbeth, N., and Haskard, D. O. 2005. Mechanisms of inflammation in gout.
     Rheumatology (Oxford) 44:1090–6.
128. Wu, X. W., et al. 1989. Urate oxidase: Primary structure and evolutionary implica-
     tions. *Proc Natl Acad Sci USA* 86:4412–6.
129. Choi, H. K., et al. 2005. Obesity, weight change, hypertension, diuretic use, and
     risk of gout in men: The health professionals follow-up study. *Arch Intern Med*
     165:742–8.
130. Choi, H. K., and Curhan, G. 2005. Gout: Epidemiology and lifestyle choices. *Curr
     Opin Rheumatol* 17:341–5.
131. Derfus, B., et al. 1998. Human osteoarthritic cartilage matrix vesicles generate both
     calcium pyrophosphate dihydrate and apatite in vitro. *Calcif Tissue Int* 63:258–62.
132. Pay, S., and Terkeltaub, R. 2003. Calcium pyrophosphate dihydrate and hydroxyapa-
     tite crystal deposition in the joint: New developments relevant to the clinician. *Curr
     Rheumatol Rep* 5:235–43.

133. Pritzker, K. P., Cheng, P. T., and Renlund, R. C. 1988. Calcium pyrophosphate crystal deposition in hyaline cartilage. Ultrastructural analysis and implications for pathogenesis. *J Rheumatol* 15:828–35.
134. Johnson, K., and Terkeltaub, R. 2005. Inorganic pyrophosphate (PPI) in pathologic calcification of articular cartilage. *Front Biosci* 10:988–97.
135. Di Giovine, F. S., et al. 1987. Interleukin 1 (IL 1) as a mediator of crystal arthritis. Stimulation of T cell and synovial fibroblast mitogenesis by urate crystal-induced IL 1. *J Immunol* 138:3213–8.
136. Inokuchi, T., et al. 2006. Plasma interleukin (IL)-18 (interferon-gamma-inducing factor) and other inflammatory cytokines in patients with gouty arthritis and monosodium urate monohydrate crystal-induced secretion of IL-18. *Cytokine* 33:21–7.
137. So, A., et al. 2007. A pilot study of IL-1 inhibition by anakinra in acute gout. *Arthritis Res Ther* 9:1228.
138. Samuels, J., and Ozen, S. 2006. Familial Mediterranean fever and the other auto-inflammatory syndromes: Evaluation of the patient with recurrent fever. *Curr Opin Rheumatol* 18:108–17.
139. Shinkai, K., McCalmont, T. H., and Leslie, K. S. 2007. Cryopyrin-associated periodic syndromes and autoinflammation. *Clin Exp Dermatol* 33:1–9.
140. Simon, A., and van der Meer, J. W. 2007. Pathogenesis of familial periodic fever syndromes or hereditary autoinflammatory syndromes. *Am J Physiol Regul Integr Comp Physiol* 292:R86–98.
141. Aksentijevich, I., et al. 2002. De novo CIAS1 mutations, cytokine activation, and evidence for genetic heterogeneity in patients with neonatal-onset multisystem inflammatory disease (NOMID): A new member of the expanding family of pyrin-associated autoinflammatory diseases. *Arthritis Rheum* 46:3340–8.
142. Dode, C., et al. 2002. New mutations of CIAS1 that are responsible for Muckle-Wells syndrome and familial cold urticaria: A novel mutation underlies both syndromes. *Am J Hum Genet* 70:1498–506.
143. Feldmann, J., et al. 2002. Chronic infantile neurological cutaneous and articular syndrome is caused by mutations in CIAS1, a gene highly expressed in polymorphonuclear cells and chondrocytes. *Am J Hum Genet* 71:198–203.
144. Hoffman, H. M., et al. 2001. Mutation of a new gene encoding a putative pyrin-like protein causes familial cold autoinflammatory syndrome and Muckle-Wells syndrome. *Nat Genet* 29:301–5.
145. Zeft, A., and Bohnsack, J. F. 2007. Cryopyrin-associated autoinflammatory syndrome: A new mutation. *Ann Rheum Dis* 66:843–4.
146. Touitou, I., et al. 2004. Infevers: An evolving mutation database for auto-inflammatory syndromes. *Hum Mutat* 24:194–8.
147. Rosengren, S., et al. 2007. Monocytes from familial cold autoinflammatory syndrome patients are activated by mild hypothermia. *J Allergy Clin Immunol* 119:991–6.
148. Hawkins, P. N., Lachmann, H. J., and McDermott, M. F. 2003. Interleukin-1-receptor antagonist in the Muckle-Wells syndrome. *N Engl J Med* 348:2583–4.
149. Duncan, J. A., et al. 2007. Cryopyrin/NALP3 binds ATP/dATP, is an ATPase, and requires ATP binding to mediate inflammatory signaling. *Proc Natl Acad Sci USA* 104:8041–6.
150. Tameling, W. I., et al. 2006. Mutations in the NB-ARC domain of I-2 that impair ATP hydrolysis cause autoactivation. *Plant Physiol* 140:1233–45.
151. Gattorno, M., et al. 2007. Pattern of interleukin-1beta secretion in response to lipopolysaccharide and ATP before and after interleukin-1 blockade in patients with CIAS1 mutations. *Arthritis Rheum* 56:3138–48.

152. Thornton, B. D., et al. 2007. Successful treatment of renal amyloidosis due to familial cold autoinflammatory syndrome using an interleukin-1 receptor antagonist. *Am J Kidney Dis* 49:477–81.

153. Arostegui, J. I., et al. 2004. Clinical and genetic heterogeneity among Spanish patients with recurrent autoinflammatory syndromes associated with the CIAS1/PYPAF1/NALP3 gene. *Arthritis Rheum* 50:4045–50.

154. Boschan, C., et al. 2006. Neonatal-onset multisystem inflammatory disease (NOMID) due to a novel S331R mutation of the CIAS1 gene and response to interleukin-1 receptor antagonist treatment. *Am J Med Genet A* 140:883–6.

155. Caroli, F., et al. 2007. Clinical and genetic characterization of Italian patients affected by CINCA syndrome. *Rheumatology* (Oxford) 46:473–8.

156. Goldbach-Mansky, R., et al. 2006. Neonatal-onset multisystem inflammatory disease responsive to interleukin-1beta inhibition. *N Engl J Med* 355:581–92.

157. Hawkins, P. N., et al. 2004. Response to anakinra in a de novo case of neonatal-onset multisystem inflammatory disease. *Arthritis Rheum* 50:2708–9.

158. Hawkins, P. N., et al. 2004. Spectrum of clinical features in Muckle-Wells syndrome and response to anakinra. *Arthritis Rheum* 50:607–12.

159. Hoffman, H. M., et al. 2004. Prevention of cold-associated acute inflammation in familial cold autoinflammatory syndrome by interleukin-1 receptor antagonist. *Lancet* 364:1779–85.

160. Leslie, K. S., et al. 2006. Phenotype, genotype, and sustained response to anakinra in 22 patients with autoinflammatory disease associated with CIAS-1/NALP3 mutations. *Arch Dermatol* 142:1591–7.

161. Ramos, E., et al. 2005. Positive clinical and biochemical responses to anakinra in a 3-yr-old patient with cryopyrin-associated periodic syndrome (CAPS). *Rheumatology* (Oxford) 44:1072–3.

162. Stack, J. H., et al. 2005. IL-converting enzyme/caspase-1 inhibitor VX-765 blocks the hypersensitive response to an inflammatory stimulus in monocytes from familial cold autoinflammatory syndrome patients. *J Immunol* 175:2630–V.

163. Onen, F. 2006. Familial Mediterranean fever. *Rheumatol Int* 26:489–96.

164. Yepiskoposyan, L., and Harutyunyan, A. 2007. Population genetics of familial Mediterranean fever: A review. *Eur J Hum Genet* 15:911–6.

165. Schaner, P., et al. 2001. Episodic evolution of pyrin in primates: Human mutations recapitulate ancestral amino acid states. *Nat Genet* 27:318–21.

166. Chae, J. J., et al. 2003. Targeted disruption of pyrin, the FMF protein, causes heightened sensitivity to endotoxin and a defect in macrophage apoptosis. *Mol Cell* 11:591–604.

167. Chae, J. J., et al. 2006. The B30.2 domain of pyrin, the familial Mediterranean fever protein, interacts directly with caspase-1 to modulate IL-1beta production. *Proc Natl Acad Sci USA* 103:9982–7.

168. Papin, S., et al. 2007. The SPRY domain of Pyrin, mutated in familial Mediterranean fever patients, interacts with inflammasome components and inhibits proIL-1beta processing. *Cell Death Differ* 14:1457–66.

169. Yu, J. W., et al. 2006. Cryopyrin and pyrin activate caspase-1, but not NF-kappaB, via ASC oligomerization. *Cell Death Differ* 13:236–49.

170. Woo, J. S., et al. 2006. Structural and functional insights into the B30.2/SPRY domain. *EMBO J* 25:1353–63.

171. Calligaris, L., et al. 2007. The efficacy of anakinra in an adolescent with colchicine-resistant familial Mediterranean fever. *Eur J Pediatr.* 167:645–696.

172. Lindor, N. M., et al. 1997. A new autosomal dominant disorder of pyogenic sterile arthritis, pyoderma gangrenosum, and acne: PAPA syndrome. *Mayo Clin Proc* 72:611–5.

173. Tallon, B., and Corkill, M. 2006. Peculiarities of PAPA syndrome. *Rheumatology* (Oxford) 45:1140–3.
174. Spencer, S., et al. 1997. PSTPIP: A tyrosine phosphorylated cleavage furrow-associated protein that is a substrate for a PEST tyrosine phosphatase. *J Cell Biol* 138:845–60.
175. Wise, C. A., et al. 2002. Mutations in CD2BP1 disrupt binding to PTP PEST and are responsible for PAPA syndrome, an autoinflammatory disorder. *Hum Mol Genet* 11:961–9.
176. Shoham, N. G., et al. 2003. Pyrin binds the PSTPIP1/CD2BP1 protein, defining familial Mediterranean fever and PAPA syndrome as disorders in the same pathway. *Proc Natl Acad Sci USA* 100:13501–6.
177. Yu, J. W., et al. 2007. Pyrin activates the ASC pyroptosome in response to engagement by autoinflammatory PSTPIP1 mutants. *Mol Cell* 28:214–27.
178. Fernandes-Alnemri, T., et al. 2007. The pyroptosome: A supramolecular assembly of ASC dimers mediating inflammatory cell death via caspase-1 activation. *Cell Death Differ* 14:1590–604.
179. Dierselhuis, M. P., et al. 2005. Anakinra for flares of pyogenic arthritis in PAPA syndrome. *Rheumatology* (Oxford) 44:406–8.
180. Podolsky, D. K. 2002. Inflammatory bowel disease. *N Engl J Med* 347:417–29.
181. Xavier, R. J., and Podolsky, D. K. 2007. Unravelling the pathogenesis of inflammatory bowel disease. *Nature* 448:427–34.
182. Sartor, R. B. 2006. Mechanisms of disease: Pathogenesis of Crohn's disease and ulcerative colitis. *Nat Clin Pract Gastroenterol Hepatol* 3:390–407.
183. Lakatos, P. L. 2006. Recent trends in the epidemiology of inflammatory bowel diseases: Up or down? *World J Gastroenterol* 12:6102–8.
184. Halme, L., et al. 2006. Family and twin studies in inflammatory bowel disease. *World J Gastroenterol* 12:3668–72.
185. Duerr, R. H., et al. 2006. A genome-wide association study identifies IL23R as an inflammatory bowel disease gene. *Science* 314:1461–3.
186. Franke, A., et al. 2007. Systematic association mapping identifies NELL1 as a novel IBD disease gene. *PLoS ONE* 2:e691.
187. Hampe, J., et al. 2007. A genome-wide association scan of nonsynonymous SNP's identifies a susceptibility variant for Crohn disease in ATG16L1. *Nat Genet* 39:207–11.
188. Hugot, J. P., et al. 2001. Association of NOD2 leucine-rich repeat variants with susceptibility to Crohn's disease. *Nature* 411:599–603.
189. Ogura, Y., et al. 2001. A frameshift mutation in NOD2 associated with susceptibility to Crohn's disease. *Nature* 411:603–6.
190. Peltekova, V. D., et al. 2004. Functional variants of OCTN cation transporter genes are associated with Crohn's disease. *Nat Genet* 36.
191. Stoll, M., et al. 2004. Genetic variation in DLG5 is associated with inflammatory bowel disease. *Nat Genet* 36:471–5.
192. Kanneganti, T. D., Lamkanfi, M., and Nunez, G. 2007. Intracellular NOD-like receptors in host defense and disease. *Immunity* 27:549–59.
193. Barnich, N., et al. 2005. Membrane recruitment of NOD2 in intestinal epithelial cells is essential for nuclear factor-{kappa}B activation in muramyl dipeptide recognition. *J Cell Biol* 170:21–6.
194. Lecine, P., et al. 2007. The NOD2-RICK complex signals from the plasma membrane. *J Biol Chem* 282:15197–207.
195. Girardin, S. E., et al. 2003. Nod2 is a general sensor of peptidoglycan through muramyl dipeptide (MDP) detection. *J Biol Chem* 278:8869–72.
196. Girardin, S. E., et al. 2003. Peptidoglycan molecular requirements allowing detection by Nod1 and Nod2. *J Biol Chem* 278:41702–8.

197. Inohara, N., et al. 2003. Host recognition of bacterial muramyl dipeptide mediated through NOD2. Implications for Crohn's disease. *J Biol Chem* 278:5509–12.

198. Ogura, Y., et al. 2001. Nod2, a Nod1/Apaf-1 family member that is restricted to monocytes and activates NF-kappaB. *J Biol Chem* 276:4812–18.

199. Ferwerda, G., et al. 2005. NOD2 and toll-like receptors are nonredundant recognition systems of *Mycobacterium tuberculosis. PLoS Pathog*, 1:279–85.

200. Kobayashi, K. S., et al. 2005. Nod2-dependent regulation of innate and adaptive immunity in the intestinal tract. *Science* 307:731–4.

201. Opitz, B., et al. 2004. Nucleotide-binding oligomerization domain proteins are innate immune receptors for internalized *Streptococcus pneumoniae. J Biol Chem* 279:36426–32.

202. Hsu, Y. M., et al. 2007. The adaptor protein CARD9 is required for innate immune responses to intracellular pathogens. *Nat Immunol* 8:198–205.

203. Park, J. H., et al. 2007. RICK/RIP2 mediates innate immune responses induced through Nod1 and Nod2 but not TLRs. *J Immunol* 178:2380–6.

204. Yang, Y., et al. 2007. NOD2 pathway activation by MDP or *Mycobacterium tuberculosis* infection involves the stable polyubiquitination of Rip2. *J Biol Chem.* 282:36223–9.

205. Underhill, D. M., and Shimada, T. 2007. A pair of 9s: It's in the CARDs. *Nat Immunol* 8:122–4.

206. Strober, W., Fuss, I., and Mannon, P. 2007. The fundamental basis of inflammatory bowel disease. *J Clin Invest* 117:514–21.

207. Li, J., et al. 2004. Regulation of IL-8 and IL-1beta expression in Crohn's disease associated NOD2/CARD15 mutations. *Hum Mol Genet* 13:1715–25.

208. Rakoff-Nahoum, S., et al. 2004. Recognition of commensal microflora by toll-like receptors is required for intestinal homeostasis. *Cell* 118:229–41.

209. Hedl, M., et al. 2007. Chronic stimulation of Nod2 mediates tolerance to bacterial products. *Proc Natl Acad Sci USA* 104:19440–5.

210. Uehara, A., et al. 2007. Various human epithelial cells express functional Toll-like receptors, NOD1 and NOD2 to produce anti-microbial peptides, but not proinflammatory cytokines. *Mol Immunol* 44:3100–11.

211. Yang, Z., et al. 2007. NOD2 transgenic mice exhibit enhanced MDP-mediated down-regulation of TLR2 responses and resistance to colitis induction. *Gastroenterology* 133:1510–12.

212. Maeda, S., et al. 2005. NOD2 mutation in Crohn's disease potentiates NF-kappaB activity and IL-1beta processing. *Science* 307:734–8.

213. Damiano, J. S., et al. 2004. Heterotypic interactions among NACHT domains: Implications for regulation of innate immune responses. *Biochem J* 381:213–9.

214. Pan, Q., et al. 2007. MDP-induced interleukin-1beta processing requires NOD2 and CIAS1/NALP3. *J Leukoc Biol* 82:177–83.

215. Watanabe, T., et al. 2006. Nucleotide binding oligomerization domain 2 deficiency leads to dysregulated TLR2 signaling and induction of antigen-specific colitis. *Immunity* 25:473–85.

216. Netea, M. G., et al. 2005. NOD2 3020insC mutation and the pathogenesis of Crohn's disease: Impaired IL-1beta production points to a loss-of-function phenotype. *Neth J Med* 63:305–8.

217. van Heel, D. A., et al. 2005. Muramyl dipeptide and toll-like receptor sensitivity in NOD2-associated Crohn's disease. *Lancet* 365:1794–6.

218. Siegmund, B. 2002. Interleukin-1beta converting enzyme (caspase-1) in intestinal inflammation. *Biochem Pharmacol* 64:1–8.

219. Siegmund, B., et al. 2001. IL-1 beta-converting enzyme (caspase-1) in intestinal inflammation. *Proc Natl Acad Sci USA* 98:13249–54.

220. Monteleone, G., et al. 1999. Bioactive IL-18 expression is up-regulated in Crohn's disease. *J Immunol* 163:143–7.
221. Pages, F., et al. 2001. Analysis of interleukin-18, interleukin-1 converting enzyme (ICE) and interleukin-18-related cytokines in Crohn's disease lesions. *Eur Cytokine Netw* 12:97–104.
222. Pizarro, T. T., et al. 1999. IL-18, a novel immunoregulatory cytokine, is up-regulated in Crohn's disease: Expression and localization in intestinal mucosal cells. *J Immunol* 162:6829–35.
223. McAlindon, M. E., et al. 1999. Investigation of the expression of IL-1beta converting enzyme and apoptosis in normal and inflammatory bowel disease (IBD) mucosal macrophages. *Clin Exp Immunol* 116:251–7.
224. McAlindon, M. E., Hawkey, C. J., and Mahida, Y. R. 1998. Expression of interleukin 1 beta and interleukin 1 beta converting enzyme by intestinal macrophages in health and inflammatory bowel disease. *Gut* 42:214–9.
225. Carter, J. D., Valeriano, J., and Vasey, F. B. 2003. Crohn disease worsened by anakinra administration. *J Clin Rheumatol* 9:276–7.
226. Bauer, C., et al. 2007. The ICE inhibitor pralnacasan prevents DSS-induced colitis in C57BL/6 mice and suppresses IP-10 mRNA but not TNF-alpha mRNA expression. *Dig Dis Sci* 52:1642–52.
227. Loher, F., et al. 2004. The interleukin-1 beta-converting enzyme inhibitor pralnacasan reduces dextran sulfate sodium-induced murine colitis and T helper 1 T-cell activation. *J Pharmacol Exp Ther* 308:583–90.
228. Bone, R. C., et al. 1992. Definitions for sepsis and organ failure and guidelines for the use of innovative therapies in sepsis. The ACCP/SCCM Consensus Conference Committee. American College of Chest Physicians/Society of Critical Care Medicine. *Chest* 101:1644–55.
229. Hotchkiss, R. S., and Karl, I. E. 2003. The pathophysiology and treatment of sepsis. *N Engl J Med* 348:138–50.
230. Levy, M. M., et al. 2003. 2001 SCCM/ESICM/ACCP/ATS/SIS International Sepsis Definitions Conference. *Intensive Care Med* 29:530–8.
231. Vincent, J. L., et al. 2006. Sepsis in European intensive care units: Results of the SOAP study. *Crit Care Med* 34:344–53.
232. Sarkar, A., et al. 2006. Caspase-1 regulates *Escherichia coli* sepsis and splenic B cell apoptosis independently of interleukin-1beta and interleukin-18. *Am J Respir Crit Care Med* 174:1003–10.
233. Fantuzzi, G., et al. 1996. Effect of endotoxin in IL-1 beta-deficient mice. *J Immunol* 157:291–6.
234. Hotchkiss, R. S., and Nicholson, D. W. 2006. Apoptosis and caspases regulate death and inflammation in sepsis. *Nat Rev Immunol* 6:813–22.
235. Fisher, C. J., Jr., et al. 1994. Initial evaluation of human recombinant interleukin-1 receptor antagonist in the treatment of sepsis syndrome: A randomized, open-label, placebo-controlled multicenter trial. *Crit Care Med* 22:12–21.
236. Zeni, F., Freeman, B., and Natanson, C. 1997. Anti-inflammatory therapies to treat sepsis and septic shock: A reassessment. *Crit Care Med* 25:1095–100.
237. Medzhitov, R. 2007. Recognition of microorganisms and activation of the immune response. *Nature* 449:819–26.
238. Kupper, T. S., and Fuhlbrigge, R. C. 2004. Immune surveillance in the skin: Mechanisms and clinical consequences. *Nat Rev Immunol* 4:211–22.
239. Companjen, A., et al. 2004. Elevated interleukin-18 protein expression in early active and progressive plaque-type psoriatic lesions. *Eur Cytokine Netw* 15:210–6.
240. Debets, R., et al. 1997. The IL-1 system in psoriatic skin: IL-1 antagonist sphere of influence in lesional psoriatic epidermis. *J Immunol* 158:2955–63.

241. Mee, J. B., et al. 2007. The psoriatic transcriptome closely resembles that induced by interleukin-1 in cultured keratinocytes: Dominance of innate immune responses in psoriasis. *Am J Pathol* 171:32–42.

242. Kong, J., Grando, S. A., and Li, Y. C. 2006. Regulation of IL-1 family cytokines IL-1alpha, IL-1 receptor antagonist, and IL-18 by 1,25-dihydroxyvitamin D3 in primary keratinocytes. *J Immunol* 176:3780–7.

243. Shepherd, J., Little, M. C., and Nicklin, M. J. 2004. Psoriasis-like cutaneous inflammation in mice lacking interleukin-1 receptor antagonist. *J Invest Dermatol* 122:665–9.

244. Antonopoulos, C., et al. 2001. Functional caspase-1 is required for Langerhans cell migration and optimal contact sensitization in mice. *J Immunol* 166:3672–7.

245. Watanabe, H., et al. 2007. Activation of the IL-1beta-processing inflammasome is involved in contact hypersensitivity. *J Invest Dermatol* 127:1956–63.

246. Antonopoulos, C., et al. 2007. IL-18 is a key proximal mediator of contact hypersensitivity and allergen-induced Langerhans cell migration in murine epidermis. *J Leukoc Biol.* 83.

247. Jin, Y., et al. 2007. NALP1 in vitiligo-associated multiple autoimmune disease. *N Engl J Med* 356:1216–25.

248. Allan, S. M., Tyrrell, P. J., and Rothwell, N. J. 2005. Interleukin-1 and neuronal injury. *Nat Rev Immunol* 5:629–40.

249. Simi, A., et al. 2007. Interleukin-1 and inflammatory neurodegeneration. *Biochem Soc Trans* 35:1122–6.

250. Lucas, S. M., Rothwell, N. J., and Gibson, R. M. 2006. The role of inflammation in CNS injury and disease. *Br J Pharmacol* 147(Suppl 1):5232–40.

251. Felderhoff-Mueser, U., et al. 2005. IL-18: A key player in neuroinflammation and neurodegeneration? *Trends Neurosci* 28:487–93.

252. Fassbender, K., et al. 1994. Proinflammatory cytokines in serum of patients with acute cerebral ischemia: Kinetics of secretion and relation to the extent of brain damage and outcome of disease. *J Neurol Sci* 122:135–9.

253. Tarkowski, E., et al. 1995. Early intrathecal production of interleukin-6 predicts the size of brain lesion in stroke. *Stroke* 26:1393–8.

254. Friedlander, R. M., et al. 1997. Expression of a dominant negative mutant of interleukin-1 beta converting enzyme in transgenic mice prevents neuronal cell death induced by trophic factor withdrawal and ischemic brain injury. *J Exp Med* 185:933–40.

255. Hara, H., et al. 1997. Attenuation of transient focal cerebral ischemic injury in transgenic mice expressing a mutant ICE inhibitory protein. *J Cereb Blood Flow Metab* 17:370–5.

256. Schielke, G. P., et al. 1998. Reduced ischemic brain injury in interleukin-1 beta converting enzyme-deficient mice. *J Cereb Blood Flow Metab* 18:180–5.

257. Rabuffetti, M., et al. 2000. Inhibition of caspase-1-like activity by Ac-Tyr-Val-Ala-Asp-chloromethyl ketone induces long-lasting neuroprotection in cerebral ischemia through apoptosis reduction and decrease of proinflammatory cytokines. *J Neurosci* 20:4398–404.

258. Mulcahy, N. J., et al. 2003. Delayed administration of interleukin-1 receptor antagonist protects against transient cerebral ischaemia in the rat. *Br J Pharmacol* 140:471–6.

259. Clark, S. R., et al. 2007. Interleukin-1 receptor antagonist penetrates human brain at experimentally therapeutic concentrations. *J Cereb Blood Flow Metab* 28:387–94.

260. Gueorguieva, I., et al. 2007. Pharmacokinetic modelling of interleukin-1 receptor antagonist in plasma and cerebrospinal fluid of patients following subarachnoid haemorrhage. *Br J Clin Pharmacol* 65:317–25.

261. Gutierrez, E. G., Banks, W. A., and Kastin, A. J. 1994. Blood-borne interleukin-1 receptor antagonist crosses the blood-brain barrier. *J Neuroimmunol* 55:153–60.

262. Touzani, O., et al. 1999. Potential mechanisms of interleukin-1 involvement in cerebral ischaemia. *J Neuroimmunol* 100:203–15.

263. Emsley, H. C., et al. 2005. A randomised phase II study of interleukin-1 receptor antagonist in acute stroke patients. *J Neurol Neurosurg Psychiatry* 76:1366–72.

264. Ross, J., et al. 2007. A selective, non-peptide caspase-1 inhibitor, VRT-018858, markedly reduces brain damage induced by transient ischemia in the rat. *Neuropharmacology* 53:638–42.

265. Adelson, P. D., and Kochanek, P. M. 1998. Head injury in children. *J Child Neurol* 13:2–15.

266. Masson, F., et al. 2001. Epidemiology of severe brain injuries: A prospective population-based study. *J Trauma* 51:481–9.

267. Bittigau, P., et al. 2003. Neuropathological and biochemical features of traumatic injury in the developing brain. *Neurotoxicol Res* 5:475–90.

268. Chan, P. H., et al. 1994. SOD-1 transgenic mice as a model for studies of neuroprotection in stroke and brain trauma. *Ann NY Acad Sci* 738:93–103.

269. Chen, Y., et al. 1996. An experimental model of closed head injury in mice: Pathophysiology, histopathology, and cognitive deficits. *J Neurotrauma* 13:557–68.

270. Jones, N. C., et al. 2004. A detrimental role for nitric oxide synthase-2 in the pathology resulting from acute cerebral injury. *J Neuropathol Exp Neurol* 63:708–20.

271. Fink, K. B., et al. 1999. Reduction of post-traumatic brain injury and free radical production by inhibition of the caspase-1 cascade. *Neuroscience* 94:1213–8.

272. Jones, N. C., et al. 2005. Antagonism of the interleukin-1 receptor following traumatic brain injury in the mouse reduces the number of nitric oxide synthase-2-positive cells and improves anatomical and functional outcomes. *Eur J Neurosci* 22:72–8.

273. Tehranian, R., et al. 2002. Improved recovery and delayed cytokine induction after closed head injury in mice with central overexpression of the secreted isoform of the interleukin-1 receptor antagonist. *J Neurotrauma* 19:939–51.

274. Yatsiv, I., et al. 2002. Elevated intracranial IL-18 in humans and mice after traumatic brain injury and evidence of neuroprotective effects of IL-18-binding protein after experimental closed head injury. *J Cereb Blood Flow Metab* 22:971–8.

275. Tasci, A., et al. 2003. Prognostic value of interleukin-1 beta levels after acute brain injury. *Neurol Res* 25:871–4.

276. Sifringer, M., et al. 2007. Activation of caspase-1 dependent interleukins in developmental brain trauma. *Neurobiol Dis* 25:614–22.

277. Vezzani, A., and Granata, T. 2005. Brain inflammation in epilepsy: Experimental and clinical evidence. *Epilepsia* 46:1724–43.

278. Nadler, J. V. 1981. Minireview. Kainic acid as a tool for the study of temporal lobe epilepsy. *Life Sci* 29:2031–42.

279. Olney, J. W., Rhee, V., and Ho, O. L. 1974. Kainic acid: A powerful neurotoxic analogue of glutamate. *Brain Res* 77:507–12.

280. Ravizza, T., et al. 2006. Inactivation of caspase-1 in rodent brain: A novel anticonvulsive strategy. *Epilepsia* 47:1160–8.

281. De Simoni, M. G., et al. 2000. Inflammatory cytokines and related genes are induced in the rat hippocampus by limbic status epilepticus. *Eur J Neurosci* 12:2623–33.

282. Vezzani, A., et al. 2000. Powerful anticonvulsant action of IL-1 receptor antagonist on intracerebral injection and astrocytic overexpression in mice. *Proc Natl Acad Sci USA* 97:11534–9.

283. Vezzani, A., et al. 2002. Functional role of inflammatory cytokines and antiinflammatory molecules in seizures and epileptogenesis. *Epilepsia* 43(Suppl 5):30–5.

284. Glenner, G. G., and Wong, C. W. 1984. Alzheimer's disease: Initial report of the puri-fication and characterization of a novel cerebrovascular amyloid protein. *Biochem Biophys Res Commun* 120:885–90.

285. Griffin, W. S., et al. 1989. Brain interleukin 1 and S-100 immunoreactivity are elevated in Down syndrome and Alzheimer disease. *Proc Natl Acad Sci USA* 86:7611–5.

286. Ojala, J., et al. 2007. Expression of interleukin-18 is increased in the brains of Alzheimer's disease patients. *Neurobiol Aging.*

287. Wyss-Coray, T. 2006. Inflammation in Alzheimer's disease: Driving force, bystander or beneficial response? *Nat Med* 12:1005–15.

288. Zhu, S. G., et al. 1999. Increased interleukin-1beta converting enzyme expression and activity in Alzheimer disease. *J Neuropathol Exp Neurol* 58:582–7.

289. Vandenabeele, P., and Fiers, W. 1991. Is amyloidogenesis during Alzheimer's disease due to an IL-1-/IL-6-mediated "acute phase response" in the brain? *Immunol Today* 12:217–9.

290. Shaftel, S. S., et al. 2007. Sustained hippocampal IL-1 beta overexpression mediates chronic neuroinflammation and ameliorates Alzheimer plaque pathology. *J Clin Invest* 117:1595–604.

291. Fahn, S. 2003. Description of Parkinson's disease as a clinical syndrome. *Ann NY Acad Sci* 991:1–14.

292. Lang, A. E., and Lozano, A. M. 1998. Parkinson's disease. First of two parts. *N Engl J Med* 339:1044–53.

293. Lang, A. E., and Lozano, A. M. 1998. Parkinson's disease. Second of two parts. *N Engl J Med* 339:1130–43.

294. Moore, D. J., et al. 2005. Molecular pathophysiology of Parkinson's disease. *Annu Rev Neurosci* 28:57–87.

295. Uhl, G. R., Hedreen, J. C., and Price, D. L. 1985. Parkinson's disease: Loss of neu-rons from the ventral tegmental area contralateral to therapeutic surgical lesions. *Neurology* 35:1215–8.

296. Tansey, M. G., McCoy, M. K., and Frank-Cannon, T. C. 2007. Neuroinflammatory mechanisms in Parkinson's disease: Potential environmental triggers, pathways, and targets for early therapeutic intervention. *Exp Neurol* 208:1–25.

297. Blum-Degen, D., et al. 1995. Interleukin-1 beta and interleukin-6 are elevated in the cerebrospinal fluid of Alzheimer's and de novo Parkinson's disease patients. *Neurosci Lett* 202:17–20.

298. Mogi, M., et al. 1994. Interleukin-1 beta, interleukin-6, epidermal growth factor and transforming growth factor-alpha are elevated in the brain from parkinsonian patients. *Neurosci Lett* 180:147–50.

299. Ferrari, C. C., et al. 2006 Progressive neurodegeneration and motor disabilities induced by chronic expression of IL-1beta in the substantia nigra. *Neurobiol Dis* 24: 183–193.

300. Mogi, M., et al. 2000. Caspase activities and tumor necrosis factor receptor R1 (p55) level are elevated in the substantia nigra from parkinsonian brain. *J Neural Transm* 107:335–41.

301. Kahns, S., et al. 2003. Caspase-1 and caspase-8 cleave and inactivate cellular parkin. *J Biol Chem* 278:23376–80.

302. Lucking, C. B., et al. 2000. Association between early-onset Parkinson's disease and mutations in the parkin gene. *N Engl J Med* 342:1560–7.

303. Shimura, H., et al. 2000. Familial Parkinson disease gene product, parkin, is a ubiq-uitin-protein ligase. *Nat Genet* 25:302–5.

304. Du, Y., et al. 2001. Minocycline prevents nigrostriatal dopaminergic neurode-generation in the MPTP model of Parkinson's disease. *Proc Natl Acad Sci USA* 98:14669–74.

305. McFarland, H. F., and Martin, R. 2007. Multiple sclerosis: A complicated picture of autoimmunity. *Nat Immunol* 8:913–9.
306. Noseworthy, J. H. 1999. Progress in determining the causes and treatment of multiple sclerosis. *Nature* 399:A40–7.
307. Pollak, Y., et al. 2002. Experimental autoimmune encephalomyelitis-associated behavioral syndrome as a model of "depression due to multiple sclerosis." *Brain Behav Immun* 16:533–43.
308. Olsson, T. 1995. Critical influences of the cytokine orchestration on the outcome of myelin antigen-specific T-cell autoimmunity in experimental autoimmune encephalomyelitis and multiple sclerosis. *Immunol Rev* 144:245–68.
309. Bauer, J., et al. 1993. Demonstration of interleukin-1 beta in Lewis rat brain during experimental allergic encephalomyelitis by immunocytochemistry at the light and ultrastructural level. *J Neuroimmunol* 48:13–21.
310. Brosnan, C. F., et al. 1995. Cytokine localization in multiple sclerosis lesions: Correlation with adhesion molecule expression and reactive nitrogen species. *Neurology* 45:516–21.
311. Losy, J., and Niezgoda, A. 2001. IL-18 in patients with multiple sclerosis. *Acta Neurol Scand* 104:171–3.
312. Maimone, D., et al. 1991. Cytokine levels in the cerebrospinal fluid and serum of patients with multiple sclerosis. *J Neuroimmunol* 32:67–74.
313. Nicoletti, F., et al. 2001. Increased serum levels of interleukin-18 in patients with multiple sclerosis. *Neurology* 57:342–4.
314. Villarroya, H., et al. 1996. Myelin-induced experimental allergic encephalomyelitis in Lewis rats: Tumor necrosis factor alpha levels in serum and cerebrospinal fluid immunohistochemical expression in glial cells and macrophages of optic nerve and spinal cord. *J Neuroimmunol* 64:55–61.
315. Furlan, R., et al. 1999. Caspase-1 regulates the inflammatory process leading to auto-immune demyelination. *J Immunol* 163:2403–9.
316. Ming, X., et al. 2002. Caspase-1 expression in multiple sclerosis plaques and cultured glial cells. *J Neurol Sci* 197:9–18.
317. Badovinac, V., et al. 1998. Interleukin-1 receptor antagonist suppresses experimental autoimmune encephalomyelitis (EAE) in rats by influencing the activation and proliferation of encephalitogenic cells. *J Neuroimmunol* 85:87–95.
318. Martin, D., and Near, S. L. 1995. Protective effect of the interleukin-1 receptor antagonist (IL-1ra) on experimental allergic encephalomyelitis in rats. *J Neuroimmunol* 61:241–5.
319. Pollak, Y., et al. 2003. The EAE-associated behavioral syndrome. II. Modulation by anti-inflammatory treatments. *J Neuroimmunol* 137:100–8.
320. Schiffenbauer, J., et al. 2000. The induction of EAE is only partially dependent on TNF receptor signaling but requires the IL-1 type I receptor. *Clin Immunol* 95:117–23.
321. Wiemann, B., et al. 1998. Combined treatment of acute EAE in Lewis rats with TNF-binding protein and interleukin-1 receptor antagonist. *Exp Neurol* 149:455–63.
322. Wildbaum, G., et al. 1998. Neutralizing antibodies to IFN-gamma-inducing factor prevent experimental autoimmune encephalomyelitis. *J Immunol* 161:6368–74.
323. Furlan, R., et al. 2007. HSV-1-mediated IL-1 receptor antagonist gene therapy ameliorates MOG(35-55)-induced experimental autoimmune encephalomyelitis in C57BL/6 mice. *Gene Ther* 14:93–8.
324. Rowland, L. P. 1998. Diagnosis of amyotrophic lateral sclerosis. *J Neurol Sci* 160(Suppl 1): 56–24.
325. Gurney, M. E., et al. 1994. Motor neuron degeneration in mice that express a human Cu,Zn superoxide dismutase mutation. *Science* 264:1772–5.

326. Ripps, M. E., et al. 1995. Transgenic mice expressing an altered murine superoxide dismutase gene provide an animal model of amyotrophic lateral sclerosis. *Proc Natl Acad Sci USA* 92:689–93.

327. Wong, P. C., et al. 1995. An adverse property of a familial ALS-linked SOD1 mutation causes motor neuron disease characterized by vacuolar degeneration of mitochondria. *Neuron* 14:1105–16.

328. Li, M., et al. 2000. Functional role of caspase-1 and caspase-3 in an ALS transgenic mouse model. *Science* 288:335–9.

329. Yoshihara, T., et al. 2002. Differential expression of inflammation- and apoptosis-related genes in spinal cords of a mutant SOD1 transgenic mouse model of familial amyotrophic lateral sclerosis. *J Neurochem* 80:158–67.

330. Pasinelli, P., et al. 1998. Caspase-1 is activated in neural cells and tissue with amyotrophic lateral sclerosis-associated mutations in copper-zinc superoxide dismutase. *Proc Natl Acad Sci USA* 95:15763–8.

331. Pasinelli, P., et al. 2000. Caspase-1 and -3 are sequentially activated in motor neuron death in Cu,Zn superoxide dismutase-mediated familial amyotrophic lateral sclerosis. *Proc Natl Acad Sci USA* 97:13901–6.

332. Ilzecka, J., Stelmasiak, Z., and Dobosz, B. 2001. Interleukin-1beta converting enzyme/caspase-1 (ICE/caspase-1) and soluble APO-1/Fas/CD 95 receptor in amyotrophic lateral sclerosis patients. *Acta Neurol Scand* 103:255–8.

333. Martin, L. J., et al. 2000. Mechanisms for neuronal degeneration in amyotrophic lateral sclerosis and in models of motor neuron death [Review]. *Int J Mol Med* 5:3–13.

334. Friedlander, R. M., et al. 1997. Inhibition of ICE slows ALS in mice. *Nature* 388:31.

3 Role of Caspases in Apoptotic-Driven Indications

Takashi Matsui

CONTENTS

3.1 INTRODUCTION

In embryonic development, controlling the number of cells is essential to generating functional organs. While proliferation and differentiation are major embryonic development processes, apoptosis, also called programmed cell death, plays an important role in regulating the number of preexisting cells by deleting them. In addition to its functional and anatomic contributions to embryo development, apoptosis is also one of the critical pathophysiological features in many diseases, including neurodegenerative diseases, autoimmune and cardiovascular diseases, as well as leukemia. The foundations for considering apoptosis-directed interventions in these diseases have been laid by elegant studies defining the molecular initiators and effectors of apoptosis. These efforts have demonstrated that inhibiting apoptosis is a feasible approach for treating a wide range of human diseases, including heart, liver, lung, neurodegenerative, and autoimmune diseases.[1–4]

Apoptosis is regulated by activation of the cell's endogenous suicide apparatus in three phases: initiation, effector, and degradation phases.[5] Extrapolation from studies of cell fate determination in the nematode *Caenorhabditis elegans* led to the identification of caspases (cysteine aspartic acid–specific proteases) as homologues of the *C. elegans* death genes (*ced*-3 and *ced*-4)[6] and an appreciation of their pivotal role in mammalian apoptosis. These molecules are synthesized as proenzymes requiring proteolytic activation.[7] There are two fundamental ways to activate caspases: the stimulation of death receptors (DRs) (the extrinsic pathway) and the release of cytochrome c from mitochondria (the intrinsic pathway).[8] In the extrinsic pathway, engagement of DRs leads to recruitment of specific adaptor molecules, including the Fas-associated death domain protein (FADD) that binds to DRs via homologous death domains (DDs).[9] Each FADD also contains a death effector domain (DED) that binds caspase-8 through a homologous DED. Overexpression of wild-type FADD is sufficient to mediate apoptosis in a way that is phenotypically indistinguishable from that mediated by death receptors in some cell types.[9] Thus, FADD is considered to be a key mediator in the induction of apoptosis in the extrinsic pathway. On the other hand, deletion of the N-terminal DED in FADD generates a dominant negative form of FADD that binds the cytoplasmic death domains of DRs but cannot activate caspase-8.[10] Taken together, the Fas/FasL receptor system, including FADD, is an important initial regulatory apparatus at the apical point of the extrinsic pathway.

The intrinsic pathway (also known as the mitochondria pathway) of apoptosis occurs when specific stimuli lead to mitochondrial release of cytochrome c and other factors, usually in association with alterations of mitochondrial membrane potential.[8] Along with dATP, cytochrome c binds Apaf-1 in the cytosol and induces its oligomerization and subsequent recruitment of procaspase-9 (Figure 3.1).[11–13] In turn, this protein complex, known as the apoptosome, cleaves and activates effector caspases, such as caspase-3, which cleave a wide variety of noncaspase cellular proteins.[11,14,15] The precise functional contribution of each substrate to programmed cell death remains poorly defined, but some targets, such as poly(ADP-ribose) polymerase (PARP), provide a useful index of caspase-mediated apoptosis.[16] Mitochondria also release Smac/DIABLO, which inhibits endogenous inhibitors of apoptosis (IAPs)[17,18] and apoptosis-inducing factor (AIF), translocates to the nucleus and induces large-scale DNA fragmentation.[19]

In addition to the original intrinsic pathways activating caspases, it has been demonstrated that the endoplasmic reticulum (ER) can sense and induce apoptotic signals caused by the accumulation of misfolded proteins.[20] Bim-1, a proapoptotic member of the Bcl-2 family, is one of the proteins that triggers ER stress.[21] Following ER stress, caspase-12 can activate caspase-9 directly without cytochrome c release and can subsequently activate caspase-3 (Figure 3.1).[22] Using knockout mice, it has been shown that caspase-12 mediates an ER-specific apoptotic pathway and may contribute to amyloid-beta neurotoxicity.[23] On the other hand, the biological effect

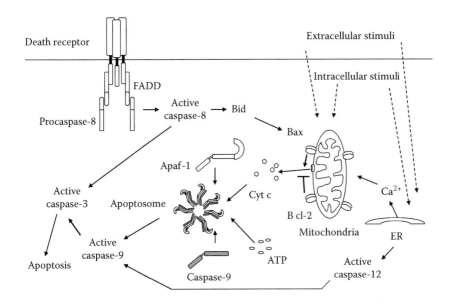

FIGURE 3.1 Overview of apoptotic pathways. Death receptors (DRs) lead to caspase-8 activation through adaptor molecules, such as FADD, with subsequent activation of caspase-3. In the mitochondrial pathway, specific stimuli induce release of cytochrome c (cyt c) from mitochondria with ATP-dependent activation of caspase-9, which then activates caspase-3. These signaling pathways are also activated by ER stress.

of human caspase-12 in apoptosis induced by ER stress is still controversial because caspase-12 has only been cloned in rodent cells.[24–26] However, it was reported that human caspase-4, a member of the caspase-1 subfamily, which includes caspase-12, may function in an ER stress-induced apoptotic pathway.[27] In other cases, ER stress induces release of cytochrome c.[28] Multidomain proapoptotic members Bax and Bak modulate apoptotic signals in both the ER and mitochondria through ER-released Ca^{2+} (Figure 3.1).[29]

Animal models of loss of function in caspases strongly suggest that caspases play an important role in many biological events modulated by apoptosis, such as organ development and ischemic injury.[30] Caspase-3-deficient mice on a 129/Sv background, born at lower frequency, are smaller than their littermates and die at 1 to 3 weeks of age.[31] Caspase-3$^{-/-}$ mice show a variety of hyperplasia and disorganized cell deployment in the brain, while no discernible histological abnormalities are observed in the heart, lung, liver, kidney, spleen, and testis.[31] Unlike on a 129/Sv background, homozygous caspase-3-deficient mice on a C57BL/6 background survive until adulthood, show minimal brain pathology,[32] and demonstrate more resistance to ischemic brain injury both *in vivo* and *in vitro*.[33] These studies confirm directly the crucial role of capase-3 not only in development but also in disease pathogenesis. While caspase-3-deficient mice exhibit profound brain defects, caspase-8-deficient mice die during fetal development due to cardiac heart failure as a result of impaired heart muscle development.[34] The phenotype of caspase-9-deficient mice resembles caspase-3-deficient mice, exhibiting embryonic lethality and defective brain development associated with decreased apoptosis.[35]

Caspases therefore play an important role in the induction of apoptosis in many settings. While inhibition of apoptosis is a pathogenic factor in most cancers, enhanced apoptosis contributes to critical cellular dysfunction in heart, liver, lung, and neuronal diseases. Here, the beneficial effects of caspase inhibitors are discussed as a novel therapeutic strategy in multiple disease settings.

3.2   CARDIOVASCULAR SYSTEM

3.2.1   MYOCARDIAL INFARCTION

3.2.1.1   Ischemic Heart Diseases and Heart Failure

Although multiple new therapeutic approaches in cardiovascular medicine have been introduced over the last decade, heart failure is still a major cause of mortality. In fact, while the number of patients with preserved cardiac function (ejection fraction) in heart failure has increased recently, mortality rates have not changed significantly in the last 15 years.[36] Despite the marked decline in acute mortality, due to improved treatments, chronic heart failure following myocardial infarction remains a major contributor to morbidity and mortality in the United States.[37] Death of heart muscle results in an irreversible decrease in cardiac function that correlates with overall morbidity and mortality in many different clinical settings. Although a number of reports in the literature describe cardiomyocyte regeneration from, in part, stem cells,[38] cardiomyocyte regeneration as a therapeutic strategy still appears inadequate to repair the injured myocardium in myocardial infarction.[39] For these reasons, there are significant clinical implications to increasing our understanding of the mechanisms that control cardiomyocyte death and identifying possible opportunities for intervention. Two major types of cell death are seen in the myocardium: necrosis and apoptosis. Necrosis is the end result of a bioenergetic catastrophe resulting from ATP depletion and is thought to be a passive form of cell death. On the other hand, apoptosis is stimulated by specific stimuli that activate intrinsic and extrinsic suicide pathways.[40] Markers of apoptosis have been identified in a wide variety of cardiac conditions in patients, including heart failure.[41–43] Cardiomyocyte apoptosis is particularly prominent in models of ischemia-reperfusion injury,[44] affecting a significant proportion of cardiomyocytes in the affected area. Also documented in human ischemia-reperfusion injury, apoptosis occurs predominantly in the border zones adjacent to infarction.[45] Effective cardioprotection would require not only cardiomyocyte survival but also preservation of function, both of which can be regulated by convergent intracellular signaling pathways in models of human cardiac diseases.[46]

3.2.1.2   Signaling Pathways and Caspase Involvement

In cardiomyocytes *in vitro*, expression of Fas and FasL are increased by hypoxia in association with an enhancement of apoptosis.[47,48] While overexpression of FADD with adenoviral gene transfer in cardiomyocytes is sufficient to induce cardiomyocyte apoptosis, expression of dominant negative FADD is also effective at inhibiting cardiomyocyte apoptosis in response to serum and oxygen deprivation.[49] FADD inhibition in cardiomyocytes blocks activation of caspase-3, -8, and -9, thereby abrogating the apoptotic response.[49] In *lpr* (lymphoproliferative) mice that lack functional Fas, both apoptosis and myocardial infarction after *in vivo* ischemia-reperfusion injury are significantly decreased.[50] The *ex vivo* ischemia-reperfusion model in the Langendorff perfused heart shows that elevated levels of Fas (CD95) ligand increase apoptosis through the receptor of CD95.[51] The *ex vivo* perfused system also demonstrated that *de novo* synthesis and release of FasL, tumor necrosis factor (TNF)-alpha, and TRAIL are increased in the reperfusion phase.[51]

In addition to the extrinsic pathway for apoptosis, a variety of strategies have shown the involvement of the intrinsic (mitochondrial) pathway in cardiac apoptosis. Antiapoptotic Bcl-family members (e.g., Bcl-2, Bcl-xL) modify apoptosis primarily at the mitochondrial membrane. Data from a murine model where Bcl-2 is overexpressed in hearts suggest that cytochrome c release from mitochondria is one of the major effectors in apoptosis induced by ischemia-reperfusion.[52,53] Acute administration of Bcl-2 by adenoviral gene transfer ameliorates ventricular remodeling and improves cardiac function for 2 to 6 weeks postinfarction.[54]

Several other proteins, such as heat shock proteins (HSPs), also regulate release of cytochrome c from mitochondria. HSPs are highly conserved chaperone proteins that are rapidly induced by environmental stresses, such as heat. Hsp27 binds to cytochrome c released by mitochondria and therefore prevents cytochrome c–mediated interaction of Apaf-1 with caspase-9 [55]; similarly, Hsp70 also suppresses apoptosis by association with Apaf-1.[56] Hsp60 appears to inhibit hypoxia-induced apoptosis by associating with and retaining the proapoptotic Bcl-family member Bax in the cytoplasm, thus preventing its insertion into the mitochondrial membrane.[57] Levels of Hsp72, the major inducible form of the HSP 70 family, increase following ischemic injury, and cardiac gene transfer of Hsp72 reduced ischemic injury at least in part through enhanced manganese superoxide dismutase (Mn-SOD) activity and improved mitochondrial respiratory function—which increased Bcl-2 levels, thus reducing cardiomyocyte apoptosis.[58]

Direct expression of caspase-3 in cardiomyocytes was found to increase infarct size.[59] Although this does not necessarily imply that inhibiting caspases would produce beneficial effects in this setting, many animal studies using caspase inhibitors have supported the beneficial effects of caspase inhibition in myocardial infarction. Two groups have used pharmacological caspase inhibitors, z-VAD-FMK and YVAD-CMK, in outbred rat models of ischemia-reperfusion injury to demonstrate significant reductions in both DNA fragmentation and infarct size at 3 and 24 hours after transient ischemia.[60,61] While the reduction in infarct size was modest (20 to 30%), these studies provide important support for the concept that apoptosis or caspase-dependent cell death contributes to clinically relevant endpoints in ischemia-reperfusion injury. In addition to the two broad-spectrum caspase inhibitors, a caspase-3/-7-selective inhibitor, MMPSI, also reduced apoptosis in a hypoxia model of cardiomyocytes *in vitro*.[62]

Coinciding with the induction of apoptosis, caspase-3 activation directly targets three components of the myofilaments: α-actin, α-actinin, and troponin T, which could also contribute to decreased cardiomyocyte function.[63] Thus, apoptotic signaling might contribute to overall cardiac dysfunction in myocardial infarction both through direct *loss* of cardiomyocytes and by *dysfunction* of surviving cardiomyocytes. Therefore, caspase inhibition is proposed to be a viable form of therapeutic intervention.

3.2.2 ATHEROSCLEROSIS

3.2.2.1 Apoptosis in Atherosclerosis

Atherosclerosis is characterized by an accumulation of lipids and fibrous elements in the arterial wall. After monocytes differentiate into macrophages at the early atherosclerotic lesion, the macrophages accumulate intracellular cholesterol via the ingestion of lipoprotein.[64,65] The rupture of the fibrous cap in the plaque plays an important role in the pathogenesis of atherosclerosis in many cardiac diseases, such as acute myocardial infarction and abdominal aortic aneurysm.[66] Therefore, reducing atherosclerotic lesions is considered an important therapeutic approach in many settings of cardiovascular diseases.

Although numerous factors are involved in the progression of atherosclerosis, multiple reports suggest that apoptosis contributes to establishing the plaque in most atherosclerotic lesions, including those in coronary, carotid, brain, and aortic arteries.[67–73] Apoptosis in macrophages is seen at both early and advanced stages in atherosclerotic lesions and plays an important role in the lesion's development,[74–76] and in humans, apoptosis is frequently identified in atherosclerotic lesions.[71–73] Additionally, terminal dUTP nick-end labeling (TUNEL) staining revealed that apoptosis of smooth muscle cells is increased in large artery aneurysms.[77] In addition to macrophage cell death, endothelial cell apoptosis is an important pathophysiological feature in atherosclerosis.[78]

3.2.2.2 Effects of Antiapoptotic Agents

The beneficial effects of antiapoptotic agents in atherosclerosis remain undefined. For instance, the apoptotic ligand TRAIL results in antiatherosclerotic effects,[79] and a recent report using z-VAD-FMK suggests that inhibiting apoptosis in atherosclerosis is detrimental in certain settings.[80] Therefore, it will be necessary to characterize the specific roles of apoptosis in each of the different cell types and at each stage of atherosclerosis during plaque development before a successful antiapoptotic approach can be considered.

3.3 LIVER DISEASES

3.3.1 APOPTOSIS IN THE LIVER

The liver has multiple biological functions that maintain whole body homeostasis in metabolism, protein synthesis, and detoxification. Although the liver can rapidly regenerate by proliferation, massive hepatocellular death caused by necrosis and apoptosis in acute and chronic liver diseases leads to liver dysfunction.[81] Although ATP-depletion-dependent necrosis is distinct from apoptosis, both types of cell death are frequently seen in liver diseases,[82,83] and it appears that apoptosis plays an important role in inflammation and fibrosis.[84,85] Given this finding, apoptosis may be a central component of pathogenesis in many liver diseases (for reviews see Malhi et al.[81] and Guicciardi and Gores[86]).

3.3.2 LIVER DISEASES AND APOPTOSIS/CASPASES

In viral hepatitis caused by hepatitis B and C viruses (HBV and HCV), expression and levels of soluble Fas correlate with disease activity and response to therapy.[87,88] In addition to this, the HCV core protein may enhance susceptibility to Fas-induced apoptosis.[89] However, it has been reported that expression of HCV proteins, including core, E1, E2, and NS2, directly or indirectly suppresses the release of cytochrome c from the mitochondria, suggesting that inhibiting the intrinsic apoptotic pathway might contribute to persistent HCV infection in the liver.[90] Apoptosis in viral hepatitis is thought to be an initial trigger of lever damage with a noninflammatory aspect. Since the subsequent inflammatory process likely results from excess apoptotic bodies, antiapoptotic agents such as caspase inhibitors, may represent a valid therapeutic strategy in liver diseases.[86] Patients with acute alcoholic hepatitis also show high levels of plasma-soluble Fas, which correlate with high expression of FADD in the liver.[91] IDN-6556, a lead compound from a class of peptidomimetic small-molecule caspase inhibitors, is being developed as a potential treatment for acute alcoholic and infectious hepatitis.[92,93] The first clinical trial of IDN-6556 involved its evaluation in normal volunteers and in patients with hepatic dysfunction,[94] and in a subsequent phase II study, IDN-6556 provided local therapeutic protection against cold-ischemia-/warm-reperfusion-mediated apoptosis and injury.[95] Information on IDN-6556 will be covered in greater detail in Chapters 8 and 10. In liver transplantation, cold ischemia/warm reperfusion is known as a major trigger of preservation injury.[96] Apoptosis induced by ischemia-reperfusion injury is observed in sinusoidal endothelial cells during early injury;[97] in a murine model of liver transplant, IDN-1965 decreased apoptosis in endothelial cells and increased survival rates.[98] Another caspase inhibitor, z-Asp-CMK, also provided protection against ischemia-reperfusion-induced apoptosis and improved liver function in animal models.[99] Even though early clinical data look very promising, more extensive studies will be required before the beneficial effects of caspase inhibitors in enhancing the success of liver transplants can be fully evaluated.

3.4 LUNG DISEASE

Cigarette smoking frequently causes chronic obstructive pulmonary disease (COPD), which is characterized by chronic inflammation of the airways and progressive destruction of lung parenchyma.[100] Apoptosis appears to be one of the critical mechanisms underlying COPD,[100,101] and a persistent increase in apoptosis occurs in COPD patients who smoke, even after smoking cessation.[102] In patients with emphysema, there is an increase in apoptosis of endothelial and mesenchymal cells, as assessed by TUNEL and DNA laddering, along with an associated increase in caspase-3 activation.[103] Mice receiving a single intratracheal injection of active caspase-3 protein show emphysematous changes with typical hallmarks of apoptosis, suggesting that caspase-induced apoptosis is sufficient for the pathogenesis observed in emphysema.[104]

Bleomycin-induced pulmonary fibrosis is a serious side effect of the chemotherapy. In a murine model of bleomycin-induced pulmonary fibrosis, Fas and FasL are over-expressed, suggesting that apoptosis plays an important role in lung pathogenesis.[105] In fact, a broad-spectrum caspase inhibitor, z-VAD-FMK, inhibits apoptosis, along with reducing caspase-1 and -3 activity, and prevents pulmonary fibrosis in a murine model of bleomycin-induced pneumopathy.[106] On the other hand, a Fas-independent pathway has also been reported in bleomycin-induced pulmonary fibrosis.[104] Therefore, studying the relative importance of apoptotic pathways in pulmonary fibrosis will be necessary.

3.5 SEPSIS

Sepsis, which is due to severe infection, is estimated to result in 751,000 cases per year in the United States, with a high mortality rate of ~29%, or approximately 215,000 deaths nationally.[107] Apoptosis and caspases strongly contribute to the pathogenesis of sepsis.[108] Hyperinflammatory reactions observed in the initial stage of sepsis are followed by an immunosuppressive state in which apoptosis plays an important role by causing depletion of immune cells.[109] During the immunosuppressive state, massive cell depletion that induces apoptosis in both the innate and adaptive immune systems (B cells and CD4+ T cells) of patients has been reported by multiple groups.[110–112] In this context, inhibiting lymphocyte apoptosis may increase patient survival rates. Supporting this, caspase inhibitors prevented lymphocyte apoptosis and improved survival rates in animal models where sepsis was induced by cecal ligation and puncture (CLP).[113,114] A recent study using short interference RNA (siRNA) against caspase-8 showed decreased apoptosis and increased survival in the CLP model, which strongly supports the model of apoptotic cell depletion during sepsis.[115] Taken together, these results suggest that sepsis may be a promising therapeutic area for a caspase inhibitor.

3.6 NEURONAL DISEASES

3.6.1 Stroke

Cerebral stroke is due to an interruption in the blood supply to the brain either by a blood clot leading to artery occlusion or from a cerebral hemorrhage. The lack of oxygen results in an ischemic response, which in turn leads to neuronal death. Animal models, such as the middle cerebral artery occlusion (MCAO) model (permanent or transient), have been used to study the effectiveness of caspase inhibitors as potential therapeutic agents for the treatment of stroke. In a murine MCAO, an irreversible caspase inhibitor, zDEVD-FMK, reduced ischemic damage as measured by infarct volume and the number of viable cells, even when administered up to 9 hours after reperfusion.[116] In a rat model, while neither z-VAD-FMK nor z-DEVD-FMK significantly reduced neuronal injury after global ischemia, both drugs significantly reduced infarction after a focal ischemia.[117] A recently developed broad-spectrum caspase inhibitor, Q-VD-OPh (quinoline-Val-Asp(Ome)-CH2-O-phenoxy), reduced cell death, resulting in significant neuroprotection in a rat model of unilateral ischemia-reperfusion brain injury.[118] Interestingly, there was a significant difference in neuroprotection between genders.[118]

Caspase-3 activity is known to be a critical downstream event in brain damage after ischemic injury.[118–120] A small, reversible caspase-3 inhibitor, M826, which selectively and potently inhibits caspase-3 enzymatic activity and apoptosis in cultured neuronal cells, significantly reduced apoptosis and brain tissue loss in the neonatal rat model of hypoxia-ischemia injury *in vivo*.[121] Although further studies are required to determine efficient timing, dose, and drug design for this therapeutic application of caspase inhibitors, these findings strongly suggest that apoptosis is a potential target for the management of stroke.

3.6.2 ALZHEIMER'S DISEASE

Alzheimer's disease is characterized by the accumulation of amyloid-β (Aβ) peptide, which is cleaved from the amyloid precursor protein (APP), as well as by extensive neuronal loss in the brain.[122,123] The Aβ peptide alone induces apoptotic cell death in cultured cells, which can be inhibited by z-VAD-FMK.[124] In addition to effectors of apoptosis, caspases (-3, -6, and -9) cleave APP at the same position to generate a C-terminal fragment (designated C31), which is found in senile plaques in Alzheimer's disease.[125] The C31 fragment is a potent inducer of apoptosis, suggesting that caspase-mediated apoptosis plays an important role in the pathogenesis of Alzheimer's.[126] The characterization of the Swedish familial mutation of APP, which results in early onset of Alzheimer's disease and in elevated production of the Aβ peptide, demonstrated that the mutation generated a consensus caspase cleavage site at a position corresponding to the β secretase cleavage site.[125] However, recent studies have suggested that Aβ is not a main pathogenic factor in Alzheimer's disease.[127] Cleavage of microtubule-associated protein tau is another consequence of caspase activation,[128] and the resulting conformational change in cleaved tau may contribute to Alzheimer's disease.[129] Although the role of apoptosis in the pathogenesis of this disease is not fully defined, caspases appear to play an important role at multiple stages. Therefore, administration of a caspase inhibitor might be a reasonable therapeutic strategy for Alzheimer's disease.

3.6.3 PARKINSON'S DISEASE

Parkinson's disease is an age-related deterioration of parts of the nervous system characterized by the loss of dopaminergic neurons and the presence of Lewy bodies in the substantia nigra. In animal models[130,131] and in patients with the disease,[132] significant apoptosis was observed in brain tissue. This suggests that caspase inhibitors may be effective treatments in Parkinson's disease, although the data are still limited.

3.7 CONCLUSIONS

Multiple lines of evidence suggest that apoptosis plays an important role in the pathogenesis of many diseases. Even though there are many unresolved questions surrounding the utility of inhibiting caspases as a therapeutic strategy, a significant body of evidence indicates that modulating apoptosis by targeting specific components of the pathway demonstrates efficacy in a variety of animal models. Furthermore, clinical data now suggest that caspase inhibition can be beneficial for certain conditions. Nevertheless, it will be important to generate both pan- and specific-caspase inhibitors that will enable a better understanding of which caspases and pathways may be optimal to inhibit in different disease settings.

REFERENCES

1. Foo, R. S., Mani, K., and Kitsis, R. N. 2005. Death begets failure in the heart. *J Clin Invest* 115:565–71.
2. Fleischer, A., et al. 2006. Modulating apoptosis as a target for effective therapy. *Mol Immunol* 43:1065–79.
3. Seibl, R., et al. 2004. Pattern recognition receptors and their involvement in the pathogenesis of arthritis. *Curr Opin Rheumatol* 16:411–18.
4. Hui, H., et al. 2004. Role of caspases in the regulation of apoptotic pancreatic islet beta-cells death. *J Cell Physiol* 200:177–200.
5. Green, D., and Kroemer, G. 1998. The central executioners of apoptosis: Caspases or mitochondria? *Trends Cell Biol* 8:267–71.
6. Yuan, J. 1996. Evolutionary conservation of a genetic pathway of programmed cell death. *J Cell Biochem* 60:4–11.
7. Alnemri, E. S., et al. 1996. Human ICE/CED-3 protease nomenclature. *Cell* 87:171.
8. Rathmell, J. C., and Thompson, C. B. 1999. The central effectors of cell death in the immune system. *Annu Rev Immunol* 17:781–828.
9. Chinnaiyan, A. M., et al. 1995. FADD, a novel death domain-containing protein, interacts with the death domain of Fas and initiates apoptosis. *Cell* 81:505–12.
10. Chinnaiyan, A., et al. 1996. FADD/MORT1 is a common mediator of CD95 (Fas/APO-1) and tumor necrosis factor receptor-induced apoptosis. *J Biol Chem* 271:4961–65.
11. Li, P., et al. 1997. Cytochrome c and dATP-dependent formation of Apaf-1/caspase-9 complex initiates an apoptotic protease cascade. *Cell* 91:479–89.
12. Jiang, X., and Wang, X. 2000. Cytochrome c promotes caspase-9 activation by inducing nucleotide binding to Apaf-1. *J Biol Chem* 275:31199–203.
13. Acehan, D., et al. 2002. Three-dimensional structure of the apoptosome: Implications for assembly, procaspase-9 binding, and activation. *Mol Cell* 9:423–32.
14. Zou, H., et al. 1997. Apaf-1, a human protein homologous to *C. elegans* CED-4, participates in cytochrome c-dependent activation of caspase-3 [see Comments]. *Cell* 90:405–13.
15. Qin, H., et al. 1999. Structural basis of procaspase-9 recruitment by the apoptotic protease-activating factor 1. *Nature* 399:549–57.
16. Tewari, M., et al. 1995. Yama/CPP32 beta, a mammalian homolog of CED-3, is a CrmA-inhibitable protease that cleaves the death substrate poly(ADP-ribose) polymerase. *Cell* 81:801–9.
17. Du, C., et al. 2000. Smac, a mitochondrial protein that promotes cytochrome c-dependent caspase activation by eliminating IAP inhibition. *Cell* 102:33–42.
18. Verhagen, A. M., et al. 2000. Identification of DIABLO, a mammalian protein that promotes apoptosis by binding to and antagonizing IAP proteins. *Cell* 102:43–53.
19. Kroemer, G., and Martin, S. J. 2005. Caspase-independent cell death. *Nat Med* 11:725–30.

20. Oyadomari, S., Araki, E., and Mori, M. 2002. Endoplasmic reticulum stress-mediated apoptosis in pancreatic beta-cells. *Apoptosis* 7:335–45.
21. Morishima, N., et al. 2004. Translocation of Bim to the endoplasmic reticulum (ER) mediates ER stress signaling for activation of caspase-12 during ER stress-induced apoptosis. *J Biol Chem* 279:50375–81.
22. Morishima, N., et al. 2002. An endoplasmic reticulum stress-specific caspase cascade in apoptosis. Cytochrome c-independent activation of caspase-9 by caspase-12. *J Biol Chem* 277:34287–94.
23. Nakagawa, T., et al. 2000. Caspase-12 mediates endoplasmic-reticulum-specific apoptosis and cytotoxicity by amyloid-beta. *Nature* 403:98–103.
24. Saleh, M., et al. 2004. Differential modulation of endotoxin responsiveness by human caspase-12 polymorphisms. *Nature* 429:75–79.
25. Rao, R. V., et al. 2001. Coupling endoplasmic reticulum stress to the cell death program. Mechanism of caspase activation. *J Biol Chem* 276:33869–74.
26. Fischer, H., et al. 2002. Human caspase 12 has acquired deleterious mutations. *Biochem Biophys Res Commun* 293:722–26.
27. Hitomi, J., et al. 2004. Involvement of caspase-4 in endoplasmic reticulum stress-induced apoptosis and Abeta-induced cell death. *J Cell Biol* 165:347–56.
28. Hacki, J., et al. 2000. Apoptotic crosstalk between the endoplasmic reticulum and mitochondria controlled by Bcl-2. *Oncogene* 19:2286–95.
29. Scorrano, L., et al. 2003. BAX and BAK regulation of endoplasmic reticulum $C_A{}^{2+}$: A control point for apoptosis. *Science* 300:135–39.
30. Los, M., Wesselborg, S., and Schulze-Osthoff, K. 1999. The role of caspases in development, immunity, and apoptotic signal transduction: Lessons from knockout mice. *Immunity* 10:629–39.
31. Kuida, K., et al. 1996. Decreased apoptosis in the brain and premature lethality in CPP32-deficient mice. *Nature* 384:368–72.
32. Leonard, J. R., et al. 2002. Strain-dependent neurodevelopmental abnormalities in caspase-3-deficient mice. *J Neuropathol Exp Neurol* 61:673–77.
33. Le, D. A., et al. 2002. Caspase activation and neuroprotection in caspase-3-deficient mice after in vivo cerebral ischemia and in vitro oxygen glucose deprivation. *Proc Natl Acad Sci USA* 99:15188–93.
34. Varfolomeev, E. E., et al. 1998. Targeted disruption of the mouse caspase 8 gene ablates cell death induction by the TNF receptors, Fas/Apo1, and DR3 and is lethal prenatally. *Immunity* 9:267–76.
35. Hakem, R., et al. 1998. Differential requirement for caspase 9 in apoptotic pathways in vivo. *Cell* 94:339–52.
36. Owan, T. E., et al. 2006. Trends in prevalence and outcome of heart failure with preserved ejection fraction. *N Engl J Med* 355:251–59.
37. Kloner, R. A., and Rezkalla, S. H. 2004. Cardiac protection during acute myocardial infarction: Where do we stand in 2004? *J Am Coll Cardiol* 44:276–86.
38. Yoon, Y. S., Lee, N., and Scadova, H. 2005. Myocardial regeneration with bone-marrow-derived stem cells. *Biol Cell* 97:253–63.
39. Rosenzweig, A. 2006. Cardiac cell therapy—Mixed results from mixed cells. *N Engl J Med* 355:1274–77.
40. Matsui, T., and Rosenzweig, A. 2003. Targeting ischemic cardiac dysfunction through gene transfer. *Curr Atheroscler Rep* 5:191–95.
41. Narula, J., et al. 1996. Apoptosis in myocytes in end-stage heart failure. *N Engl J Med* 335:1182–89.
42. Mallat, Z., et al. 1996. Evidence of apoptosis in arrhythmogenic right ventricular dysplasia. *N Engl J Med* 335:1190–96.

43. Abbate, A., et al. 2002. Persistent infarct-related artery occlusion is associated with an increased myocardial apoptosis at postmortem examination in humans late after an acute myocardial infarction. *Circulation* 106:1051–54.

44. Gottlieb, R. A., et al. 1994. Reperfusion injury induces apoptosis in rabbit cardiomyocytes. *J Clin Invest* 94:1621–28.

45. Saraste, A., et al. 1997. Apoptosis in human acute myocardial infarction. *Circulation* 95:320–23.

46. Matsui, T., and Rosenzweig, A. 2005. Convergent signal transduction pathways controlling cardiomyocyte survival and function: The role of PI 3-kinase and Akt. *J Mol Cell Cardiol* 38:63–71.

47. Tanaka, M., et al. 1994. Hypoxia induces apoptosis with enhanced expression of Fas antigen messenger RNA in cultured neonatal rat cardiomyocytes. *Circ Res* 75:426–33.

48. Stephanou, A., et al. 2001. Induction of apoptosis and Fas receptor/Fas ligand expression by ischemia/reperfusion in cardiac myocytes requires serine 727 of the STAT-1 transcription factor but not tyrosine 701. *J Biol Chem* 276:28340–47.

49. Chao, W., et al. 2002. Importance of FADD signaling in serum deprivation- and hypoxia-induced cardiomyocyte apoptosis. *J Biol Chem* 277:31639–45.

50. Lee, P., et al. 2003. Fas pathway is a critical mediator of cardiac myocyte death and MI during ischemia-reperfusion in vivo. *Am J Physiol Heart Circ Physiol* 284:H456–63.

51. Jeremias, I., et al. 2000. Involvement of CD95/Apo1/Fas in cell death after myocardial ischemia. *Circulation* 102:915–20.

52. Brocheriou, V., et al. 2000. Cardiac functional improvement by a human Bcl-2 transgene in a mouse model of ischemia/reperfusion injury. *J Gene Med* 2:326–33.

53. Chen, Z., et al. 2001. Overexpression of Bcl-2 attenuates apoptosis and protects against myocardial I/R injury in transgenic mice. *Am J Physiol Heart Circ Physiol* 280:H2313–20.

54. Chatterjee, S., et al. 2002. Viral gene transfer of the antiapoptotic factor Bcl-2 protects against chronic postischemic heart failure. *Circulation* 106(Suppl 1):I212–17.

55. Bruey, J. M., et al. 2000. Hsp27 negatively regulates cell death by interacting with cytochrome c. *Nat Cell Biol* 2:645–52.

56. Beere, H. M., et al. 2000. Heat-shock protein 70 inhibits apoptosis by preventing recruitment of procaspase-9 to the Apaf-1 apoptosome. *Nat Cell Biol* 2:469–75.

57. Gupta, S., and Knowlton, A. A. 2002. Cytosolic heat shock protein 60, hypoxia, and apoptosis. *Circulation* 106:2727–33.

58. Suzuki, K., et al. 2002. Heat shock protein 72 enhances manganese superoxide dismutase activity during myocardial ischemia-reperfusion injury, associated with mitochondrial protection and apoptosis reduction. *Circulation* 106(Suppl 1):I270–76.

59. Condorelli, G., et al. 2001. Heart-targeted overexpression of caspase-3 in mice increases infarct size and depresses cardiac function. *Proc Natl Acad Sci USA* 98:9977–82.

60. Yaoita, H., et al. 1998. Attenuation of ischemia/reperfusion injury in rats by a caspase inhibitor. *Circulation* 97:276–81.

61. Holly, T. A., et al. 1999. Caspase inhibition reduces myocyte cell death induced by myocardial ischemia and reperfusion in vivo. *J Mol Cell Cardiol* 31:1709–15.

62. Chapman, J. G., et al. 2002. A novel nonpeptidic caspase-3/7 inhibitor, (S)-(+)-5-[1-(2-methoxymethylpyrrolidinyl)sulfonyl]isatin reduces myocardial ischemic injury. *Eur J Pharmacol* 456:59–68.

63. Communal, C., et al. 2002. Functional consequences of caspase activation in cardiac myocytes. *Proc Natl Acad Sci USA* 99:6252–56.

64. Brown, M. S., and Goldstein, J. L. 1983. Lipoprotein metabolism in the macrophage: Implications for cholesterol deposition in atherosclerosis. *Annu Rev Biochem* 52:223–61.

65. Glass, C. K., and Witztum, J. L. 2001. Atherosclerosis. The road ahead. *Cell* 104:503–16.

66. Daugherty, A., and Cassis, L. A. 2002. Mechanisms of abdominal aortic aneurysm formation. *Curr Atheroscler Rep* 4:222–27.

67. Isner, J. M., et al. 1995. Apoptosis in human atherosclerosis and restenosis. *Circulation* 91:2703–11.

68. Bjorkerud, S., and Bjorkerud, B. 1996. Apoptosis is abundant in human atherosclerotic lesions, especially in inflammatory cells (macrophages and T cells), and may contribute to the accumulation of gruel and plaque instability. *Am J Pathol* 149:367–80.

69. Hara, A., Yoshimi, N., and Mori, H. 1998. Evidence for apoptosis in human intracranial aneurysms. *Neurol Res* 20:127–30.

70. Holmes, D. R., et al. 1996. Smooth muscle cell apoptosis and p53 expression in human abdominal aortic aneurysms. *Ann NY Acad Sci* 800:286–87.

71. Lopez-Candales, A., et al. 1997. Decreased vascular smooth muscle cell density in medial degeneration of human abdominal aortic aneurysms. *Am J Pathol* 150:993–1007.

72. Henderson, E. L., et al. 1999. Death of smooth muscle cells and expression of mediators of apoptosis by T lymphocytes in human abdominal aortic aneurysms. *Circulation* 99:96–104.

73. Rowe, V. L., et al. 2000. Vascular smooth muscle cell apoptosis in aneurysmal, occlusive, and normal human aortas. *J Vasc Surg* 31:567–76.

74. Kockx, M. M., and Herman, A. G. 2000. Apoptosis in atherosclerosis: Beneficial or detrimental? *Cardiovasc Res* 45:736–46.

75. Geng, Y. J., and Libby, P. 2002. Progression of atheroma: A struggle between death and procreation. *Arterioscler Thromb Vasc Biol* 22:1370–80.

76. Tabas, I. 2005. Consequences and therapeutic implications of macrophage apoptosis in atherosclerosis: The importance of lesion stage and phagocytic efficiency. *Arterioscler Thromb Vasc Biol* 25:2255–64.

77. Satta, J., Mennander, A., and Soini, Y. 2002. Increased medial TUNEL-positive staining associated with apoptotic bodies is linked to smooth muscle cell diminution during evolution of abdominal aortic aneurysms. *Ann Vasc Surg* 16:462–66.

78. DeKroon, R., et al. 2006. APOE4-VLDL inhibits the HDL-activated phosphatidylinositol 3-kinase/Akt pathway via the phosphoinositol phosphatase SHIP2. *Circ Res* 99:829–36.

79. Secchiero, P., et al. 2006. Systemic tumor necrosis factor-related apoptosis-inducing ligand delivery shows antiatherosclerotic activity in apolipoprotein E-null diabetic mice. *Circulation* 114:1522–30.

80. Martinet, W., et al. 2006. z-VAD-fmk-induced non-apoptotic cell death of macrophages: Possibilities and limitations for atherosclerotic plaque stabilization. *Autophagy* 2:312–14.

81. Malhi, H., Gores, G. J., and Lemasters, J. J. 2006. Apoptosis and necrosis in the liver: A tale of two deaths? *Hepatology* 43(Suppl 1):S31–44.

82. Lemasters, J. J. 2005. Dying a thousand deaths: Redundant pathways from different organelles to apoptosis and necrosis. *Gastroenterology* 129:351–60.

83. Jaeschke, H., and Lemasters, J. J. 2003. Apoptosis versus oncotic necrosis in hepatic ischemia/reperfusion injury. *Gastroenterology* 125:1246–57.

84. Bantel, H., et al. 2004. Detection of apoptotic caspase activation in sera from patients with chronic HCV infection is associated with fibrotic liver injury. *Hepatology* 40:1078–87.

85. Lauber, K., et al. 2003. Apoptotic cells induce migration of phagocytes via caspase-3-mediated release of a lipid attraction signal. *Cell* 113:717–30.

86. Guicciardi, M. E., and Gores, G. J. 2005. Apoptosis: A mechanism of acute and chronic liver injury. *Gut* 54:1024–33.

87. Lapinski, T. W., et al. 2004. Serum concentration of sFas and sFasL in healthy HBsAg carriers, chronic viral hepatitis B and C patients. *World J Gastroenterol* 10:3650–53.

88. Song, L. H., et al. 2004. Variations in the serum concentrations of soluble Fas and soluble Fas ligand in Vietnamese patients infected with hepatitis B virus. *J Med Virol* 73:244–49.

89. Ruggieri, A., et al. 1997. Sensitization to Fas-mediated apoptosis by hepatitis C virus core protein. *Virology* 229:68–76.

90. Machida, K., et al. 2001. Inhibition of cytochrome c release in Fas-mediated signaling pathway in transgenic mice induced to express hepatitis C viral proteins. *J Biol Chem* 276:12140–46.

91. Tagami, A., et al. 2003. Fas-mediated apoptosis in acute alcoholic hepatitis. *Hepatogastroenterology* 50:443–48.

92. Poordad, F. F. 2004. IDN-6556 Idun Pharmaceuticals Inc. *Curr Opin Invest Drugs* 5:1198–204.

93. Hoglen, N. C., et al. 2004. Characterization of IDN-6556 (3-[2-(2-tert-butyl-phenyl-aminooxalyl)-amino]-propionylamino)-4-oxo-5-(2,3,5,6-tetrafluoro-phenoxy)-pentanoic acid): A liver-targeted caspase inhibitor. *J Pharmacol Exp Ther* 309:634–40.

94. Valentino, K. L., et al. 2003. First clinical trial of a novel caspase inhibitor: Anti-apoptotic caspase inhibitor, IDN-6556, improves liver enzymes. *Int J Clin Pharmacol Ther* 41:441–49.

95. Baskin-Bey, E. S., et al. 2007. Clinical trial of the pan-caspase inhibitor, IDN-6556, in human liver preservation injury. *Am J Transplant* 7:218–25.

96. Ploeg, R. J., et al. 1993. Risk factors for primary dysfunction after liver transplantation—A multivariate analysis. *Transplantation* 55:807–13.

97. Gao, W., et al. 1998. Apoptosis of sinusoidal endothelial cells is a critical mechanism of preservation injury in rat liver transplantation. *Hepatology* 27:1652–60.

98. Natori, S., et al. 1999. Apoptosis of sinusoidal endothelial cells occurs during liver preservation injury by a caspase-dependent mechanism. *Transplantation* 68:89–96.

99. Cursio, R., et al. 1999. A caspase inhibitor fully protects rats against lethal normothermic liver ischemia by inhibition of liver apoptosis. *FASEB J* 13:253–61.

100. Demedts, I. K., et al. 2006. Role of apoptosis in the pathogenesis of COPD and pulmonary emphysema. *Respir Res* 7:53.

101. Voelkel, N. F., and Cool, C. D. 2003. Pulmonary vascular involvement in chronic obstructive pulmonary disease. *Eur Respir J Suppl* 46:28s–32s.

102. Hodge, S., et al. 2005. Increased airway epithelial and T-cell apoptosis in COPD remains despite smoking cessation. *Eur Respir J* 25:447–54.

103. Imai, K., et al. 2005. Correlation of lung surface area to apoptosis and proliferation in human emphysema. *Eur Respir J* 25:250–58.

104. Aoshiba, K., Yokohori, N., and Nagai, A. 2003. Alveolar wall apoptosis causes lung destruction and emphysematous changes. *Am J Respir Cell Mol Biol* 28:555–62.

105. Kuwano, K., et al. 1999. Essential roles of the Fas-Fas ligand pathway in the development of pulmonary fibrosis. *J Clin Invest* 104:13–19.

106. Kuwano, K., et al. 2001. Attenuation of bleomycin-induced pneumopathy in mice by a caspase inhibitor. *Am J Physiol Lung Cell Mol Physiol* 280:L316–25.

107. Angus, D. C., et al. 2001. Epidemiology of severe sepsis in the United States: Analysis of incidence, outcome, and associated costs of care. *Critical Care Med* 29:1303–10.

108. Hotchkiss, R. S., and Nicholson, D. W. 2006. Apoptosis and caspases regulate death and inflammation in sepsis. *Nat Rev Immunol* 6:813–22.

109. Hotchkiss, R. S., et al. 1999. Apoptotic cell death in patients with sepsis, shock, and multiple organ dysfunction. *Critical Care Med* 27:1230–51.

110. Hotchkiss, R. S., et al. 2001. Sepsis-induced apoptosis causes progressive profound depletion of B and CD4+ T lymphocytes in humans. *J Immunol* 166:6952–63.

111. Toti, P., et al. 2004. Spleen depletion in neonatal sepsis and chorioamnionitis. *Am J Clin Pathol* 122:765–71.

112. Felmet, K. A., et al. 2005. Prolonged lymphopenia, lymphoid depletion, and hypoprolactinemia in children with nosocomial sepsis and multiple organ failure. *J Immunol* 174:3765–72.

113. Hotchkiss, R. S., et al. 1999. Prevention of lymphocyte cell death in sepsis improves survival in mice. *Proc Natl Acad Sci USA* 96:14541–46.

114. Hotchkiss, R. S., et al. 2000. Caspase inhibitors improve survival in sepsis: A critical role of the lymphocyte. *Nature Immunol* 1:496–501.

115. Wesche-Soldato, D. E., et al. 2005. In vivo delivery of caspase-8 or Fas siRNA improves the survival of septic mice. *Blood* 106:2295–301.

116. Fink, K., et al. 1998. Prolonged therapeutic window for ischemic brain damage caused by delayed caspase activation. *J Cereb Blood Flow Metab* 18:1071–76.

117. Li, H., et al. 2000. Caspase inhibitors reduce neuronal injury after focal but not global cerebral ischemia in rats. *Stroke* 31:176–82.

118. Renolleau, S., et al. 2007. Specific caspase inhibitor Q-VD-OPh prevents neonatal stroke in P7 rat: A role for gender. *J Neurochem* 100:1062–71.

119. Schulz, J. B., et al. 1998. Extended therapeutic window for caspase inhibition and synergy with MK-801 in the treatment of cerebral histotoxic hypoxia. *Cell Death Differ* 5:847–57.

120. Asai, A., et al. 1999. High level calcineurin activity predisposes neuronal cells to apoptosis. *J Biol Chem* 274:34450–58.

121. Han, B. H., et al. 2002. Selective, reversible caspase-3 inhibitor is neuroprotective and reveals distinct pathways of cell death after neonatal hypoxic-ischemic brain injury. *J Biol Chem* 277:30128–36.

122. Katzman, R. 1986. Alzheimer's disease. *N Engl J Med* 314:964–73.

123. Kang, J., et al. 1987. The precursor of Alzheimer's disease amyloid A4 protein resembles a cell-surface receptor. *Nature* 325:733–36.

124. Xu, J., et al. 2001, Amyloid beta peptide-induced cerebral endothelial cell death involves mitochondrial dysfunction and caspase activation. *J Cereb Blood Flow Metab* 21:702–10.

125. Gervais, F. G., et al. 1999. Involvement of caspases in proteolytic cleavage of Alzheimer's amyloid-beta precursor protein and amyloidogenic A beta peptide formation. *Cell* 97:395–406.

126. Lu, D. C., et al. 2000. A second cytotoxic proteolytic peptide derived from amyloid beta-protein precursor. *Nat Med* 6:397–404.

127. Dickson, D. W. 2004. Apoptotic mechanisms in Alzheimer neurofibrillary degeneration: Cause or effect? *J Clin Invest* 114:23–27.

128. Rissman, R. A., et al. 2004. Caspase-cleavage of tau is an early event in Alzheimer's disease tangle pathology. *J Clin Invest* 114:121–30.

129. Weaver, C. L., et al. 2000. Conformational change as one of the earliest alterations of tau in Alzheimer's disease. *Neurobiol Aging* 21:719–27.

130. He, X. J., et al. 2006. Evidence of apoptosis in the subventricular zone and rostral migratory stream in the MPTP mouse model of Parkinson's disease. *J Neuropathol Exp Neurol* 65:873–82.

131. Tatton, N. A., and Kish, S. J. 1997. In situ detection of apoptotic nuclei in the substantia nigra compacta of 1-methyl-4-phenyl-1,2,3,6-tetrahydropyridine-treated mice using terminal deoxynucleotidyl transferase labelling and acridine orange staining. *Neuroscience* 77:1037–48.

132. Mochizuki, H., et al. 1996. Histochemical detection of apoptosis in Parkinson's disease. *J Neurol Sci* 137:120–23.

4 Kinetics and Catalytic Activity of Caspases

Kip A. Nalley

CONTENTS

4.1 INTRODUCTION

The mechanism by which caspases cleave substrates is a multistep process that involves several distinct steps. First, the caspase must be converted from the inactive precursor protein (the procaspase) to the active form, which then binds a protein substrate and catalyzes the subsequent cleavage of a peptide bond. This highly coordinated process results in the activation of caspases when required, yet keeps the proteins inactive to prevent aberrant cell death.

The proteolytic conversion from an inactive precursor to an active state presents several discrete points for development of caspase inhibitors or activators. Additionally, small molecules or proteins that bind to each of the states will potentially behave very differently *in vivo*. Compounds that inhibit the catalytic activity of the protein in the active state have been the most widely studied and therefore have given us the most information about the caspase active sites. However, compounds that bind to the caspase zymogen and prevent or enhance activation of the apoptotic signal are also potentially interesting. There are many examples of interacting proteins that either directly inhibit catalytic activity or indirectly inhibit caspase activity by preventing maturation of the inactive (or procaspase) zymogen to the active state. The active site inhibitory serpin-like molecules CrmA, XIAP, and p35 are competitive, active site binding proteins that behave much like small molecule active site inhibitors. In contrast, other members of the inhibitor of apoptosis (IAP) family of proteins do not directly bind to the active site but prevent activation of the caspase by interactions with and modifications of the zymogen or adaptor proteins (see reviews in Eckelman et al.[1] and Callus and Vaux[2]). This chapter will cover the mechanisms of caspase activation, their catalytic mechanism, and features that determine substrate specificity. Additionally, how each of these states can be modulated to regulate apoptosis will be discussed.

4.2 CASPASE STRUCTURE

Much of the information regarding the catalytic mechanism, as well as the determinants for substrate binding, has been determined by analysis of the structures of the caspase family members. To date, structures of caspases-1,[3–6] -2,[7] -3,[8–13] -7,[14–18] -8,[19–22] and -9[23,24] have been submitted to the protein data bank (PDB) and are accessible from the Molecular Modeling Database (MMDB) at www.ncbi.nlm.nih.gov/entrez/query.fcgi?db=Structure. Active caspases are heterodimers consisting of two large, 17–20 kDa, and two small, 10–12 kDa, subunits. The overall fold of the caspases consists of six parallel and antiparallel β-sheets intertwined with five α-helices. The topological features as well as the crystal structure of the caspase-3 catalytic domain are shown in Figure 4.1. The contacts between the individual monomers

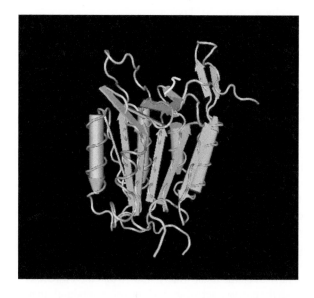

FIGURE 4.1 (See color insert.) Structure of caspase-3 monomer showing the large subunit (purple), the small subunit (blue), and DEVD-CHO (yellow) bound to the active site.

occur through interactions between the $\beta6$ strands of the polypeptides, resulting in a twelve-strand β-sheet throughout the dimer.

4.3 ZYMOGEN ACTIVATION

Caspases are prevented from aberrant activation by regulation of the inactive zymogen. Zymogen activation of proteases has been reviewed extensively by several groups.[25–27] The effector caspases-3 and -7 show an exceptional difference between the activities of the procaspase versus the active caspase of greater than 10,000-fold. In contrast, the initiator caspase zymogens are only less active by 10-fold for caspase-9 and 100-fold for caspase-8.[28–32] Although there are similarities among the mechanisms by which the various caspases are activated, there are significant differences between activation of the initiator caspases and the effector caspases. The effector caspases-3, -6, and -7 exist as inactive homodimers that are activated by limited proteolysis, freeing the active dimer from the prodomain.[10,17,33,34] In contrast, activation of the initiator and pro-inflammatory caspases occurs through dimerization of the inactive monomers to form an active multiprotein assembly that frequently contains other partner proteins.[24,28,35]

4.3.1 EFFECTOR CASPASES

Procaspase-3 monomers and presumably procaspase-7 monomers associate to form tightly bound, inactive heterodimers. The crystal structures of procaspase-7, uninhibited caspase-7, and inhibited caspase-7 show remarkable similarity in the overall fold of the enzyme (Figure 4.2a).[14–18] The conformation of residues forming the catalytic dyad are almost identical in the active form versus the zymogen. However, displacement of the active site cysteine appears to disrupt the oxyanion hole. The most evident differences between the inactive and active forms of caspase-7 are the positions of the 341- and 381-loops, which are major determinants for substrate specificity (Figure 4.2b). For consistency in discussing differences among the caspases, the caspase-1 numbering convention for labeling amino acids will be used,[36] which allows for discussion of similar and dissimilar residues among the caspases based on the relative position of the residue in the structure of the caspase. In the zymogen, the loops are disordered and extend into the solvent in contrast to the inhibitor bound enzyme, where the loops are well defined, which severely affects the position of Arg-341, the critical residue for defining the positively charged P1 and P3 pockets. In addition, other residues necessary for tight binding of substrate are disordered in the procaspase-7 structure, including Trp-340, Gln-381b, and Phe-381h (due to the larger size of the 381-loop in caspase-7, each of the inserted residues is annotated with a letter designation from Ser-381a to His-381i). The disordered structure of the zymogen disrupts the substrate binding pocket of each monomer, resulting in an inactive enzyme. The displacement of the 341-loop is due to the insertion of the intersubunit linker peptides into the active site cavity, sterically blocking the 341-loop from adopting the conformation that forms an ordered substrate binding pocket. The inserted loop occupies space that would be occupied by residues Val-334–Tyr-337 from the adjacent monomer in the active enzyme.[17,18] In addition to removing the steric interference from intersubunit loops, cleavage of procaspase-7 also allows

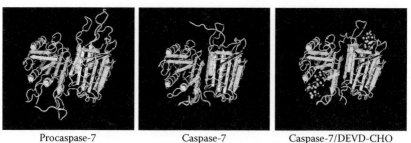

Procaspase-7   Caspase-7   Caspase-7/DEVD-CHO

(a)

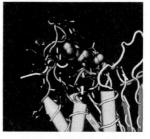

Procaspase-7   Caspase-7/DEVD-CHO

(b)

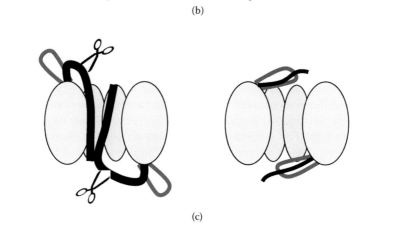

(c)

FIGURE 4.2 (See color insert.) (a) Structure of procaspase-7, apocaspase-7, and caspase 7 bound to Ac-DEVD-CHO. (b) Close-up view of the conformational change induced near the active site of caspase-7 by binding Ac-DEVD-CHO. (c) Cartoon representation of the conformational change induced by cleavage of the prodomain.

for interactions between the newly formed N- and C-termini that stabilize the active enzyme conformation.[17] These ionic interactions in combination with the 381-loop form a stable complex that does not allow the N-termini of the small subunits to fold back into the center cavity of the active enzyme (Figure 4.2c).

Since the catalytic domains of caspases-3, -6, and -7 have a high degree of sequence identity and the folds of the active caspases are nearly identical, it is reasonable to assume that the activation of caspases-3 and -6 is accomplished in a manner similar to that of caspase-7. In fact, the residues in the intersubunit linkers are conserved or are replaced with residues with similar characteristics (see sequence alignments in Fuentes-Prior and Salvesen[27] and Earnshaw et al.[37]). Therefore, it is quite likely that the substrate binding loops of procaspases-3 and -6 are also disordered due to insertion of the intersubunit linker into the central cavity. Spectroscopic investigations into the local environment around Trp-348 in procaspase-6 before and after proteolytic cleavage confirmed that Trp-348 is in a different environment after activation.[34] Additionally, active caspases-3 and -6 have been produced by changing the order of the large and small subunits in the expression constructs, which alleviates the steric clashes among the intersubunit linkers and allows spontaneous maturation to the active form.[38,39]

4.3.2 INITIATOR CASPASES

Although many of the initiator caspases can be activated by proteolytic cleavage in a manner similar to that of the effector caspases, it appears that the mechanism of activation is very different in cells. Additionally, it has been determined that proteolytic cleavage of caspase-8 or -9 is neither necessary nor sufficient for their activation. Procaspases-8 and 9 exist primarily as monomers and require dimerization to form their active states;[40] thus, a key factor regulating their activation is the relative concentration of monomer versus dimer at equilibrium.[23,28,41] Therefore, activation of caspases-8 and -9 is accomplished by forcing the monomers together, thereby increasing the amount of dimeric active enzyme in the cell.[31,42] Although cleavage of the procaspase is not required for activation, proteolytic processing may stabilize the active dimer.

Activation of procaspase-9 is mediated by interactions between its caspase recruitment domain (CARD) and its interacting partner APAF-1, which together form the apoptosome complex.[43,44] The details of the activation mechanism of the initiator caspases have been elucidated from biochemical characterization of active initiator caspase complexes as well as from the structures of procaspase-9 and procaspase-9 bound to the CARD domains of APAF-1. Analysis of the procaspase-9 crystal structure shows that the important 341-loop can adopt an active conformation without the necessity of proteolytic processing.[23] Moreover, hydrophobic interactions at the interface between the two monomers stabilize the dimers without the need for the N- and C-termini and the 341-loop that stabilizes the caspase-7 active structure. In fact, a version of caspase-9 containing a mutation at the cleavage sites that prevents proteolytic processing can still activate procaspase-3 in complex with the apoptosome in vitro,[45–47] and can cleave small fluorescent substrates with similar activity to wild-type enzyme.[48] Electron micrographs of APAF-1 show that it forms a heptameric complex with a central core composed of the CARD domains.[49] Upon the addition of caspase-9 to APAF-1, the electron micrographs now contain additional density, which presumably represents a tetrameric complex of caspase-9 molecules flanked by two APAF-1 heptamers. The appearance of the caspase-9 tetramer suggests that the tetrameric complex formed in the caspase-9 crystal[23] (Figure 4.3) is the activated form of the enzyme.

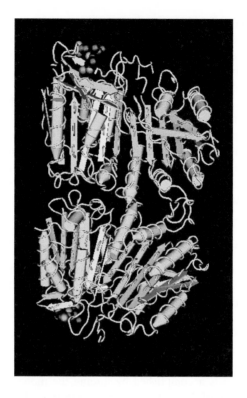

FIGURE 4.3 (See color insert.) Structure of the caspase-9 tetramer.

The crystal structure of caspase-9 also confirmed the finding that only one of the caspase-9 monomers bound the irreversible inhibitor VAD-fluoromethyl ketone,[23] whereas the other caspase-9 monomer contained a disordered substrate binding peptide, similar to what is found in the inactive structure of caspase-7. This suggests that the caspase-9 dimer contains one active and one inactive monomer and that both forms exist within the apoptosome. Mutational analyses of residues that stabilize the dimeric interface further support the hypothesis that these hydrophobic interactions are important determinants of enzyme activity.[28,50] For example, mutation of Phe-390 to the hydrophobic Val residue did not compromise the ability of the enzyme to be activated; however, replacement of Phe-390 with a charged Asp residue reduced enzyme activity. Additionally, charge reversal at residues Lys-396/Lys-397 to Asp/Asp prevented enzyme activation by APAF-1 and failed to induce cell death in 293 and HeLa cells.[51]

Although the crystal structures of procaspases-8 and -10 have not been reported, biochemical analysis of procaspase-8 as well as the Fas/FADD/FLIP (FLICE [FADD-Like ICE] inhibitory protein) complex have provided a model for procaspase-8 activation. Procaspase-8 subunits, like procaspase-9, appear to exist as monomers in solution.[28,52] and it has been reported that activation of procaspase-8 occurs near the cell membrane in a transient manner. In cells the oligomerization and subsequent activation of procaspase-8 occurs at the death-inducing signaling complex (DISC) through interactions between the death effector domain (DED) and the death domain (DD), and the DISC complex appears to contain trimerized Fas DD or FADD DD molecules.[53] A similar heterohexameric complex containing the FADD DED domains and procaspase-8 has also been described.[54] Mutations at the caspase-8 dimerization interface severely restrict its ability to activate the apoptotic pathway through activation with Fas ligand.[55] Further support of the forced dimerization hypothesis for activation of procaspase-8 is the observation that mutations preventing cleavage of the intersubunit linker do not inhibit cell death when these variants are transfected into cells.[56,57] The procaspase-8 homologue c-FLIP$_L$, which does not contain protease activity, was identified as a *bona fide* activator of apoptosis and has led to the discovery of a novel mechanism for activation of procaspase-8.[58–61] It was found that c-FLIP$_L$ is recruited to the DISC complex and binds procaspase-8 to form a heterodimer; however, the intersubunit linker of procaspase-8 is not cleaved in this heterodimer.[60,61] *In vitro* the c-FLIP$_L$/procaspase-8 heterodimer has similar activity and substrate specificity as the caspase-8 homodimer.[62] This interesting result suggests a novel mode of activation where formation of a complex containing an enzymatically active and an enzymatically inactive caspase homologue enhances the activity of the caspase.

4.3.3 INFLAMMATORY CASPASES

Activation of the inflammatory procaspases is very similar to activation of the initiator caspases. Several CARD-containing complexes have been shown to bind to the CARD of procaspase-1, including Rip2/RICK/CARDIAK,[63–66] CARD2/Ipaf/CLAN,[67,68] and the ASC/PyCARD and NALP1 (inflammasome) complex.[69–71] The Ser/Thr kinase Rip2 interacts with the CARD domain of procaspase-1, yet its involvement in activation of procaspase-1 has not been rigorously demonstrated. In contrast, Ipaf (ICE protein activating factor) oligomerizes with procaspase-1 in a CARD-dependent manner, and this oligomerization is required for activation of procaspase-1.[72] Ipaf-dependent oligomerization is reminiscent of the apoptosome and the DISC complexes, suggesting a similar method for activation of procaspase-1. In a similar manner, addition of lipopolysaccharide (LPS) nucleates heterodimerization of the ASC and NALP1 proteins, allowing the CARD domains to interact with the CARD domains of caspases-1 and -5.[69] Formation of this complex forces procaspase-1 dimerization, followed by cleavage of the intersubunit linker. This method of activation by forced proximity is supported by the fact that overexpression of procaspase-1 results in autoactivation due to increased protein concentration effects.[73,74]

In summary, caspase activation and maturation occurs via two major mechanisms. The effector caspases are activated by cleavage of the prodomain from an established protein dimer, which allows the enzyme to fold into a stable molecule. The initiator and pro-inflammatory caspases are activated by what has become known as the induced proximity model where monomeric procaspases are forced into a multiprotein complex promoting dimerization. By forcing the subunits into proximity, they are able to fold into their active forms allowing them to cleave their substrates *in vivo*.

4.4 CATALYTIC MECHANISM

Catalysis of peptide cleavage by caspases is very similar to the mechanism utilized by other cysteine proteases as well as serine proteases. Small molecule drug discovery efforts with other proteases helped the development of very effective, reversible aldehyde–based caspase inhibitors as well as irreversible inhibitors that covalently modify the active site cysteine. Confirmation of the active site cysteine was determined by analysis of the caspase-1 crystal structure showing the inhibitor bound to Cys-285,[3,75] which was shown to be located at the junction between the large and small subunits of a single caspase monomer. In addition to the active site cysteine, a conserved histidine residue near the active site forms a catalytic dyad increasing the nucleophilicity of the active site cysteine, acting as a general base, as well as activating a water molecule to complete product release as seen in Figure 4.4. Mutational studies have confirmed the importance of Cys-285 and His-237 residues in the catalytic cycle,[3] and mutations in the active site cysteine act as dominant negative regulators of apoptosis when overexpressed.

The catalytic activity of the caspases can vary greatly among family members. Even caspases with identical or nearly identical substrate preferences, such as caspases-3 and -7, have k_{cat}/K_m values that can vary by approximately eightfold. The catalytic activity of the recombinantly expressed caspases has been determined using a variety of substrates,[76–79] and data for the proposed optimal substrate for each caspase are summarized in Table 4.1. It is difficult to compare the activities of the caspases to each other due to wide differences in preferred substrates as well as the fact that some caspases exist as multiprotein complexes, which may enhance their activity against naturally occurring substrates. Therefore, it is more informative to discuss the factors that determine the substrate preferences for each of the caspases. This has been gleaned from analysis of crystal structures of the caspases bound to inhibitors as well as the studies that determined K_m and k_{cat}/K_m for different substrates.

4.5 SUBSTRATE SPECIFICITY

Since all caspases use a very similar catalytic mechanism to cleave peptide bonds, the major determinant in substrate selection is the binding between enzyme and substrate. As the substrate binding pocket is well ordered in the published caspase crystal structures, it is possible to determine which residues contribute to substrate binding and selectivity. For illustration, the substrate binding pocket of caspase-7 with the residues known to make contact with the substrate is shown in Figure 4.5. Although the substrate binding sites are highly conserved, there are remarkable differences in the makeup of the binding pockets among the family members that help enhance selectivity.

4.5.1 SPECIFICITY AT THE S1 POCKET

Caspases are exquisitely selective for aspartic acid at the P1 position. The P1 binding pocket consists of two highly conserved arginine residues (Arg-179 and Arg-341) and a conserved glutamine (Gln-283). The positively charged arginines form strong ionic interactions with the negatively charged aspartate, and generate an ideal binding

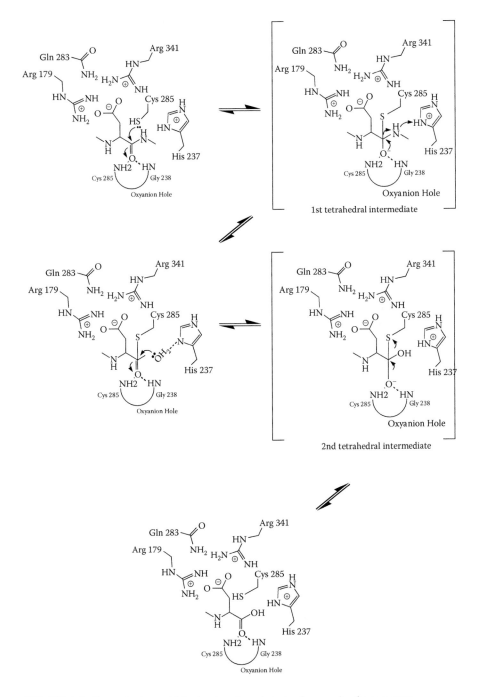

FIGURE 4.4 (See color insert.) Mechanism for substrate cleavage by the caspases.

TABLE 4.1

Kinetic Parameters for Caspases with Optimal Substrates

	Substrate (P4(5) to P1)	k_{cat}/K_m $(M^1 s^{-1})$	k_{cat} (s^{-1})	K_m (μM)
Caspase-1	WEHD[78]	330,000	13	4.0
Caspase-1	YVAD[79]	79,000	0.57	7.3
Caspase-2	VDVAD[79]	84,000	4.5	53
Caspase-3	DQMD[79]	262,000	11	44
Caspase-3	DEVD[79]	218,000	2.4	11
Caspase-4	LEVD[79]	3,200	0.14	44
Caspase-5	WEHD[78]	15,000	2.3	15
Caspase-6	VEID[79]	168,000	5.0	30
Caspase-7	DEVD[79]	37,000	0.43	12
Caspase-8	DEVD[79]	9,300	0.37	4.0
Caspase-9	VEHD[78]	490	0.20	408
Caspase-10	VEHD[78]	n.d	n.d	42

pocket for the negatively charged aspartic amino acid. Consistent with this ideal fit, catalytic efficiency is reduced up to four orders of magnitude when the P1 residue is changed to glutamate.[80] Although the selectivity for aspartic acid at P1 is very high, it should be noted that the transcription factor Max is cleaved by caspase-5 at an Ile-Glu-Val-Glu/Ser site. However, caspase-5 does not cleave a peptide substrate with an identical sequence, showing that additional contacts between caspase-5 and the protein are necessary to facilitate cleavage of the glutamate-containing substrate.[81]

4.5.2 SPECIFICITY AT THE S2 POCKET

Binding of P2 in the S2 pocket contributes to some of the differences in affinity determined for substrate binding.[14,21,82] The S2 pocket for the effector caspases-3

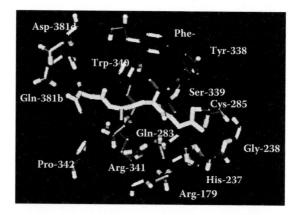

FIGURE 4.5 (See color insert.) Detailed view of the interaction between caspase-7 and the inhibitor Ac-DEVD-CHO.

and -7 contains a trio of aromatic side chains from residues Tyr-338, Trp-340, and Phe-381. Therefore, the pocket is best suited to accommodate small, hydrophobic residues such as valine or alanine. In the crystal structure of apocaspase-3, the side chain of Tyr-338 occupies the S2 pocket, preventing substrate binding. However, upon substrate binding, the side chain rotates ~90° around the $C\alpha$-$C\beta$ bond to allow access of the bulky P2 residue. The consequence of this rotation on binding has been confirmed by analysis of inhibitor binding to caspase-3.[8,12] Although there are differences in activity among substrates with different P2 residues, there appears to be remarkable plasticity in the S2 pocket for binding different substrates. For instance, it has been shown that caspase-6 can accommodate Ile, Thr, Trp, and Leu residues with nearly equal catalytic efficiencies.[83] In the pro-inflammatory and initiator caspases, Tyr-338 is replaced by the smaller valine (or alanine in caspase-2), allowing for the binding of bulkier residues such as histidine or isoleucine.

4.5.3 SPECIFICITY AT THE S3 POCKET

The residues flanking the S3 pocket are fairly well conserved throughout the caspases. Most caspases have a strong preference for glutamate at the P3 position due to ionic interactions with the Arg-341 side chain.[14,20,84–86] In addition, Arg-341 contributes to main chain/main chain hydrogen bonding with the substrate backbone, which stabilizes substrates in the active site. Caspases-8 and -9 have additional interactions between Arg-177 and the glutamate side chain that further stabilize substrate binding.[19,87] Other caspases without the additional arginine do not show the same degree of selectivity for glutamate as caspases-1, -2, -3, and -7, and therefore can accommodate a wider range of residues, including Ala, Gln, Thr, and Val.[88]

4.5.4 SPECIFICITY AT THE S4 POCKET

Differences among the S4 pockets appear to be the major determinants of substrate specificity and are used to classify the caspases into groups based on the best accommodated P4 residue. The S4 pocket of the pro-inflammatory caspases-1, -4, and -5 is a hydrophobic pocket that best accommodates large aromatic residues. In particular, the tryptophan indole group forms interactions with multiple residues within the S4 pocket.[89] In contrast, the apoptotic caspases contain a tryptophan at position 348 that prevents bulkier residues from occupying the S4 pocket. This tryptophan also contributes hydrophobic interactions that stabilize the small aliphatic residues preferred by the initiator caspases and forms a hydrogen bond with the aspartate residue preferred by the effector caspases.[7,10,11,21] In fact, although caspase-8 is classified as an initiator caspase, it can cleave substrates with an aliphatic residue or aspartate at P4 with nearly equal efficiency. Additional interactions between the P4 side chain and the S4 pocket are responsible for the differences in the side chain preferences. For instance, in caspase-8, hydrophobic interactions between the P4 residue and Tyr-340 enhance substrate binding,[20] whereas in caspases-2,[7] -3,[10,11] and -8,[19] hydrogen bonding between the P4 aspartate and Asn-342, or Gln -81b in caspase-7,[14] contributes to the substrate binding energy. Additionally, in caspases-3 and -7 steric interactions between the bigger loop, which constricts the S4 pocket, and the substrate result in a strong preference for the smaller aspartate versus a glutamate at P4.[90]

4.5.5 THE S5 POCKET AND CASPASE-2

Most of the caspases do not have any substrate specificity beyond the P4 residue. A notable exception to this is caspase-2, which requires occupancy of the S5 subsite for efficient cleavage. In this case the presence of a P5 residue results in a 35-fold increase in the catalytic efficiency of caspase-2.[91] The crystal structure of caspase-2 bound to an inhibitor containing a P5 leucine residue revealed hydrophobic inter-actions between the leucine and two residues of the 381-loop, consistent with the known preference for small hydrophobic residues at the P5 position. In addition, the presence of a P5 residue may result in tighter interactions between the P4 aspartate and the enzyme, resulting in a more tightly bound substrate.[7] The requirement for a P5 residue may make caspase-2 particularly difficult to inhibit reversibly with small molecules. Occupying the entire S1-to-S5 binding pocket would require an extended inhibitor approximately the size of the preferred substrate VDVAD that may have difficulties in cell permeability as well as poor pharmacokinetic properties. At this point, there is a low level of therapeutic interest in caspase-2 given that it is not a critical player in many cell death pathways.[92]

4.5.6 SPECIFICITY AT THE S1′ POCKET

The prime binding sites of the caspases do not show significant differences among the caspase families. All of the caspases tested cleave peptides with small residues (glycine, alanine, and serine) at the S1′ pocket efficiently,[93] and surprisingly, phenyl-alanine and tyrosine were also well tolerated. However, polar residues and proline were not well tolerated and significantly reduced the catalytic efficiency of peptide cleavage. The advantage of promiscuity of the S1′ pocket for the development of irreversible caspase inhibitors is that it permits a large variety of leaving groups to be utilized. However, the disadvantage is that it does not allow for discrimination among the members of the caspase family, and therefore requires specificity to be determined by the S1-S4(5) binding pockets. Although there is little specificity among the different caspases at the S1′ pocket, the nature of inhibitor binding to the S1′ pocket can determine whether the compound binds reversibly or irreversibly. In a series of studies examining the effects of different leaving groups on caspase-1 inhibitors, aspartic acid ketone–containing inhibitors were shown to interact with the active site cysteine, forming a reversible thiohemiketal intermediate similar to the tetrahedral transition state proposed for substrate cleavage (see Figure 4.4). The active site cysteine could then potentially displace the leaving group via an S_N2 mechanism, forming the irreversible enzyme/inhibitor thioether complex (Figure 4.6). This class of inhibitors demonstrated varied kinetic profiles and was classified as reversible, irreversible, or bimodal. Bimodal inhibitors were found to rapidly bind tightly but reversibly to the enzyme, and then more slowly to irrevers-ibly modify the enzyme ($K_i > k_3$). In contrast with calpain inhibitors, irreversibility of caspase-1 inhibitors did not depend significantly on the pK_a of the leaving group but on the angle of the S-C-Cα′-LG bonds. All of the inhibitors tested were found to have a S-C-Cα′-LG bond angle near 60° or 180°. Inhibitors with a bond angle of 180° were likely to be bimodal or irreversible, while those with bond angles of 60° were always reversible.

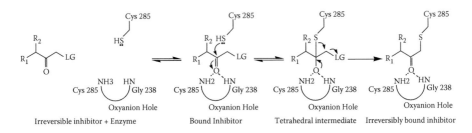

FIGURE 4.6 Mechanism for binding and irreversible inhibition by common irreversible and bimodal inhibitors.

To fully characterize the extent of the bimodal inhibition, it is necessary to examine both the reversible binding events and the rate of the reversible step. To accomplish this task, Wu and Fritz fit the inhibition curves at a variety of concentrations of inhibitors to equation (4.1).[94]

$$[P]_t = [E]_O^T \left(\frac{[S] K_i}{[I] K_s} \right) \left(\frac{k_s}{k_3} \right) [1 - e^{-k_3 t / \{1 + K_i / [I](1 + [S]/[K_s])\}}] \tag{4.1}$$

By fitting the data to this equation, the parameters for K_i and k_3 can be determined for each bimodal inhibitor.

4.6 SUMMARY

Analysis of the mechanism of activation and the kinetic parameters for different substrates among the caspases has led to significant advances in the understanding of caspase biology. Each of the steps, from zymogen activation through substrate binding and cleavage, presents a potential target for intervention by small molecule inhibitors. Compounds that target the zymogen and prevent activation of the caspase could have great potential as prophylactic treatments to prevent cell loss in cases where there is a known risk of the apoptotic cell death. Although this is a limited set of circumstances, there are many cases where such a compound would be useful in a medical setting. Once caspases have been activated, active site inhibitors would be useful to prevent the progression of the apoptotic pathway. All of the caspases follow a similar catalytic mechanism with the active site cysteine as the nucleophile to hydrolyze the peptide bond. Covalent modification of the active site cysteine results in irreversible inhibition of the caspase. Similarly, transition-state analogues such as the peptide aldehydes are also potent inhibitors due to interactions with the active site cysteine. The common catalytic mechanism of the caspases may be very useful for the development of inhibitors that target many of the caspases. In addition, each of the caspases has a very similar three-dimensional core structure leading to a highly homologous active site, and this structural similarity makes it possible to develop inhibitors that can bind to nearly all of the caspases. However, there are significant differences among the caspases and particularly among the initiator, effector, and inflammatory classes of caspases. These differences are evident in the different substrate preferences for each of the caspases and could be exploited to make small molecules that bind specifically to a single caspase, or to a class of caspases. Thus, the caspases are a family of enzymes that have intriguing properties for drug discovery. In many ways all of the caspases are similar, yet they can be broken down into distinct classes by activation mechanism, substrate preference, and biological activity. The similarities and differences among the caspases make them prime targets for development of both broad-spectrum and specific inhibitors.

REFERENCES

1. Eckelman, B. P., Salvesen, G. S., and Scott, F. L. 2006. Human inhibitor of apoptosis proteins: Why XIAP is the black sheep of the family. *EMBO Rep.* 7:988.
2. Callus, B. A., and Vaux, D. L. 2007. Caspase inhibitors: Viral, cellular and chemical. *Cell Death Differ.* 14:73.
3. Wilson, K. P., et al. 1994. Structure and mechanism of interleukin-1 beta converting enzyme. *Nature* 370:270.
4. Rano, T. A., et al. 1997. A combinatorial approach for determining protease specificities: Application to interleukin-1beta converting enzyme (ICE). *Chem. Biol.* 4:149.
5. Okamoto, Y. et al. 1999. Peptide based interleukin-1 beta converting enzyme (ICE) inhibitors: Synthesis, structure activity relationships and crystallographic study of the ICE-inhibitor complex. *Chem. Pharm. Bull.* (Tokyo) 47:11.
6. Romanowski, M. J., et al. 2004. Crystal structures of a ligand-free and malonate-bound human caspase-1: Implications for the mechanism of substrate binding. *Structure* 12:1361.
7. Schweizer, A., Briand, C., and Grutter, M. G. 2003. Crystal structure of caspase-2, apical initiator of the intrinsic apoptotic pathway. *J. Biol. Chem.* 278:42441.
8. Ni, C. Z., et al. 2003. Conformational restrictions in the active site of unliganded human caspase-3. *J. Mol. Recognit.* 16:121.
9. Lee, D., et al. 2000. Potent and selective nonpeptide inhibitors of caspases 3 and 7 inhibit apoptosis and maintain cell functionality. *J. Biol. Chem.* 275:16007.
10. Mittl, P. R., et al. 1997. Structure of recombinant human CPP32 in complex with the tetrapeptide acetyl-Asp-Val-Ala-Asp fluoromethyl ketone. *J. Biol. Chem.* 272:6539.
11. Rotonda, J., et al. 1996. The three-dimensional structure of apopain/CPP32, a key mediator of apoptosis. *Nat. Struct. Biol.* 3:619.
12. Becker, J. W., et al. 2004. Reducing the peptidyl features of caspase-3 inhibitors: A structural analysis. *J. Med. Chem.* 47:2466.
13. Riedl, S. J., et al. 2001. Structural basis for the inhibition of caspase-3 by XIAP. *Cell* 104:791.
14. Wei, Y., et al. 2000. The structures of caspases-1, -3, -7 and -8 reveal the basis for substrate and inhibitor selectivity. *Chem. Biol.* 7:423.
15. Chai, J., et al. 2001. Structural basis of caspase-7 inhibition by XIAP. *Cell* 104:769.
16. Huang, Y., et al. 2001. Structural basis of caspase inhibition by XIAP: Differential roles of the linker versus the BIR domain. *Cell* 104:781.
17. Chai, J., et al. 2001. Crystal structure of a procaspase-7 zymogen: Mechanisms of activation and substrate binding. *Cell* 107:399.
18. Riedl, S. J., et al. 2001. Structural basis for the activation of human procaspase-7. *Proc. Natl. Acad. Sci. USA* 98:14790.
19. Blanchard, H., et al. 1999. The three-dimensional structure of caspase-8: An initiator enzyme in apoptosis. *Structure* 7:1125.

20. Watt, W., et al. 1999. The atomic-resolution structure of human caspase-8, a key activator of apoptosis. *Structure* 7:1135.
21. Blanchard, H., et al. 2000. Caspase-8 specificity probed at subsite S(4): Crystal structure of the caspase-8-Z-DEVD-cho complex. *J. Mol. Biol.* 302:9.
22. Xu, G., et al. 2001. Covalent inhibition revealed by the crystal structure of the caspase-8/p35 complex. *Nature* 410:494.
23. Renatus, M., et al. 2001. Dimer formation drives the activation of the cell death protease caspase 9. *Proc. Natl. Acad. Sci. USA* 98:14250.
24. Shiozaki, E. N., et al. 2003. Mechanism of XIAP-mediated inhibition of caspase-9. *Mol. Cell* 11:519.
25. Donepudi, M., and Grutter, M. G. 2002. Structure and zymogen activation of caspases. *Biophys. Chem.* 101–102:145.
26. Lazure, C. 2002. The peptidase zymogen proregions: Nature's way of preventing undesired activation and proteolysis. *Curr. Pharm. Des.* 8:511.
27. Fuentes-Prior, P., and Salvesen, G. S. 2004. The protein structures that shape caspase activity, specificity, activation and inhibition. *Biochem. J.* 384:201.
28. Boatright, K. M., et al. 2003. A unified model for apical caspase activation. *Mol. Cell* 11:529.
29. Stennicke, H. R., and Salvesen, G. S. 1999. Catalytic properties of the caspases. *Cell Death Differ.* 6:1054.
30. Donepudi, M., et al. 2003. Insights into the regulatory mechanism for caspase-8 activation. *Mol. Cell* 11:543.
31. Salvesen, G. S., and Dixit, V. M. 1999. Caspase activation: The induced-proximity model. *Proc. Natl. Acad. Sci. USA* 96:10964.
32. Muzio, M., et al. 1998. An induced proximity model for caspase-8 activation. *J. Biol. Chem.* 273:2926.
33. Pop, C., et al. 2001. Removal of the pro-domain does not affect the conformation of the procaspase-3 dimer. *Biochemistry* 40:14224.
34. Kang, B. H., et al. 2002. The structure of procaspase 6 is similar to that of active mature caspase 6. *Biochem. J.* 364:629.
35. Talanian, R. V., et al. 1996. Stability and oligomeric equilibria of refolded interleukin-1beta converting enzyme. *J. Biol. Chem.* 271:21853.
36. Alnemri, E. S., Livingston, D. J., Nicholson, D. W., Salvesen, G., Thornberry, N. A., Wong, W. W., and Yuan, J. 1996. Human ICE/CED-3 protease nomenclature. *Cell* 87:171.
37. Earnshaw, W. C., Martins, L. M., and Kaufmann, S. H. 1999. Mammalian caspases: Structure, activation, substrates, and functions during apoptosis. *Annu. Rev. Biochem.* 68:383.
38. Srinivasula, S. M., et al. 1998. Generation of constitutively active recombinant caspases-3 and -6 by rearrangement of their subunits. *J. Biol. Chem.* 273:10107.
39. Suzuki, Y., Nakabayashi, Y., and Takahashi, R. 2001. Ubiquitin-protein ligase activity of X-linked inhibitor of apoptosis protein promotes proteasomal degradation of caspase-3 and enhances its anti-apoptotic effect in Fas-induced cell death. *Proc. Natl. Acad. Sci. USA* 98:8662.
40. Chang, D. W., et al. 2003. Oligomerization is a general mechanism for the activation of apoptosis initiator and inflammatory procaspases. *J. Biol. Chem.* 278:16466.
41. Donepudi, M., et al. 2003. Insights into the regulatory mechanism for caspase-8 activation. *Mol. Cell* 11:543.
42. Muzio, M., et al. 1998. An induced proximity model for caspase-8 activation. *J. Biol. Chem.* 273:2926.
43. Qin, H., et al. 1999. Structural basis of procaspase-9 recruitment by the apoptotic protease-activating factor 1. *Nature* 399:549.

44. Zhou, P., et al. 1999. Solution structure of Apaf-1 CARD and its interaction with caspase-9 CARD: A structural basis for specific adaptor/caspase interaction. *Proc. Natl. Acad. Sci. USA* 96:11265.

45. Stennicke, H. R., et al. 1999. Caspase-9 can be activated without proteolytic processing. *J. Biol. Chem.* 274:8359.

46. Srinivasula, S. M., et al. 2001. A conserved XIAP-interaction motif in caspase-9 and Smac/DIABLO regulates caspase activity and apoptosis. *Nature* 410:112.

47. Bratton, S. B., et al. 2001. Recruitment, activation and retention of caspases-9 and -3 by Apaf-1 apoptosome and associated XIAP complexes. *EMBO J.* 20:998.

48. Srinivasula, S. M., et al. 2001. A conserved XIAP-interaction motif in caspase-9 and Smac/DIABLO regulates caspase activity and apoptosis. *Nature* 410:112.

49. Acehan, D., et al. 2002. Three-dimensional structure of the apoptosome: Implications for assembly, procaspase-9 binding, and activation. *Mol. Cell* 9:423.

50. Shiozaki, E. N., et al. 2003. Mechanism of XIAP-mediated inhibition of caspase-9. *Mol. Cell* 11:519.

51. Chang, D. W., et al. 2003. Oligomerization is a general mechanism for the activation of apoptosis initiator and inflammatory procaspases. *J. Biol. Chem.* 278:16466.

52. Donepudi, M., et al. 2003. Insights into the regulatory mechanism for caspase-8 activation. *Mol. Cell* 11:543.

53. Weber, C. H., and Vincenz, C. 2001. A docking model of key components of the DISC complex: Death domain superfamily interactions redefined. *FEBS Lett.* 492:171.

54. Kaufmann, M., et al. 2002. Identification of a basic surface area of the FADD death effector domain critical for apoptotic signaling. *FEBS Lett.* 527:250.

55. Chang, D. W., et al. 2003. Interdimer processing mechanism of procaspase-8 activation. *EMBO J.* 22:4132.

56. Muzio, M., et al. 1998. An induced proximity model for caspase-8 activation. *J. Biol. Chem.* 273:2926.

57. Martin, D. A., et al. 1998. Membrane oligomerization and cleavage activates the caspase-8 (FLICE/MACHalpha1) death signal. *J. Biol. Chem.* 273:4345.

58. Irmler, M., et al. 1997. Inhibition of death receptor signals by cellular FLIP. *Nature* 388:190.

59. Yeh, W. C., et al. 2000. Requirement for Casper (c-FLIP) in regulation of death receptor-induced apoptosis and embryonic development. *Immunity* 12:633.

60. Micheau, O., et al. 2002. The long form of FLIP is an activator of caspase-8 at the Fas death-inducing signaling complex. *J. Biol. Chem.* 277:45162.

61. Chang, D. W., et al. 2002. c-FLIP(L) is a dual function regulator for caspase-8 activation and CD95-mediated apoptosis. *EMBO J.* 21:3704.

62. Boatright, K. M., et al. 2004. Activation of caspases-8 and -10 by FLIP(L). *Biochem. J.* 382:651.

63. Inohara, N., et al. 1998. RICK, a novel protein kinase containing a caspase recruitment domain, interacts with CLARP and regulates CD95-mediated apoptosis. *J. Biol. Chem.* 273:12296.

64. McCarthy, J. V., Ni, J., and Dixit, V. M. 1998. RIP2 is a novel NF-kappaB-activating and cell death-inducing kinase. *J. Biol. Chem.* 273:16968.

65. Kobayashi, K., et al. 2002. RICK/Rip2/CARDIAK mediates signalling for receptors of the innate and adaptive immune systems. *Nature* 416:194.

66. Thome, M., et al. 1998. Identification of CARDIAK, a RIP-like kinase that associates with caspase-1. *Curr. Biol.* 8:885.

67. Poyet, J. L., et al. 2001. Identification of Ipaf, a human caspase-1-activating protein related to Apaf-1. *J. Biol. Chem.* 276:28309.

68. Franchi, L., et al. 2006. Cytosolic flagellin requires Ipaf for activation of caspase-1 and interleukin 1beta in salmonella-infected macrophages. *Nat. Immunol.* 7:576.

69. Mariathasan, S., et al. 2004. Differential activation of the inflammasome by caspase-1 adaptors ASC and Ipaf. *Nature* 430:213.

70. Srinivasula, S. M., et al. 2002. The PYRIN-CARD protein ASC is an activating adaptor for caspase-1. *J. Biol. Chem.* 277:21119.

71. Martinon, F., Burns, K., and Tschopp, J. 2002. The inflammasome: A molecular platform triggering activation of inflammatory caspases and processing of proIL-beta. *Mol. Cell* 10:417.

72. Poyet, J. L., et al. 2001. Identification of Ipaf, a human caspase-1-activating protein related to Apaf-1. *J. Biol. Chem.* 276:28309.

73. Van Criekinge, W., et al. 1996. Functional characterization of the prodomain of interleukin-1beta-converting enzyme. *J. Biol. Chem.* 271:27245.

74. Ramage, P., et al. 1995. Expression, refolding, and autocatalytic proteolytic processing of the interleukin-1 beta-converting enzyme precursor. *J. Biol. Chem.* 270:9378.

75. Walker, N. P., et al. 1994. Crystal structure of the cysteine protease interleukin-1 beta-converting enzyme: A (p20/p10)2 homodimer. *Cell* 78:343.

76. Stennicke, H. R., et al. 2000. Internally quenched fluorescent peptide substrates disclose the subsite preferences of human caspases 1, 3, 6, 7 and 8. *Biochem. J.* 350:563.

77. Talanian, R. V., et al. 1997. Substrate specificities of caspase family proteases. *J. Biol. Chem.* 272:9677.

78. Garcia-Calvo, M., et al. 1999. Purification and catalytic properties of human caspase family members. *Cell Death Differ.* 6:362.

79. Talanian, R. V., et al. 1997. Substrate specificities of caspase family proteases. *J. Biol. Chem.* 272:9677.

80. Stennicke, H. R., et al. 2000. Internally quenched fluorescent peptide substrates disclose the subsite preferences of human caspases 1, 3, 6, 7 and 8. *Biochem. J.* 350:563.

81. Krippner-Heidenreich, A., et al. 2001. Targeting of the transcription factor Max during apoptosis: Phosphorylation-regulated cleavage by caspase-5 at an unusual glutamic acid residue in position P1. *Biochem. J.* 358:705.

82. Chereau, D., et al. 2003. Structural and functional analysis of caspase active sites. *Biochemistry* 42:4151.

83. Thornberry, N. A., et al. 1997. A combinatorial approach defines specificities of members of the caspase family and granzyme B. Functional relationships established for key mediators of apoptosis. *J. Biol. Chem.* 272:17907.

84. Thornberry, N. A., et al. 1997. A combinatorial approach defines specificities of members of the caspase family and granzyme B. Functional relationships established for key mediators of apoptosis. *J. Biol. Chem.* 272:17907.

85. Talanian, R. V., et al. 1997. Substrate specificities of caspase family proteases. *J. Biol. Chem.* 272:9677.

86. Garcia-Calvo, M., et al. 1998. Inhibition of human caspases by peptide-based and macromolecular inhibitors. *J. Biol. Chem.* 273:32608.

87. Blanchard, H., et al. 2000. Caspase-8 specificity probed at subsite S(4): Crystal structure of the caspase-8-Z-DEVD-cho complex. *J. Mol. Biol.* 302:9.

88. Thornberry, N. A., et al. 1997. A combinatorial approach defines specificities of members of the caspase family and granzyme B. Functional relationships established for key mediators of apoptosis. *J. Biol. Chem.* 272:17907.

89. Rano, T. A., et al. 1997. A combinatorial approach for determining protease specificities: application to interleukin-1beta converting enzyme (ICE). *Chem. Biol.* 4:149.

90. Stennicke, H. R., et al. 2000. Internally quenched fluorescent peptide substrates disclose the subsite preferences of human caspases 1, 3, 6, 7 and 8. *Biochem. J.* 350:563.

91. Talanian, R. V., et al. 1997. Substrate specificities of caspase family proteases. *J. Biol. Chem.* 272:9677.
92. Bergeron, L., et al. 1998. Defects in regulation of apoptosis in caspase-2-deficient mice. *Genes Dev.* 12:1304.
93. Stennicke, H. R., et al. 2000. Internally quenched fluorescent peptide substrates disclose the subsite preferences of human caspases 1, 3, 6, 7 and 8. *Biochem. J.* 350:563.
94. Wu, J. C., and Fritz, L. C. 1999. Irreversible caspase inhibitors: Tools for studying apoptosis. *Methods* 17:320.

5 Nonpeptide Small Molecule Inhibitors of Caspases

Alexandre V. Ivachtchenko, Ilya Okun,
Sergey E. Tkachenko, Alex S. Kiselyov,
Yan A. Ivanenkov, and Konstantin V. Balakin

CONTENTS

5.1 INTRODUCTION

Cellular apoptosis, a basic form of programmed cell death, is critical for various physiological processes, including maintenance of homeostasis in multicellular organisms. Apoptosis plays causative or contributing roles in various human conditions. For instance, excessive accumulation of aberrant cells may cause tumors and autoimmune disorders. Conversely, abnormal cell death was associated with heart failure, stroke, AIDS, neurodegenerative diseases, and liver injury.[1] Key players in the regulation of cellular apoptosis include death receptors and their ligands, transcriptional regulators (p53, IAP), Bcl-2 proteins, caspases, and endogenous caspase inhibitors.

Proteolytic enzymes such as caspases are important effector molecules in apoptosis.[2] Caspases represent a subfamily of cysteinyl aspartate–specific proteases that are involved in initiation, execution, and regulatory phases of cellular apoptosis. Twelve different members of this family have been identified in humans.[3] In healthy cells, caspases exist as dormant proenzymes. Upon receipt of a death-inducing signal, they undergo two (or more) cleavage events to liberate a large subunit and a small subunit that heterodimerize into the active enzyme. Caspases can be divided into three groups with distinct substrate specificities.[4] The group I caspases (caspase-1, -4, and -5) recognize preferentially the tetrapeptide WEHD. The group I caspases do not participate directly in the apoptotic pathways; rather, they have been shown to play an important role in cytokine maturation. The group II caspases (caspase-2, -3, and -7) recognize the tetrapeptide DEXD (where X is any amino acid) and are termed the executioner caspases because they are directly responsible for the cleavage and disabling of the homeostatic and structural proteins during apoptosis. The group III caspases (caspase-6, -8, -9, and -10) recognize tetrapeptides with the sequence I/L/VEXD. They are termed the activator caspases because their substrate specificity corresponds to the cleavage sites between the large and small subunits of most but not all group II and group III caspases. Both *in vitro* and *in vivo* data support a critical role for caspase-8 and -9 in the initiation of the apoptotic proteolytic cascade leading to cell death.

Although a plethora of signals initiate apoptosis, its phenotype is similar for different cell types. The typical morphological and biochemical hallmarks of apoptosis include cell shrinkage, nuclear DNA fragmentation, and membrane blebbing. Indeed, the final stages of apoptotic death are highly conserved and regulated by caspases. This discovery has boosted the search for caspase inhibitors as promising therapeutic agents. Inhibition of caspases is considered to have a potential in treating inflammatory, autoimmune diseases, rheumatoid arthritis, liver injury, myocardial infarction, neurodegenerative disorders,[5] including Alzheimer's,[6] Huntington's,[7] and Parkinson's disease,[8] and viral infections. The caspase inhibitors described to-date include peptides, peptidomimetics, and nonpeptide small molecules. Natural molecules preventing either caspase activation or its activity have also been described. Some of these compounds are proteins derived from viruses (p35, CrmA) or endogenously expressed in mammals (inhibitor of apoptosis protein [IAPs]).

It should be noted that the concept of caspase-dependent apoptosis as the major mechanism by which cells are eliminated may not universally apply, and caspase-independent apoptosis or other modes of cell death also have to be considered in the development of anti-apoptotic therapeutics.

This review covers recent advances in the development of small molecule caspase inhibitors of a nonpeptide structure. Perceived advantages of such synthetic molecules over peptide or peptidomimetic therapies include tunable activity and selectivity along with favorable pharmacokinetics, namely, better bioavailability, oral activity, and enhanced *in vivo* stability.

During the last decade, major progress has been made to further understand caspase structure and function, providing a unique basis for drug design and development.[9,10] Structure-based design of inhibitors of caspases yielded potent peptide and peptidomimetic compounds. Therapeutic potential of small molecule peptidomimetics has been demonstrated in multiple preclinical studies. VX-740 (pralnacasan), VX-765, and IDN-6556/PF-3,491,390 (emricasan) are examples of such inhibitors undergoing clinical trials as anti-inflammatory and anti-apoptotic agents. However, these inhibitors typically possess pharmacokinetic and physicochemical properties that significantly limit their utility in the clinic. Therapeutic efficacy of peptides has been limited due to poor cell permeability, limited activity in intact cells, and the lack of stability.

The development of nonpeptide small molecule inhibitors of caspases could overcome these limitations, and therefore, extensive studies of such agents have been performed in the past decade. In the next sections, we will describe the main chemical classes of such agents and discuss their role in the current strategies for the design of anti-apoptotic therapeutic agents.

5.2 NITRIC OXIDE DONORS

In the past few years, nitric oxide (NO) has become one of the most studied entities in biochemistry.[11–19] It is a highly reactive molecule displaying dichotomous regulatory roles under physiological and pathological conditions. As a biological messenger, it was found to be involved in diverse physiological processes such as vasodilatory and anti-platelet effects, macrophage-induced cytotoxicity, and neurotransmission.[11] Moreover, nitric oxide emerged as an important endogenous regulator of apoptosis.[12] This chemical prevented apoptosis in hepatocytes induced by trophic factor withdrawal, exposure to tumor necrosis factor alpha (TNF-α), anti-Fas antibody, and lipopolysaccharide.[13] It was recently found that angiotensin II–induced apoptosis of human umbilical venous endothelial cells (HUVECs) and caspase-3 activity could by completely inhibited by both molecular NO and various nitric oxide donors, such as sodium nitroprusside or S-nitrosopenicillamine.[14] *In vitro* studies demonstrated an inhibitory effect of NO on caspase-1, -2, -3, -4, -6, -7, and -8,[15] and the mechanism of inhibition was suggested to be due to direct S-nitrosylation of the cysteine (Cys) residue within the active side of enzyme, an effect that was reversible by the addition of dithiothreitol.[16] Inhibition of the caspase-1 inhibitory pathway was established by studies on iNOS knockout mice.[17] However, as communicated by Brune et al.,[18] enzymatic assay conditions do not fully reflect the real situation in a living cell, and therefore it remains unclear if the anti-apoptotic action of nitric oxide is due to reversible caspase inhibition. Accordingly, several other mechanisms of cytoprotection were considered. Recent studies on ischemia-reperfusion injury in cardiomyocytes confirmed that inhibition of NOS increased apoptosis and endogenous NO synthesis might protect cells against apoptotic death.[19]

Nitric oxide reacts with molecular oxygen, reactive oxygen species, and free thiols, such as glutathione and proteins, to form nitrosothiols. Nitrosothiols may functionally mimic NO *in vivo*.[20] Enzymes containing Cys (such as ornithine decarboxylase, vacuolar H^+-ATPase, and protein tyrosine phosphatase [PTPase]) were modified by NO-donors via an S-nitrosylation of the thiol moiety.

Several derivatives of N-nitrosoanilines with longer half-lives have been recently tested for their inhibitory activity against recombinant caspase-3 (Figure 5.1, Table 5.1).[21] The most potent compounds from this series were itrosoanilines **1**, **2**, and **3**. Compounds **4–7** exhibited less potent inhibition.[22] N-Nitroso-1H-indole **8** was also synthesized and tested for its activity against caspases. This compound inhibited

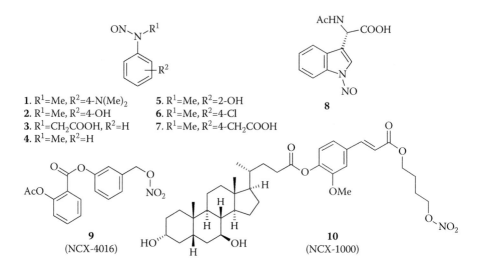

1. R^1=Me, R^2=4-N(Me)$_2$ 5. R^1=Me, R^2=2-OH
2. R^1=Me, R^2=4-OH   6. R^1=Me, R^2=4-Cl
3. R^1=CH$_2$COOH, R^2=H   7. R^1=Me, R^2=4-CH$_2$COOH
4. R^1=Me, R^2=H

9
(NCX-4016)

10
(NCX-1000)

FIGURE 5.1 Nitric oxide donors, inhibitors of caspases.

caspase-1 activity with a pseudo-first-order rate constant $k_{(obsd)}$ of 0.076 min^{-1}. The *N*-nitrosoaniline warhead has also been introduced into peptide-based inhibitors.

A new class of NO-releasing nonsteroidal anti-inflammatory drugs has been introduced that might target caspase-1 in a cyclooxygenase-independent pathway, including, e.g., NCX-4016 (**9**) (Figure 5.1).[23] NCX-4016 is a nitric oxide NO-donating derivative of acetyl salicylic acid that effectively inhibits proinflammatory cytokine release from endotoxin (LPS)-challenged monocytes. NCX-4016 caused a 40–80% inhibition of release of IL-1-β, IL-8, IL-12, IL-18, interferon (IFN)-γ, and TNF-α from stimulated human monocytes, with an EC$_{50}$ of 10–20 μM for IL-1β and IL-18 without affecting cytokine messenger RNA expression. Inhibition of caspase-1 by NCX-4016 was reversed by the addition of dithiothreitol (DTT), confirming

TABLE 5.1
Pseudo-First-Order Rate Constants
for N-Nitrosoanilines 1–8

Compound	Caspase-3, $k_{(obsd)}$ (min^{-1})
1	0.195
2	0.129
3	0.109
4	0.038
5	0.023
6	0.009
7	0.023
8	0.076

S-nitrosylation of caspase-1 as the likely mechanism of inhibition.[24] This compound was tested using a mitochondria-dependent model of apoptosis in HUVECs. It was shown that treatment with NCX-4016 protected HUVECs against the apoptotic action of staurosporine, but did not protect against aspirin (a non-NO-releasing analogue of NCX-4016 did not protect against staurosporine). In contrast to some other NO-releasing analogues, NCX-4016 did not increase mitochondrial oxidative stress.

Several steroid-based NO-donors were recently obtained and tested for their activity against a range of caspases. For example, NO-ursodeoxycholic acid **10** (NCX-1000) (Figure 5.1) represents the prototype of a family of NO-releasing derivatives of ursodeoxycholic acid (UDCA).[25] Discovered by NicOx, this compound is currently in phase II clinical development as a treatment for chronic liver disease, including portal hypertension. NCX-1000 inhibited activity of caspase-3, -8, and -9 in the micromolar range[25] and was also effective in protecting against N-$a$cetyl-$p$ara-$a$mino$p$henol acetaminophen (APAP)-induced hepatotoxicity when administered in a therapeutic manner.[26]

In summary, the present data indicate that the addition of an NO-releasing moiety to both steroid-based and nonsteroidal pharmacophores confers effective immunoregulatory and anti-apoptotic activities to initial compounds, and that these effects are caused mainly by the S-nitrosylation/inhibition of both the pro-apoptotic and proinflammatory branches of the caspase superfamily. Given the limited efficacy of currently used drugs in treating immunomediated inflammatory diseases, assessment of the effects of such agents in a clinical setting appears to be warranted.

5.3 METAL IONS, ARSINES, AND DITHIOCARBAMATES

While nitric oxide and NO-donors have been intensively studied, the underlying mechanisms of nitrosative stress with resulting pathology remain unclear. Previous studies have demonstrated that NO exposure increases free zinc ions, Zn^{2+}, within cells. However, the resulting effects on endothelial cell survival have not been adequately resolved. It has been suggested that zinc cations (Zn^{2+}) act as endogenous inhibitors of apoptosis and that depletion of intracellular zinc induces protein synthesis-dependent neuronal apoptosis in cortical cultures.[27] *In vitro* studies showed that activity of caspase-3, -6, -7, and -8 was abolished by Zn^{2+} in the submicromolar range.[27–30] Micromolar Zn^{2+} inhibited apoptosis in tubular cells and was protective in ischemic kidneys in rats.[30]

Other metal ions can inhibit caspase enzymes and thus act as anti-apoptotic agents. For example, tellurium (IV) tartrate significantly inhibited caspase-1 activity in peripheral blood mononuclear cells (PBMCs)[31] and cadmium salts blocked apoptosis of CHO K1-BH4 cells induced by hygromycin B and actinomycin D.[32] Furthermore, cadmium salts were effective inhibitors of caspase-3 activity, with nontoxic levels of these agents resulting in up to approximately 60% inhibition of the apoptotic caspase pathway.[33]

These studies have documented the potent protective effects of metal ions against cell apoptosis, suggesting their potential clinical use as protective agents. For example, Zn^{2+} might be considered a useful agent in renal protection during *in vivo* ischemia-reperfusion.[30] However, for *in vivo* use, one needs to keep in mind that all the mentioned metal ions are toxic at relatively low concentrations. For example,

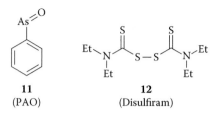

11   **12**
(PAO)   (Disulfiram)

FIGURE 5.2 Phenylarsine oxide and disulfiram.

Zn^{2+} induced cell injury and death at concentrations of 100 μM or above.[30] Therefore, for *in vivo* studies, the doses of metal ions need to be carefully controlled.

Phenylarsine oxide (PAO) **11** (Figure 5.2), a small arsine-organic molecule, inhibited recombinant human caspases as well as endogenous caspases that were active in chicken DU249 cells (S/M extracts).[34] At 100 μM concentration, this compound blocked morphological changes of nuclear apoptosis *in vitro* and internucleosomal DNA fragmentation in S/M extracts without interfering with poly (ADP-ribose) polymerse (PARP) or lamin A cleavage.[35] Similar to arsenic trioxide (As_2O_3), PAO was found to be a potent vicinal thiol-binding agent. PAO inhibited proliferation of both OCIM2 and OCI/AML3 cells in a dose-dependent fashion (0.01 to 0.1 μM).[36] This agent almost completely abolished production of active caspases and suppressed caspase-induced activation of NF-kappaB in acute myelogenous leukemia cells.

Dithiocarbamates (DCs) have been recently reported to be potent inhibitors of apoptosis in several different model systems, including thymocytes,[37] leukemic cells,[38] and fibroblasts.[39] These compounds efficiently blocked proteolytic activation of several caspases.[40] As expected, addition of a copper chelator compromised the inhibitory activity of reduced DCs.[40] DC disulfides were introduced as more potent inhibitors of T cell apoptosis than their reduced counterparts. Disulfide-related inhibition of apoptosis was correlated with the inhibition of caspase-3 proenzyme processing and its activation, particularly in a cell-free model system. The inhibitory effect of DC disulfides was abolished by adding DTT, suggesting the formation of a disulfide moiety between dithiocarbamates and the enzyme. Experiments using ^{35}S-labeled disulfiram (DSF) **12**, which is currently used clinically for the treatment of alcoholism, have shown that DSF strongly inhibited caspase-1 and -3 through this mechanism of action.[41] Alternative mechanisms of anti-apoptotic activity of **12** may include stabilization of mitochondrial membrane potential and suppression of reactive oxygen species.[42] These findings reveal the potential for other thiol-oxidizing toxicants to inhibit apoptosis by preventing the proteolytic activation of caspases.

5.4 MICHAEL ACCEPTORS AND QUINONE-CONTAINING COMPOUNDS

5.4.1 QUINONES AND THEIR BIOISOSTERIC ANALOGUES

Michael acceptors are known to inhibit cysteine-containing enzymes,[43] and a possible mechanism of inhibition of caspases by these molecules is shown in Figure 5.3. Reaction between SH-group of the active site cysteine of a caspase and activated

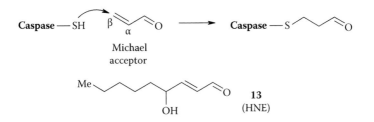

FIGURE 5.3 Mechanism of inhibition of caspase activity by Michael acceptors.

electrophilic double C=C bond of a Michael acceptor results in irreversible inactivation of the enzyme. For example, 4-hydroxynonenal (HNE) **13** (Figure 5.3), a metabolite generated by lipid peroxidation, is thiol reactive. It exhibited oxidative stress-linked pathological events in humans, such as cellular growth inhibition, apoptosis inhibition, and induction.[44] This compound blocked the release of lipopolysaccharide (LPS)-stimulated IL-1β (EC$_{50}$ = 5 µM) and IL-10 (EC$_{50}$ = 2 µM) in a dose-dependent manner in human peripheral blood mononuclear cells.[45,46] At low concentrations, the agent significantly enhanced the proliferation index, whereas at high concentrations it blocked cell proliferation.[47,48]

Several cyclic α,β-unsaturated systems were recently synthesized and tested for their activity against caspases (Figure 5.4). For example, quinone derivatives were found to inhibit caspases with IC$_{50}$ values in the micromolar range. p-Benzoquinone (BQ) **14**, a benzene metabolite, inhibits caspase-1[49] and -3.[50] Anti-apoptotic activity of hydroquinone (HQ) **15** in myeloblasts was explained by inhibition of caspase-1 and -3 by BQ, which is formed by oxidation of hydroquinone.[51] BQ and HQ inhibited apoptotic death of NIH3T3 cells induced by both serum starvation and lack of extracellular matrix (ECM), and cells treated with BQ and HQ showed greater proliferation than normal cells under low-serum conditions and anchorage-independent growth in soft agar.[52]

Bioisosteric analogues of cyclic quinones were developed and tested for their anti-apoptotic potential and caspase inhibitory activity. Xylaric acid or γ-pyrone-3-acetic acid (L-741,494) **16**, a water-soluble Michael acceptor-based substance, isolated from the fungal genus Xylaria, was found to be a competitive, irreversible inhibitor of caspase-1. At the same time, it was inactive against papain and trypsin.[53] Benzoquinones **17–20** (Figure 5.4) inhibited caspase-3 with an IC$_{50}$ in the low nanomolar range.[54] Quinone derivative **21** was reported as a novel potent caspase inhibitor produced by *Streptomyces* sp.[55] and inhibited both caspase-1 activity (IC$_{50}$ = 3.9 µM) and mature IL-1β secretion from THP-1 cells induced by LPS with an IC$_{50}$ value of 5.2 µM.

Minocycline **22**, a second-generation tetracycline, displays topological organization comparable to cyclic quinones. This compound demonstrated broad neuroprotective properties in experimental models of ischemic stroke, Huntington's disease (HD), amyotrophic lateral sclerosis (ALS), traumatic brain injury, multiple sclerosis, and Parkinson's disease.[56] Minocycline inhibited release of cytochrome c from mitochondria in a transgenic mouse model of ALS. In addition, this compound affected caspase-1 and caspase-3 mRNA upregulation and activation of p38MAP kinase.[56]

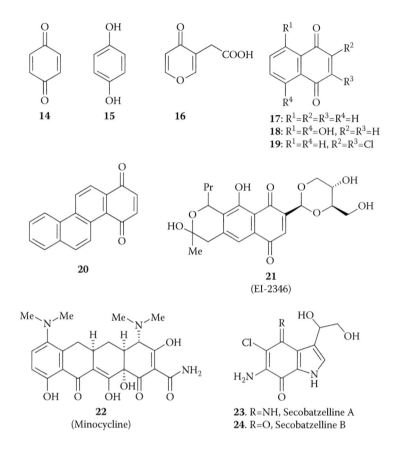

14   **15**   **16**

17: $R^1=R^2=R^3=R^4=H$
18: $R^1=R^4=OH$, $R^2=R^3=H$
19: $R^1=R^4=H$, $R^2=R^3=Cl$

20

21
(EI-2346)

22
(Minocycline)

23. R=NH, Secobatzelline A
24. R=O, Secobatzelline B

FIGURE 5.4 Quinones and quinones-like compounds as caspase inhibitors.

Quinones secobatzelline-A **23** and -B **24** have been isolated from marine sponges of the *Batzella* genus.[57] Secobatzelline-A inhibited caspase-3 ($IC_{50} = 0.02$ µg/mL) as well as proliferation of P388 murine leukemia ($IC_{50} = 0.06$ µg/mL) and A549 human lung adenocarcinoma ($IC_{50} = 0.04$ µg/mL) cells.

Despite potent *in vitro* caspase inhibitory activity, quinone-containing compounds usually show poor selectivity. In addition to reaction between the caspase active site cystein thiol and the activated double C=C bonds (Figure 5.3), interaction of quinone-containing compounds with caspases may also include redox reactions leading to oxidation of the cysteine moiety, and X-philic or nucleophilic substitution in the case of haloquinones. Such a high chemical reactivity clearly limits the clinical utility of quinone-based anti-apoptotic agents. To prevent some of these undesirable chemical transformations, additional functionalities have been described as alternative electrophiles. For example, azidomethylene derivatives inhibited caspase-1 in a competitive fashion ($IC_{50} < 10$ nM)[58] and featured high stability *in vitro* to thiols. Thus, several compounds containing this core fragment can be designed as selective inhibitors of caspase family enzymes. Quinone-based epoxides and their structural analogues represent another useful modification with increased selectivity and

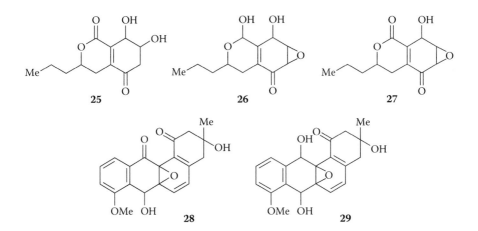

FIGURE 5.5 Quinone and quinone-based epoxides as caspase inhibitors.

clinical potential. Unfortunately, Michael acceptors may not have *in vivo* activity against caspases.

5.4.2 QUINONE-BASED EPOXIDES AND THEIR STRUCTURAL ANALOGUES

A series of potent quinone-based epoxides was recently obtained and tested for their anti-apoptotic activity (Figure 5.5). Epoxides **26** and **27** isolated from the culture broths of *Farrowia* sp. selectively inhibited human recombinant caspase-1 activity with IC_{50} values of 0.086 and 0.006 μM, respectively, without inhibiting elastase or cathepsin B.[59,60] Their close structural analogue **25**, lacking the epoxide function, inhibited the same enzyme with an IC_{50} value of 10 μM. Compounds **26** and **27** also inhibited mature IL-1β secretion from THP-1 cells induced by LPS with IC_{50} values of 5.0 and 10.3 μM, respectively (Table 5.2).

Epoxides **28** (El-1507-1) and **29** (El-1507-2) were isolated from the culture broths of *Streptomyces* sp. These compounds selectively inhibited recombinant human caspase-1 activity with IC_{50} values of 0.23 and 0.42 μM, and inhibited mature IL-1β secretion from THP-1 cells with IC_{50} values of 1.1 and 1.4 μM, respectively.[61] The epoxide functionality also appeared in relatively potent manumycin-related caspase-1 inhibitors **30–32** and its oxidation product **33** (IC_{50} of 0.07 μM) (Figure 5.6).[62]

Compounds **30–32**, caspase-1 inhibitors isolated from the culture broths of *Streptomyces* sp., selectively inhibited recombinant human caspase activity with IC_{50} values of 0.2, 0.38, and 0.09 μM, respectively, without inhibiting elastase or cathepsin B. Compounds **31–33** also inhibited mature caspase-1 secretion from THP-1 cells with IC_{50} values of 2.2, 3.6, and 5.4 μM, respectively.[63]

NEPP-11 **34** and its analogue **35** (Figure 5.6), members of neurite outgrowth-promoting prostaglandins (NEPPs), prevented manganese-induced apoptosis of PC12 cells and neuronal cell death of HT22 cells.[64] In preclinical studies NEPP-11 significantly decreased the number of apoptotic neurons in primary cortical cultures exposed to NMDA, via inhibition of NMDA-mediated caspase-3 activation.

TABLE 5.2
***In Vitro* Activity of Compounds 25–35 against Caspase-1**

Compound	IC$_{50}$ (μM)	
	Caspase-1	THP-1 Cells
25	0.086	5
26	0.006	10.3
27	10	—
28	0.23	1.1
29	0.42	1.4
30	0.2	2.2
31	0.38	3.6
34	—	1[a]
35	0.59	0.28

[a] In primary cortical cultures exposed to NMDA.

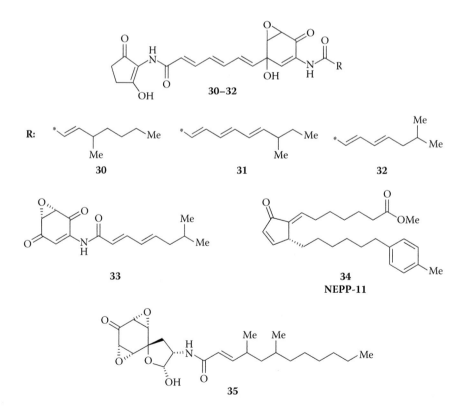

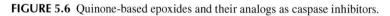

FIGURE 5.6 Quinone-based epoxides and their analogs as caspase inhibitors.

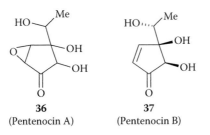

36
(Pentenocin A)

37
(Pentenocin B)

FIGURE 5.7 Pentenocins A and B isolated from Trichoderma hamatum.

It has been shown that NMDA increased caspase-3 activity and NEPP-11 significantly suppressed this increase with an IC_{50} value of 1 µM. Additionally, this compound (5 µM) attenuated manganese-induced DNA fragmentation by approximately 50%. Compound **35** was isolated from the culture broths of *Penicillium* sp. E-2128 and selectively inhibited human recombinant caspase-1 activity with an IC_{50} value of 0.59 µM, without inhibiting elastase or cathepsin B. This compound also inhibited mature IL-1β secretion from THP-1 cells induced by LPS with an IC_{50} value of 0.28 µM.[65]

Another epoxide-containing caspase-1 inhibitor, pentenocin A **36** (IC_{50} = 0.575 µM), has been isolated together with its α,β-carbonyl analogue, pentenocin B **37** (IC_{50} = 0.25 µM), from the broth of *Trichoderma hamatum* FO-6903 (Figure 5.7).[66] More recently, four possible diastereomers of pentenocin B were synthesized in a stereocontrolled manner, and the first total synthesis of a natural enantiomer of (+)-pentenocin B unequivocally established the absolute stereochemistry to be 4S,5R,6R.[67]

These studies demonstrated that quinone-based epoxides and their structural analogues represent a useful modification of quinone-containing agents described in Section 5.4.1 due to increased selectivity. At the same time, these compounds also contain highly reactive electrophilic moieties, and their interaction with proteins can lead to undesired reactions. Thus, apart from being potential Michael acceptors, the reviewed compounds may also undergo nucleophic epoxide ring opening (Figure 5.8), and in a hydrophobic environment, the thiol-capture function of the

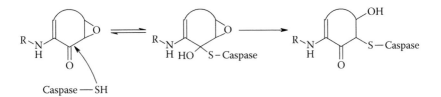

FIGURE 5.8 Thiophilic epoxide ring opening via the addition of cysteine thiol to the carbonyl group followed by irreversible 1,2-shift.

ketone moiety may facilitate an electrophilic attack.[68] High reactivity of these molecules still severely limits their utility as therapeutic agents.

5.5 CEPHEM OXIDES, HALOMETHYL AMIDES, PYRIMIDOTRIAZINEDIONES, AND THEIR STRUCTURAL ANALOGUES

Additional caspase-1 inhibitors have appeared in the patent literature. These include cephem oxides **38–47**,[69] halomethyl amides, e.g., **48** ($IC_{50} < 1\ \mu M$),[70] and pyrimidotriazinediones, e.g., **49** ($IC_{50} = 74\ \mu M$) (Figure 5.9).[71] All these compounds are very reactive, with the pyrimido[5,4-*e*][1,2,4]triazine-5,7(6*H*,8*H*)-dione **49** apparently being the most reactive compound from this series. Nontrivial topological analogues of pyrimidotriazinediones were described, such as YM-215438 **50**[72] (caspase-3

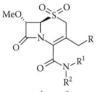

38. R=OAc, R^1=H, R^2=3,4-*di*-Cl-CH$_2$C$_6$H$_3$
39. R=OAc, R^1=Me, R^2=3,4-*di*-Cl-CH$_2$C$_6$H$_3$
40. R=OAc, R^1=H, R^2=3-I-CH$_2$C$_6$H$_4$
41. R=OAc, R^1=Me, R^2=3-I-CH$_2$C$_6$H$_4$

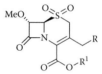

42. R=OAc, R^1=CH$_2$C$_6$H$_5$
43. R=OAc, R^1=4-NO$_2$-CH$_2$C$_6$H$_4$
44. R=OAc, R^1=3-I-4-Me-CH$_2$C$_6$H$_3$
45. R=OAc, R^1=3,4-*di*-Cl-C$_6$H$_3$
46. R= -SO$_2$-C$_6$H$_5$, 3,4-*di*-Cl-CH$_2$C$_6$H$_3$
47. R= -5S-(1-methyl-1*H*-tetrazole), R^1=3,4-*di*-Cl-CH$_2$C$_6$H$_3$

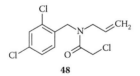

48

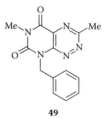

49

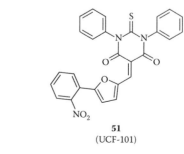

50
(YM-215438)

51
(UCF-101)

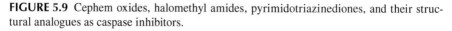

FIGURE 5.9 Cephem oxides, halomethyl amides, pyrimidotriazinediones, and their structural analogues as caspase inhibitors.

inhibitor) and UCF-101 **51**, which are Omi/HtrA2 (pro-apoptotic mitochondrial serine protease) inhibitors.[73,74] Omi/HtrA2 induces cell death in a caspase-dependent manner (caspase-3, -7, and -9) by interacting with the inhibitor of apoptosis protein (XIAP) as well as in a caspase-independent manner that relies on its protease activity. Additionally, it was recently shown that UCF-101 induces cellular responses independently of its known target, Omi/HtrA2.[75]

It can be suggested that the clinical potential of these agents is compromised by possible adverse reactions of these highly electrophilic species with different proteins. At the same time, compounds **50** and **51** are currently undergoing preclinical studies for the treatment of Alzheimer's, Parkinson's, and Huntington's diseases, ischemic stroke, dermatologic disorders, and cardiovascular diseases.[72–74]

5.6 QUINOLINES, QUINAZOLINES, AND PYRIDAZINES

Quinolines **52–55** (Figure 5.10, Table 5.3) have been recently synthesized and tested for their activity against recombinant caspases. The most potent compound from this series **54** inhibited caspase-3 activity with an IC_{50} value of 13.8 μM (and 92% inhibition was observed at 20 μM).[76] The design, synthesis, and biological activity of quinazolinone-based caspase-3 inhibitors **56–58** have been reported.[77] 6,8-*Di*-substituted 2-(phenylamino)quinazolinones **59–67** were found to be inhibit human

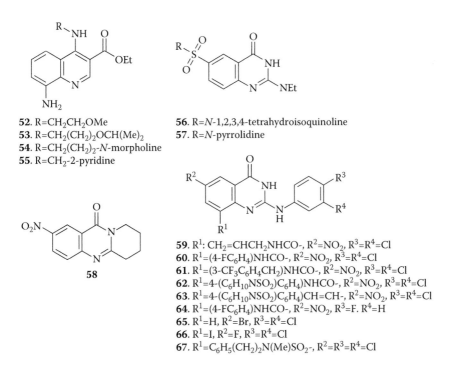

52. R=CH$_2$CH$_2$OMe
53. R=CH$_2$(CH$_2$)$_2$OCH(Me)$_2$
54. R=CH$_2$(CH$_2$)$_2$-*N*-morpholine
55. R=CH$_2$-2-pyridine

56. R=*N*-1,2,3,4-tetrahydroisoquinoline
57. R=*N*-pyrrolidine

58

59. R^1: CH$_2$=CHCH$_2$NHCO-, R^2=NO$_2$, R^3=R^4=Cl
60. R^1=(4-FC$_6$H$_4$)NHCO-, R^2=NO$_2$, R^3=R^4=Cl
61. R^1=(3-CF$_3$C$_6$H$_4$CH$_2$)NHCO-, R^2=NO$_2$, R^3=R^4=Cl
62. R^1=4-(C$_6$H$_{10}$NSO$_2$)C$_6$H$_4$)NHCO-, R^2=NO$_2$, R^3=R^4=Cl
63. R^1=4-(C$_6$H$_{10}$NSO$_2$)C$_6$H$_4$)CH=CH-, R^2=NO$_2$, R^3=R^4=Cl
64. R^1=(4-FC$_6$H$_4$)NHCO-, R^2=NO$_2$, R^3=F. R^4=H
65. R^1=H, R^2=Br, R^3=R^4=Cl
66. R^1=I, R^2=F, R^3=R^4=Cl
67. R^1=C$_6$H$_5$(CH$_2$)$_2$N(Me)SO$_2$-, R^2=R^3=R^4=Cl

FIGURE 5.10 Caspase inhibitors based on quinoline and quinazoline scaffolds.

TABLE 5.3
***In Vitro* Activity of Compounds 52–55 against Caspase-3**

Compound	Caspase-3 IC_{50} (µM)	Inhibition(%)[a]
52	16.5	88.8
53	13.5	92.5
54	13.8	92.3
55	13.2	90.1

[a] Percent inhibition of caspase-3 activity by compounds **52–54** (20 µM).

caspase-3 with K_i values of 88–793 nM (Table 5.4).[78] Moreover, synthesized aminoquinazolines (AQZs) were evaluated for their inhibitory activity against several other caspase family members. As shown in Table 5.4, exemplified compounds did not appreciably inhibit caspase-1, possessing K_i values above 10 µM. Additionally, the AQZs also did not have significant activity against caspase-2 and -7, the two enzymes with high substrate similarity to caspase-3. Forty percent of the tested AQZs showed partial inhibition of caspase-6, with K_i values ranging from 199 nM to 3.7 µM. Although caspase-8 has a substrate specificity similar to that of caspase-6, this isoform was not inhibited by the tested AQZs. Thus, among the six caspases

TABLE 5.4
Activity and Selectivity Profile of AQZs (59–64) against the Caspase Family Enzymes

Compound	SH-SY5Y Cells IC_{50} (µM)[a]	$K_i^b \pm$ S.E.M.[c] (nM)					
		Caspase-1	Caspase-2	Caspase-3	Caspase-6	Caspase-7	Caspase-8
59	14.9[†]	>10.000	>10.000	247 ± 81	>10.000	>10.000	>10.000
60	5.7[†]–8.2[††]	>10.000	>10.000	560 ± 59	567 ± 44	>10.000	>10.000
61	3[††]	>10.000	>10.000	793 ± 354	>10.000	>10.000	>10.000
62	—	>10.000	>10.000	260 ± 86	>10.000	>10.000	>10.000
63	—	>10.000	>10.000	154 ± 41	3.710 ± 414	>10.000	>10.000
64	—	>10.000	>10.000	88 ± 9	199 ± 44	>10.000	>10.000

[a] SH-SY5Y cells: †, inhibition of staurosporine-induced intracellular DEVDase activity; ††, inhibition of staurosporine-induced DNA fragmentation.

[b] Random screening for inhibitors of recombinant human caspase-3.

[c] S.E.M. = standard error of the mean.

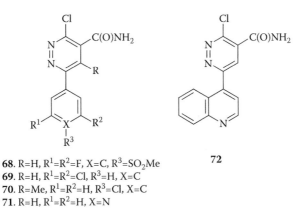

68. R=H, R^1=R^2=F, X=C, R^3=SO$_2$Me
69. R=H, R^1=R^2=Cl, R^3=H, X=C
70. R=Me, R^1=R^2=H, R^3=Cl, X=C
71. R=H, R^1=R^2=H, X=N

72

FIGURE 5.11 Pyridazine-based caspase inhibitors.

evaluated in this assay, only caspase-3 was potently and noticeably inhibited by the tested AQZs. Compounds **59–61** are examples of selective caspase-3 inhibitors and were further evaluated in cellular assays of caspase-3 activity.

AQZs inhibited endogenous caspase-3 activity as assessed using a cell-permeable, exogenously added substrate in staurosporine-treated SH-SY5Y cells. These agents also reduced biochemical and cellular features of apoptosis that are thought to be a consequence of caspase-3 activation, including DNA fragmentation and various morphological features that define the terminal stages of apoptotic cell death. Moreover, aminoquinazolines **59–67** inhibited apoptosis induced by nerve growth factor withdrawal from differentiated PC12 cells. Based on these observations, quinazoline-based compounds **56–67** represent a structurally novel class of inhibitors, some of which selectively inhibit caspase-3 and thereby allow evaluation of the role of caspase-3 activity in various cellular models of apoptosis. Several AQZs are equipotent against caspase-6, although most lack activity against this isoform and caspase-1, -2, -7, and -8. Additionally, it was recently reported that the 2-phenyl-4-quinozolinone derivative HMJ-38 stimulates the cleavage of pro-caspase-9 and -3. The significant reduction of caspase activity and apoptosis indicated that the HMJ-38-induced apoptosis was mainly mediated by activation of caspase-9 and -3.[79]

3-Chloro-4-carboxamido-6-arylpyridazines **68–72** (Figure 5.11) were identified as time-dependent inhibitors of caspase-1,[80] with the most potent compound from this series inhibiting caspase-1 with an IC$_{50}$ in the low micromolar range. The mechanism of this inactivation is believed to involve alkylation of the enzyme's active site thiol by displacement of the 3-chloro group.

5.7 α-KETOAMIDES, INDOLONES, AND ISATINS

α-Ketoamides, e.g., **73**, along with cyclic 5-aminosulfonyl-2-indolones **74–82**[81] and 1,5-substituted isatins **83–97**[82-85] were identified as caspase inhibitors (Figures 5.12 and 5.13). For example, 5-nitroisatins **83** and **84** developed by Smith Kline Beecham

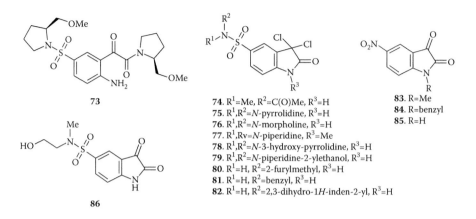

FIGURE 5.12 α-Ketoamide, indolone, and isatin-based caspase inhibitors.

featured IC$_{50}$ values of 1 and 0.25 μM against caspase-3, respectively, whereas a related analogue **85** was less active (IC$_{50}$ = 3 μM). The electrophilicity of the reactive carbonyl moiety in isatins was critical for their inhibitory activity, suggesting a mechanism of action that involves nucleophilic attack of the catalytic cysteine residue of the enzyme (Figure 5.14). Further development of this series yielded 5-sulfamoylisatins **86–97**.[83] Compound **90** inhibited caspase-3 and -7 with IC$_{50}$ values of 15 and 47 nM, respectively. The importance of the chiral methoxymethylpyrrolidine group for caspase binding potential was emphasized,[84] and it was found that the S-enantiomer and the open-ring derivative **86** were significantly less active. Increasing the ring size from five to seven atoms had little effect on activity, but decreasing it to an azetidine resulted

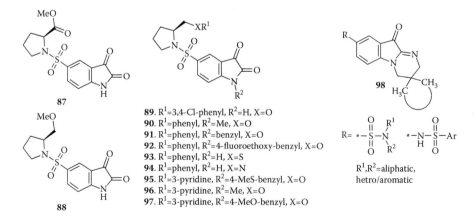

FIGURE 5.13 Isatin-like caspase inhibitors.

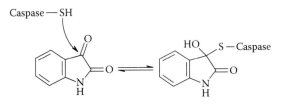

FIGURE 5.14 Possible mechanism of caspase inactivation via SH-group interaction with active carbonyl group in the isatins.

in a tenfold increase in potency relative to that of unsubstituted pyrrole. A co-crystal structure of the complex between **90** and caspase-3 demonstrated that the catalytic cysteine thiol and isatin ketone carbonyl group form a tetrahedral intermediate, and the S1 subsite was occupied only by a water molecule.[85] The pyrrolidine ring in **90** is involved in hydrophobic interactions with the S2 pocket of the enzyme, which are responsible for the observed high selectivity. WC-II-89 **92** demonstrated potent inhibition of caspase-3 and -7, and high selectivity against caspase-1, -6, and -8. Biodistribution studies using ^{18}F-labeled WC-II-89[86] revealed higher uptake in liver and spleen of cycloheximide-treated rats, an animal model of apoptosis, than in control animals, and MicroPET imaging studies confirmed the high uptake of the radiotracer in the liver of a cycloheximide-treated rat compared to the untreated control pattern. The presence of activated caspase-3 in the liver and spleen of cycloheximide-treated animals was confirmed by Western blot analysis. Therefore, ^{18}F-labeled WC-II-89 could be used as a potential radiotracer for imaging caspase-3 activation in tissues undergoing apoptosis. Several cyclic bioisosteric analogues were recently developed by Wyeth Pharmaceuticals,[87] for example, 3,4-dihydropyrimido[1,2-a]indol-10(2H)-one **98** selectively inhibited caspase-3 and -7 activity at micromolar concentrations.

In general, these studies demonstrated impressive *in vitro* efficacy and selectivity of substituted isatins. However, despite the high *in vitro* potency, their efficacy in cell-based models of apoptosis was moderate; for example, the anti-apoptotic activity of **90** in a cell-based model of caspase-3-dependent apoptosis was very low (ED$_{50}$ of about 10 μM). The attenuated potency of the inhibitor has been attributed to the reversible binding between **90** and cytosolic constituents.[83]

5.8 ISOQUINOLINE-1,3,4-TRIONES

A series of isoquinoline-1,3,4-trione derivatives were identified as potent inhibitors of caspases and promising anti-apoptotic agents through structural modification of the original isoquinoline **99** (IC$_{50}$ value of 0.15 μM for caspase-3) (Figure 5.15, Table 5.5).[88] Based on high-throughput screening of a diverse small molecule library of 8,000 compounds using recombinant caspase-3, various *N*-substituted derivatives of the starting molecule **99** were identified as potent caspase-2, -3, -6, -7, and -8 inhibitors. It was also found that introduction of an *N*-acyl group improved activity against caspase-3 (compounds **100–103**). Some of these agents

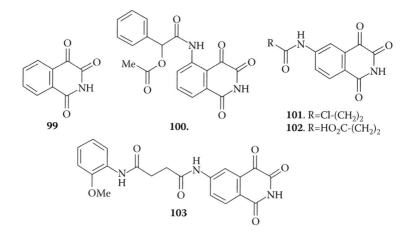

FIGURE 5.15 Isoquinoline-like triones as caspase inhibitors.

showed significant protection against apoptosis in cells and exhibited a dose-dependent decrease in infarct volume (up to 77% compared to vehicle control) in a transient middle cerebral artery occlusion (MCAO) stroke model. For example, compound **103** entered preclinical studies as an agent for the treatment of diseases involving abnormally upregulated apoptosis.[88] Furthermore, it was recently shown that these inhibitors attenuated apoptosis induced by beta-amyloid in PC12 cells and primary neuronal cells.[89] It was also reported that compounds containing this core fragment inhibit caspase-1 activity in an irreversible and slow-binding manner, and inhibit cellular caspase-1 activity and the maturation of interleukin-1β in U-937 cells.[90]

TABLE 5.5
***In Vitro* Activity of Isoquinoline-1,3,4-Triones 99–103 against Caspase Family Enzymes**

Compound	IC$_{50}$ (μM)				
	Caspase-2	Caspase-3	Caspase-6	Caspase-7	Caspase-8
99	1.53 ± 0.24	0.149 ± 0.015	0.474 ± 0.083	0.386 ± 0.034	1.91 ± 0.15
100	0.233 ± 0.027	0.040 ± 0.003	0.216 ± 0.014	0.063 ± 0.007	0.425 ± 0.055
101	0.53 ± 0.18	0.083 ± 0.010	0.283 ± 0.061	0.139 ± 0.015	0.575 ± 0.003
102	0.859 ± 0.073	0.068 ± 0.006	0.201 ± 0.006	0.136 ± 0.014	1.12 ± 0.043
103	0.537 ± 0.035	0.113 ± 0.01	0.137 ± 0.006	0.218 ± 0.024	0.835 ± 0.016

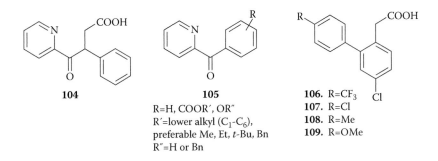

FIGURE 5.16 Pyridine- and 1,1′-biphenyl-containing carboxylic acids as caspase inhibitors.

5.9 PYRIDINE-BASED CARBOXYLIC ACIDS AND THEIR ANALOGUES

Several pyridine-based carboxylic acids were recently found to indirectly inhibit caspase-1. Based on x-ray crystallographic analysis and a right-handed cartesian coordinate system, Albrecht et al. have identified twenty-two binding sites within caspase-1 and developed several pyridine derivatives, each containing a carboxylic functionality, e.g., compounds **104** and **105** (Figure 5.16), as novel compounds that indirectly inhibit caspase-1.[91] The tested compounds commonly alkylated caspase-1 hydroxy or mercapto groups to generate an irreversible covalent bond, and this was shown to occur at three separate binding sites on caspase-1. Notably, one of these binding sites was established as a known site that was involved in inhibition of caspase-1. Therefore, the tested compounds can be broadly classified as potential inhibitors of caspase-1 activity. For example, phenyl(pyridin-2-yl)methanone was shown to significantly block caspase-1 activity at 10^{-6} mol/l. Lardy et al. designed a 1,1′-biphenyl-2-ylacetic acid template that is structurally related to pyridine-based acids **104** and **105**,[92] and compounds **106–109** selectively inhibited caspase-10 activity (IC$_{50}$ ca. 9–23 µM, Table 5.6). These compounds were claimed as being

TABLE 5.6
In Vitro Activity of Compounds 106-109 against a Panel of Recombinant Caspases

Compound	IC$_{50}$ (µM)					
	Caspase-2	Caspase-3	Caspase-6	Caspase-8	Caspase-9	Caspase-10
106	>200	>200	>200	>200	>200	23.6
107	>200	>200	>200	>200	>200	10.1
108	>200	>200	>200	>200	>200	13.1
109	>200	>200	>200	>200	>200	9.0

potentially useful for the treatment of diabetic retinopathy. It is worthy to note that compounds **104–109** are topologically related to NCX-4016 **9** (see Figure 5.1).

5.10 PYRROLO[3,4-C]QUINOLINE CASPASE INHIBITORS

Ivachtchenko et al. have described a promising class of capase-3 inhibitors discovered using a focused library of protease inhibitors (Figure 5.17).[93–95] For example, 1,3-dioxo-2,3-dihydro-1H-pyrrolo[3,4-c]quinolines **110** and **111a–j** inhibit caspase-3 catalyzed proteolytic breakdown of its fluorogenic substrate, Ac-DEVD-AMC. The most active compounds featured alkyl, aryl, and heteroaryl 4-substituents displaying activity in the 20–60 nM range. Potency of these agents depended on the nature of the substituent in position 4 of this heterocyclic system, with 4-methyl and 4-phenyl derivatives being the most potent molecules (IC_{50} = 23–27 nM, Table 5.7).[93] A selectivity profile against a panel of recombinant human caspases (1–10) determined that the exemplified compounds were more potent against caspase-3 ($logIC_{50}$ = 7.53), but the level of selectivity versus the remaining caspases was modest (Table 5.8).

Pyrrolo[3,4-c]quinolines **112a–c** were also active against caspase-3,[94] and the most active compounds within this series, **112b** and **112c**, inhibited caspase-3 activity with an IC_{50} = 6 nM. Kinetic data indicated that these were noncompetitive and reversible inhibitors of caspase-3.

Compounds of general structure **113a–g** (Figure 5.17) have also been tested for their ability to inhibit caspase-3.[95] For a group of 2-unsubstituted compounds

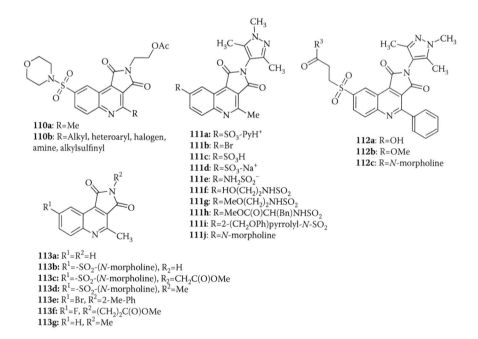

110a: R=Me
110b: R=Alkyl, heteroaryl, halogen, amine, alkylsulfinyl

111a: R=SO$_3$-PyH$^+$
111b: R=Br
111c: R=SO$_3$H
111d: R=SO$_3$-Na$^+$
111e: R=NH$_2$SO$_2^-$
111f: R=HO(CH$_2$)$_2$NHSO$_2$
111g: R=MeO(CH$_2$)$_2$NHSO$_2$
111h: R=MeOC(O)CH(Bn)NHSO$_2$
111i: R=2-(CH$_2$OPh)pyrrolyl-N-SO$_2$
111j: R=N-morpholine

112a: R=OH
112b: R=OMe
112c: R=N-morpholine

113a: R^1=R^2=H
113b: R^1=-SO$_2$-(N-morpholine), R$_2$=H
113c: R^1=-SO$_2$-(N-morpholine), R$_2$=CH$_2$C(O)OMe
113d: R^1=-SO$_2$-(N-morpholine), R^2=Me
113e: R^1=Br, R^2=2-Me-Ph
113f: R^1=F, R^2=(CH$_2$)$_2$C(O)OMe
113g: R^1=H, R^2=Me

FIGURE 5.17 Pyrrolo[3,4-c]quinoline-based caspase inhibitors.

TABLE 5.7
In Vitro **Caspase-3 Inhibitory Potency**
of Substituted Pyrrolo [3,4-c]Quinoline-
1,3-Diones 110–115

Compound	Caspase-3 IC$_{50}$ (μM)
110	0.023–27.9
111a–j	0.004–0.36
112a	0.020
112b	0.006
112c	0.006
113a	>0.1
113b	0.00021
113c	0.000016
113d	0.000044
113e	0.00254
113f	0.00108
113g	0.00636

with R^1 = H, the IC$_{50}$ value exceeded 100 μM; 8-fluoro derivatives (R^1 = F) and 8-bromo derivatives (R^1 = Br) have IC$_{50}$ values of 62.8 and 37.1 μM, respectively; and an 8-(morpholine-4-sulfonyl)-substituted compound had an IC$_{50}$ = 0.21 μM. In this series, the activity changed by three orders of magnitude. Similar dependencies were observed within all other congeneric series with identical 2-substituents. The observed correlations between the electron-withdrawing ability of the 8-substituent (R^1) and potency suggested that electrophilicity of the imide carbonyls is important for activity. The nature of 2-substituents also influenced the activity of the synthesized compounds against caspase-3, and in all the studied congeneric series with identical 8-substituents, minimal activity was observed for 2-unsubstituted compounds.

The mechanism of inhibition has been studied for a large variety of compounds possessing electrophilic carbonyls, such as peptide aldehydes, isatins, homophthalimides, quinazolinones, etc.[96–98] In characterized cases, the mechanism involved addition of the enzyme's catalytic cysteine residue to the carbonyl moiety (Figure 5.18). The ability of thiols to reversibly interact with phthalimide-like compounds in a similar manner[99] suggests that caspases could be reversibly inactivated by the electrophilic carbonyls of compounds **110–113**, assuming appropriate substituents R, R^1, and R^2 are present.

Based on these studies, pyrrolo[3,4-c]quinoline caspase inhibitors represent a novel and promising structural class of caspase inhibitors potentially useful for the treatment of diseases involving abnormally upregulated apoptosis. Thus, to assess the anti-apoptotic activity of 2-[4-methyl-8-(morpholin-4-ylsulfonyl)-1,3-dioxo-1,3-dihydro-2H-pyrrolo[3,4-c]quinolin-2-yl]ethyl acetate (IC$_{50}$ = 23 $\pm$ 2 nM) in a

TABLE 5.8
In Vitro Activity of Compound 110a against a Panel of Recombinant Caspases

Compound	Caspase-1	Caspase-2	Caspase-3	Caspase-4	Caspase-5	Caspase-6	Caspase-7	Caspase-8	Caspase-9	Caspase-10
					$logIC_{50}$ (nM)					
100a	6.67	5.57	7.53	6.64	6.47	6.24	6.48	6.46	5.74	5.76

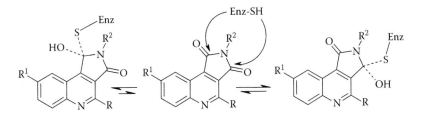

FIGURE 5.18 The mechanism of inhibition of caspases by pyrriodione-containing compounds.

cell-based assay, the viability of human Jurkat T cells treated with 10 µM staurosporin was measured. The tested compound demonstrated a higher level of protection than a cell-permeable z-VAD-FMK peptide inhibitor.

5.11 CASPASE INHIBITORS OF MISCELLANEOUS STRUCTURAL TYPES

Several other small molecule compounds were recently identified as potent caspase inhibitors (Figure 5.19). For example, dihydropyrazolo[1,5-*a*][1,3,5]triazin **114** inhibited caspase-1 activity with micromolar activity.[100] berkeleydione **115** and berkeleytrione **116**, new bioactive metabolites from an acid mine organism, were developed by Harvard Medical School and the University of Montana as potent caspase inhibitors.[101] Finally, the natural compound discorhabdin P **117** was discovered by Harbor Branch Oceanographic Institution to be a potent caspase-3 inhibitor, with possible use as a potential anti-apoptotic agent.[102]

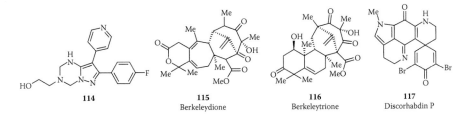

FIGURE 5.19 Caspase inhibitors of miscellaneous structural types.

5.12 CONCLUSIONS

Apoptosis, or programmed cell death, is a common property of all multicellular organisms. It can be triggered by a number of factors, including ultraviolet or γ-irradiation, growth factor withdrawal, chemotherapeutic drugs, or signaling via death receptors. Caspases play a central role in the regulation and execution of apoptotic cell death. Increased levels of apoptosis and caspase activity are frequently observed at sites of cellular damage in both acute (e.g., myocardial infarction, stroke, sepsis) and chronic (e.g., Alzheimer's, Parkinson's, and Huntington's disease) disorders. Caspase inhibition has been demonstrated to be therapeutically effective in moderating excessive programmed cell death.[9,10]

Caspase inhibitors that displayed efficacy in preclinical tests include peptides, peptidomimetics, and nonpeptide small molecules. However, the peptide-based and peptidomimetic caspase inhibitors often demonstrate pharmacokinetic and physicochemical liabilities restricting their utility *in vivo*. Therapeutic efficacy of these inhibitors has also been limited due to poor cell permeability, marginal activity in intact cells, and metabolic instability. In addition, the complex structure of many peptides and peptidomimetics poses challenges to their structural optimization. As a result, small molecule caspase inhibitors that could address these deficiencies are being pursued by many pharmaceutical companies.

Development of nonpeptide small molecule inhibitors of caspases could overcome these limitations, and therefore, extensive studies of such agents have been performed in the past decade. As a result, several preclinical and at least one clinical candidate have been developed, which are exemplified in Table 5.9. The available *in vitro* and *in vivo* data suggest many potential benefits associated with their use in the treatment of many diseases involving abnormally upregulated apoptosis. Although

TABLE 5.9
Preclinical and Clinical Development of Nonpeptide Small Molecule Inhibitors

Name	Target Caspases	Pathologies
NO-Ursodeoxy-cholic acid, (NCX-1000)[25,26] (**10**, Figure 5.1), phase II	Caspase-3, -8, and -9	Chronic liver diseases
NEPP-11[64] (**35**, Figure 5.6), preclinical	Caspase-3	Neurodegenerative diseases, stroke
YM-215438[72] (**51**, Figure 5.9), preclinical	Caspase-3	Alzheimer's, Parkinson's, and Huntington's diseases, ischemic stroke, dermatologic disorders, cardiovascular diseases
UCF-101[72] (**52**, Figure 5.9), preclinical	Caspase-3 and -9	Alzheimer's, Parkinson's, and Huntington's diseases, ischemic stroke, dermatologic disorders, cardiovascular diseases
WC-II-89[88] (**92**, Figure 5.13), preclinical	Caspase-1, -2, -3, -6, -7, and -8	Diseases involving abnormally upregulated apoptosis

significant progress toward this goal has been achieved, clinical efficacy of synthetic nonpeptide caspase inhibitors remains to be demonstrated. Despite impressive *in vitro* enzyme inhibition data, the majority of such agents are still in early stages of development. The clinical effectiveness of small molecule caspase inhibitors may be compromised by non-drug-like properties, e.g., the presence of highly reactive electrophilic warheads, and a possible lack of sufficient selectivity.

In many cases, nonpeptide inhibitors of caspases can represent new tools to further explore the role of different caspase enzymes in various cellular models of apoptosis. This should facilitate a better understanding of *in vivo* settings where these enzymes play a critical role in the execution of the apoptotic death response. Such studies will help elucidate both the therapeutic value and potential mechanism-based liabilities of inhibition by particular caspase enzymes.

Because of the potential detrimental effects on cell survival due to inappropriate activation or inhibition of apoptotic events, apoptotic pathways have to be tightly controlled in human cells. Thus, the enhanced expression of endogenous inhibitors of apoptotic mechanisms (such as IAP) is often observed in various cancer cells.[103] Therefore, the development of an anti-apoptotic therapy for chronic administration may be associated with potential toxic effects caused by the abrogation of normal apoptosis in the human adult, which accounts for over 10^{11} cell deaths per day.[104] A major goal in the development of small molecule therapeutic agents for treatment of chronic degenerative diseases is the prevention of unwanted cell death while not promoting proliferative diseases (e.g., cancer, rheumatoid arthritis–associated synovial hypertrophy).

In this context, understanding the influence of anti-apoptotic agents on specific apoptotic proteases and their function in highly proliferative cells is an important challenge. Nonselective nonpeptide caspase inhibitors described in this article are likely to be used only in acute settings such as salvaging cells destined to die by apoptosis following stroke, myocardial ischemia, or acute liver degeneration. In these cases, the therapeutic window is short, and so global apoptosis inhibition is unlikely to be an issue. However, if cancer cell survival is indeed enhanced by nonselective caspase inhibition, then therapeutic approaches will need to be selective for the target cell or tissue type so that general cellular hypertrophy is not stimulated. More tissue-specific apoptosis inhibitors, in terms of activity against a specific caspase or directed tissue distribution, are probably required for chronic neurodegenerative diseases or osteoarthritis, which will ultimately make these more difficult targets. Overcoming this challenge will probably be required for the successful clinical development of nonpeptide caspase inhibitors.

REFERENCES

1. Jin, Z., and El-Deiry, W. S. 2005. Overview of cell death signaling pathways. *Cancer Biol. Ther.* 4:139.
2. Philchenkov, A. 2004. Caspases: Potential targets for regulating cell death. *J. Cell. Mol. Med.* 8:432.
3. Fan, T. J., et al. 2005. Caspase family proteases and apoptosis. *Acta Biochim. Biophys Sin.* 37:719.

4. Nicholson, D. W. 1999. Caspase structure, proteolytic substrates and function during apoptotic cell death. *Cell Death Differ.* 6:1028.

5. Schwerk, C., and Schulze-Osthoff, K. 2003. Non-apoptotic functions of caspases in cellular proliferation and differentiation. *Biochem. Pharmacol.* 66:1453.

6. Mattson, M. P., Partin, J., and Begley, J. G. 1998. Amyloid beta-peptide induces apoptosis-related events in synapses and dendrites. *Brain Res.* 807:167.

7. Gutekunst, C. A., Norflus, F., and Hersch, S. M. 2000. Recent advances in Huntington's disease. *Curr. Opin. Neurol.* 13:445.

8. Hartley, J. M., et al. 1994. Complex I inhibitors induce dose-dependent apoptosis in PC12 cells: Relevance to Parkinson's disease. *J. Neurochem.* 63:1987.

9. O'Brien, T., and Lee, D. 2004. Prospects for caspase inhibitors. *Mini Rev. Med. Chem.* 4:153.

10. Linton, S. D. 2005. Caspase inhibitors: A pharmaceutical industry perspective. *Curr. Top. Med. Chem.* 5:1697.

11. Bonavida, B., et al. 2006. Therapeutic potential of nitric oxide in cancer. *Drug Resist. Updat.* 9:157.

12. Chung, H. T., et al. 2001. Nitric oxide as a bioregulator of apoptosis. *Biochem. Biophys. Res. Commun.* 282:1075.

13. Kim, Y. M., et al. 2000. Nitric oxide prevents tumor necrosis factor alpha-induced rat hepatocyte apoptosis by the interruption of mitochondrial apoptotic signaling through S-nitrosylation of caspase-8. *Hepatology* 32:770.

14. Dimmeler, S., et al. 1997. Angiotensin II induces apoptosis of human endothelial cells. Protective effect of nitric oxide. *Circ. Res.* 81:970.

15. Ascenzi, P., et al. 2001. Inhibition of cysteine protease activity by NO-donors. *Curr. Protein Pept. Sci.* 2:137.

16. Rossig, L., et al. 1999. Nitric oxide inhibits caspase-3 by S-nitrosation *in vivo. J. Biol. Chem.* 274:6823.

17. Kim, Y. M., et al. 1998. Nitric oxide prevents IL-1beta and IFN-gamma-inducing factor (IL-18) release from macrophages by inhibiting caspase-1 (IL-1beta-converting enzyme). *J. Immunol.* 161:4122.

18. Brune, B., von Knethen, A., and Sandau, K. B. 1998. Nitric oxide and its role in apoptosis. *Eur. J. Pharmacol.* 351:261.

19. Weiland, U., et al. 2000. Inhibition of endogenous nitric oxide synthase potentiates ischemia-reperfusion-induced myocardial apoptosis via a caspase-3 dependent pathway. *Cardiovasc. Res.* 45:671.

20. Stamler, J. S., et al. 1992. Nitric oxide circulates in mammalian plasma primarily as an S-nitroso adduct of serum albumin. *Proc. Natl. Acad. Sci. USA* 89:7674.

21. Zhengmao, G., et al. 2001. *N*-Nitrosoanilines: A new class of caspase-3 inhibitors. *Bioorg. Med. Chem.* 9:99.

22. Guo, Z.-M., et al. 1996. *S*-Nitrosylation of proteins by *N*-alkyl-*N*-nitrosoanilines. *Bioorg. Med. Chem. Lett.* 6:573.

23. Fiorucci, S., et al. 2000. IL-1 beta converting enzyme is a target for nitric oxide-releasing aspirin: New insights in the antiinflammatory mechanism of nitric oxide-releasing nonsteroidal antiinflammatory drugs. *J. Immunol.* 165:5245.

24. Fiorucci, S., et al. 2002. NCX-4016, a nitric oxide-releasing aspirin, protects endothelial cells against apoptosis by modulating mitochondrial function. *FASEB J.* 16:1645.

25. Fiorucci, S., et al. 2004. Treatment of portal hypertension with NCX-1000, a liver-specific NO donor. *Cardiovasc. Drug Rev.* 22:135.

26. Fiorucci, S., et al. 2004. Liver delivery of NO by NCX-1000 protects against acute liver failure and mitochondrial dysfunction induced by APAP in mice. *Br. J. Pharmacol.* 143:1.

27. Ahn, Y. H., et al. 1998. Depletion of intracellular zinc induces protein synthesis-dependent neuronal apoptosis in mouse cortical culture. *Exp. Neurol.* 154:47.

28. Stennicke, H. R., and Salvesen, G. S. 1997. Biochemical characteristics of caspases-3, -6, -7, and -8. *J. Biol. Chem.* 272:25719.

29. Perry, D. K., et al. 1997. Zinc is a potent inhibitor of the apoptotic protease, caspase-3. A novel target for zinc in the inhibition of apoptosis. *J. Biol. Chem.* 272:18530.

30. Wei, Q., et al. 2004. Inhibition of apoptosis by Zn^{2+} in renal tubular cells following ATP depletion. *Am. J. Physiol. Renal Physiol.* 287:F492.

31. Albeck, M., et al. 1998. Tellurium compounds: Selective inhibition of cysteine proteases and model reaction with thiols. *Inorg. Chem.* 37:1704.

32. Yuan, C., et al. 2000. Possible role of caspase-3 inhibition in cadmium-induced blockage of apoptosis. *Toxicol. Appl. Pharmacol.* 164:321.

33. Coutant, A., et al. 2006. Cadmium-induced apoptosis in lymphoblastoid cell line: Involvement of caspase-dependent and -independent pathways. *Biochimie* 88:1815.

34. Takahashi, A., et al. 1997. Inhibition of ICE-related proteases (caspases) and nuclear apoptosis by phenylarsine oxide. *Exp. Cell. Res.* 231:123.

35. Gurr, J. R., et al. 1999. Dithiothreitol enhances arsenic trioxide-induced apoptosis in NB4 cells. *Mol. Pharmacol.* 56:102.

36. Estrov, Z., et al. 1999. Phenylarsine oxide blocks interleukin-1beta-induced activation of the nuclear transcription factor NF-kappaB, inhibits proliferation, and induces apoptosis of acute myelogenous leukemia cells. *Blood* 94:2844.

37. Wolfe, J. T., Ross, D., and Cohen, G. M. 1994. A role for metals and free radicals in the induction of apoptosis in thymocytes. *FEBS Lett.* 352:58.

38. Bessho, R., et al. 1994. Pyrrolidine dithiocarbamate, a potent inhibitor of nuclear factor kappa B (NF-kappa B) activation, prevents apoptosis in human promyelocytic leukemia HL-60 cells and thymocytes. *Biochem. Pharmacol.* 48:1883.

39. Albrecht, H., Tschopp, J., and Jongeneel, C. V. 1994. Bcl-2 protects from oxidative damage and apoptotic cell death without interfering with activation of NF-kappa B by TNF. *FEBS Lett.* 351:45.

40. Nobel, C. S., et al. 1997. Mechanism of dithiocarbamate inhibition of apoptosis: Thiol oxidation by dithiocarbamate disulfides directly inhibits processing of the caspase-3 proenzyme. *Chem. Res. Toxicol.* 10:636.

41. Nobel, C. S., et al. 1997. Disulfiram is a potent inhibitor of proteases of the caspase family. *Chem. Res. Toxicol.* 10:1319.

42. Zhao, A., et al. 2000. Disulfiram inhibits TNF-alpha-induced cell death. *Cytokine* 12:1356.

43. Dragovich, P. S., et al. 1998. Structure-based design, synthesis, and biological evaluation of irreversible human rhinovirus 3C protease inhibitors. 1. Michael acceptor structure-activity studies. *J. Med. Chem.* 41:2806.

44. Nakashima, I., et al. 2003. 4-Hydroxynonenal triggers multistep signal transduction cascades for suppression of cellular functions. *Mol. Aspects Med.* 24:231.

45. Davis, D. W., Hamilton, R. F., Jr., and Holian, A. 1997. 4-Hydroxynonenal inhibits interleukin-1 beta converting enzyme. *J. Interferon Cytokine Res.* 17:205.

46. Larini, A., Bianchi, L., and Bocci, V. 2004. Effect of 4-hydroxynonenal on antioxidant capacity and apoptosis induction in Jurkat T cells. *Free Radic. Res.* 38:509.

47. Awasthi, Y. C., et al. 2003. Role of 4-hydroxynonenal in stress-mediated apoptosis signaling. *Mol. Aspects Med.* 24:219.

48. Rabacchi, S. A., et al. 2004. Divergence of the apoptotic pathways induced by 4-hydroxynonenal and amyloid beta-protein. *Neurobiol. Aging* 25:1057.

49. Niculescu, R., et al. 1995. Inhibition of the conversion of pre-interleukins-1 alpha and 1 beta to mature cytokines by p-benzoquinone, a metabolite of benzene. *Chem. Biol. Interact.* 98:211.

50. Kokel, D., and Xue, D. 2006. A class of benzenoid chemicals suppresses apoptosis in *C. elegans. Chembiochemistry* 7:2010.
51. Hazel, B., Baum, G., and Kalf, G. 1996. Hydroquinone, a bioreactive metabolite of benzene, inhibits apoptosis in myeloblasts. *Stem. Cells* 14:730.
52. Ibuki, Y., and Goto, R. 2004. Dysregulation of apoptosis by benzene metabolites and their relationships with carcinogenesis. *Biochim. Biophys. Acta* 1690:11.
53. Salvatore, M. J., et al. 1994. L-741,494, a fungal metabolite that is an inhibitor of interleukin-1 beta converting enzyme. *J. Nat. Prod.* 57:755.
54. Graczyk, P. P. 2002. Caspase inhibitors as anti-inflammatory and antiapoptotic agents. *Prog. Med. Chem.* 39:1.
55. Koizumi, F., et al. 2004. EI-2346, a novel interleukin-1beta converting enzyme inhibitor produced by *Streptomyces* sp. E-2346. II. Structure elucidation. *J. Antibiot.* 57:289.
56. Chen, M., et al. 2000. Minocycline inhibits caspase-1 and caspase-3 expression and delays mortality in a transgenic mouse model of Huntington disease. *Nat. Med.* 6:797.
57. Gunasekera, S. P., et al. 1999. Secobatzellines A and B, two new enzyme inhibitors from a deep-water Caribbean sponge of the genus Batzella. *J. Nat. Prod.* 62:1208.
58. Thanh, G. L., et al. 2006. Organic azide inhibitors of cysteine proteases. *J. Am. Chem. Soc.* 128:12396.
59. Koizumi, F., Matsuda, Y., and Nakanishi, S. 2003. EI-1941-1 and -2, novel interleukin-1beta converting enzyme inhibitors produced by *Farrowia* sp. E-1941. I. Biochemical characterization of EI-1941-1 and -2. *J. Antibiot.* 56:464.
60. Shoji, M., et al. 2005. Enantio- and diastereoselective total synthesis of EI-1941-1, -2, and -3, inhibitors of interleukin-1beta converting enzyme, and biological properties of their derivatives. *J. Org. Chem.* 70:9905.
61. Tsukuda, E., et al. 1996. EI-1507-1 and -2, novel interleukin-1 beta converting enzyme inhibitors produced by *Streptomyces* sp. E-1507. *J. Antibiot.* 49:333.
62. Tanaka, T., et al. 1996. EI-1511-3, -5 and EI-1625-2, novel interleukin-1 beta converting enzyme inhibitors produced by *Streptomyces* sp. E-1511 and E-1625. III. Biochemical properties of EI-1511-3, -5 and EI-1625-2. *J. Antibiot.* 49:1085.
63. Tanaka, T., et al. 1996. EI-1511-3, -5 and EI-1625-2, novel interleukin-1 beta converting enzyme inhibitors produced by *Streptomyces* sp. E-1511 and E-1625. I. Taxonomy of producing strain, fermentation and isolation. *J. Antibiot.* 49:1073.
64. Hirata, Y., et al. 2004. Anti-apoptotic and pro-apoptotic effect of NEPP11 on manganese-induced apoptosis and JNK pathway activation in PC12 cells. *Brain. Res.* 1021:241.
65. Koizumi, F., et al. 2003. EI-2128-1, a novel interleukin-1beta converting enzyme inhibitor produced by *Penicillium* sp. E-2128. *J. Antibiot.* 56:891.
66. Matsumoto, T., et al. 1999. Novel cyclopentanone derivatives pentenocins A and B with interleukin-1beta converting enzyme inhibitory activity, produced by *Trichoderma hamatum* FO-6903. *J. Antibiot.* 52:754.
67. Sugahara, T., Fukuda, H., and Iwabuchi, Y. 2004. Total synthesis and absolute stereochemistry of pentenocin B, a novel interleukin-1 beta converting enzyme inhibitor. *J. Org. Chem.* 69:1744.
68. Wipf, P., Jeger, P., and Kim, Y. 1998. Thiophilic ring-opening and rearrangement reactions of epoxyketone natural products. *Bioorg. Med. Chem. Lett.* 8:351.
69. Levy, M. A., and Gleason, J. G. 1997. Interleukin converting enzyme and apoptosis. WO Patent 007805.
70. Dolle, R. E., and Rinker, J. M. 1995. Halomethyl amides as IL-1β protease inhibitors, WO Patent 9529672, 1995. *Chem. Abstr.* 124:76524.
71. Albreht, H.-P., et al. 1996. Medicinal use of pyrimidotriazine-dione derivatives, DE Patent 19534164, 1996. *Chem. Abstr.* 125:336889.

72. Ohmori, J., et al. 2001. Discovery, synthesis, and SAR of novel non-peptide, small molecule and zinc(II)-potentiated caspase-3 inhibitors. In *21st Med. Chem. Symp.*, November 28–30, Kyoto, Japan, abstract 1P-30.

73. Cilenti, L., et al. 2003. Characterization of a novel and specific inhibitor for the pro-apoptotic protease Omi/HtrA2. *J. Biol. Chem.* 278:11489.

74. Liu, H. R., et al. 2005. Role of Omi/HtrA2 in apoptotic cell death after myocardial ischemia and reperfusion. *Circulation* 111:90.

75. Klupsch, K., and Downward, J. 2006. The protease inhibitor Ucf-101 induces cellular responses independently of its known target, HtrA2/Omi. *Cell Death Differ.* 13:2157.

76. Kim, S.-G., et al. 2006. Quinoline derivatives as caspase-3 inhibitor, preparation for producing the same and pharmaceutical composition comprising the same. U.S. Patent 7009053.

77. Greenfield, A., et al. 2005. Design, synthesis and biological activity of novel quinazolinone-based caspase 3 inhibitors. In *230th ACS National Meeting*, Washington, DC, August 28–September 1, abstract MEDI 318.

78. Scott, C. V., et al. 2003. Novel small molecule inhibitors of caspase-3 block cellular and biochemical features of apoptosis. *J. Pharm. Exp. Ther.* 304:433.

79. Yang, J. S., et al. 2004. Selective induction of G2/M arrest and apoptosis in HL-60 by a potent anticancer agent, HMJ-38. *Anticancer Res.* 24:1769.

80. Dolle, R. E., et al. 1997. 3-Chloro-4-carboxamido-6-arylpyridazines as a non-peptide class of interleukin-1β converting enzyme inhibitor. *Bioorg. Med. Chem. Lett.* 7:1003.

81. Brady, K. D., et al. 1999. A catalytic mechanism for caspase-1 and for bimodal inhibition of caspase-1 by activated aspartic ketones. *Bioorg. Med. Chem.* 7:621.

82. Wenhua, C., et al. 2005. *N*-Benzylisatin sulfonamide analogues as potent caspase-3 inhibitors: Synthesis, *in vitro* activity, and molecular modeling studies. *J. Med. Chem.* 48:7637.

83. Lee, D., et al. 2001. Potent and selective nonpeptide inhibitors of caspases 3 and 7. *J. Med. Chem.* 44:2015.

84. Kopka, K., et al. 2006. 5-Pyrrolidinylsulfonyl isatins as a potential tool for the molecular imaging of caspases in apoptosis. *J. Med. Chem.* 49:6704.

85. Lee, D., et al. 2000. Potent and selective nonpeptide inhibitors of caspases 3 and 7 inhibit apoptosis and maintain cell functionality. *J. Biol. Chem.* 275:16007.

86. Zhou, D., et al. 2006. Synthesis, radiolabeling, and *in vivo* evaluation of an ^{18}F-labeled isatin analog for imaging caspase-3 activation in apoptosis. *Bioorg. Med. Chem. Lett.* 16:5041.

87. Dollings, P. J., Dietrich, A. J., and Havran, S. M. 2005. Pyrimidoindolones and methods for using same. U.S. Patent 20050250798A1.

88. Chen, Y.-H., et al. 2006. Design, synthesis, and biological evaluation of isoquinoline-1,3,4-trione derivatives as potent caspase-3 inhibitors. *J. Med. Chem.* 49:1613.

89. Zhang, Y. H., et al. 2006. Isoquinoline-1,3,4-trione and its derivatives attenuate beta-amyloid-induced apoptosis of neuronal cells. *FEBS J.* 273:4842.

90. Ma, X. Q., et al. 2007. Novel irreversible caspase-1 inhibitor attenuates the maturation of intracellular interleukin-1beta. *Biochem. Cell Biol.* 85:56.

91. Albrecht, H.-P., et al. 1998. Organic-chemical compound with ice-inhibitory action, U.S. Patent 5798247, 1998. *Chem. Abstr.* 129:186149.

92. Lardy, C., et al. 2006. Selective caspase 10 inhibitors potentially useful for the treatment of diabetic retinopathy. WO Patent 2006058592.

93. Kravchenko, D. V., et al. 2005. Synthesis and structure-activity relationship of 4-substituted 2-(2-acetyloxyethyl)-8-(morpholine-4-sulfonyl)pyrrolo[3,4-c]quinoline-1,3-diones as potent caspase-3 inhibitors. *J. Med. Chem.* 48:3680.

94. Kravchenko, D. V., et al. 2005. Synthesis and caspase-3 inhibitory activity of 8-sulfonyl-1,3-dioxo-2,3-dihydro-1*H*-pyrrolo[3,4-*c*]quinolines. *Farmaco* 60:804.

95. Kravchenko, D. V., et al. 2005. 1,3-Dioxo-4-methyl-2,3-dihydro-1*H*-pyrrolo[3,4-*c*]quinolines as potent caspase-3 inhibitors. *Bioorg. Med. Chem. Lett.* 15:1841.

96. Webber, S. E., et al. 1998. Tripeptide aldehyde inhibitors of human rhinovirus 3C protease: Design, synthesis, biological evaluation, and cocrystal structure solution of P1 glutamine isosteric replacements. *J. Med. Chem.* 41:2786.

97. Webber, S. E., et al. 1996. Design, synthesis, and evaluation of nonpeptidic inhibitors of human rhinovirus 3C protease. *J. Med. Chem.* 39:5072.

98. Wang, Q. M., et al. 1998. Dual inhibition of human rhinovirus 2A and 3C proteases by homophthalimides. *Antimicrob. Agents Chemother.* 42:916.

99. Oswaldo, N., Jose, R., and Larry, A. 1994. Kinetic study of the formation and rupture of stable tetrahedral intermediates. C-O, C-N and C-S bond formation. *J. Phys. Org. Chem.* 7:80.

100. Oku, T., et al. 1994. New heterocyclic derivatives. JP Patent 239864.

101. Stierle, D. B., et al. 2004. Berkeleydione and berkeleytrione, new bioactive metabolites from an acid mine organism. *Org. Lett.* 6:1049.

102. Gunasekera, S. P., et al. 1999. Discorhabdin P, a new enzyme inhibitor from a deep-water Caribbean sponge of the genus *Batzella*. *J. Nat. Prod.* 62:173.

103. LaCasse, E. C., et al. 1998. The inhibitors of apoptosis (IAPs) and their emerging role in cancer. *Oncogene* 17:3247.

104. Nuttall, M. E. et al. 2001. Selective inhibitors of apoptotic caspases: Implications for novel therapeutic strategies. *Drug Disc. Today* 6:85.

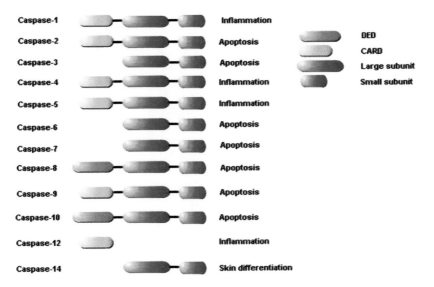

Function

Caspase-1		Inflammation
Caspase-2		Apoptosis
Caspase-3		Apoptosis
Caspase-4		Inflammation
Caspase-5		Inflammation
Caspase-6		Apoptosis
Caspase-7		Apoptosis
Caspase-8		Apoptosis
Caspase-9		Apoptosis
Caspase-10		Apoptosis
Caspase-12		Inflammation
Caspase-14		Skin differentiation

DED
CARD
Large subunit
Small subunit

FIGURE 1.1

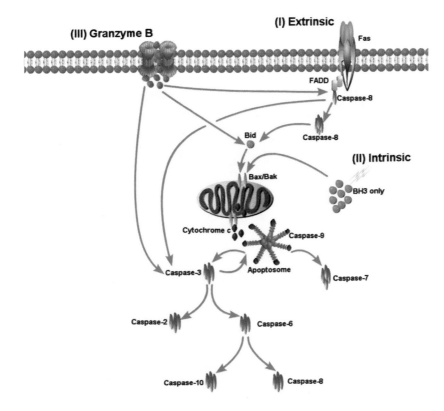

FIGURE 1.2

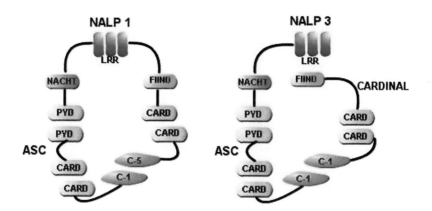

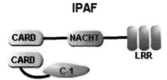

FIGURE 1.3

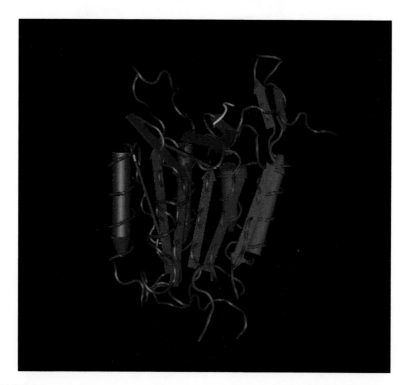

FIGURE 4.1

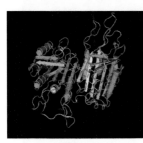
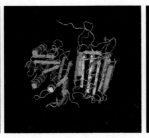

Procaspase-7   Caspase-7   Caspase-7 / DEVD-CHO

(a)

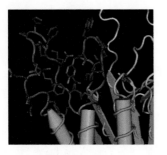
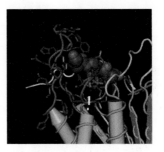

Procaspase-7   Caspase-7 / DEVD-CHO

(b)

(c)

FIGURE 4.2

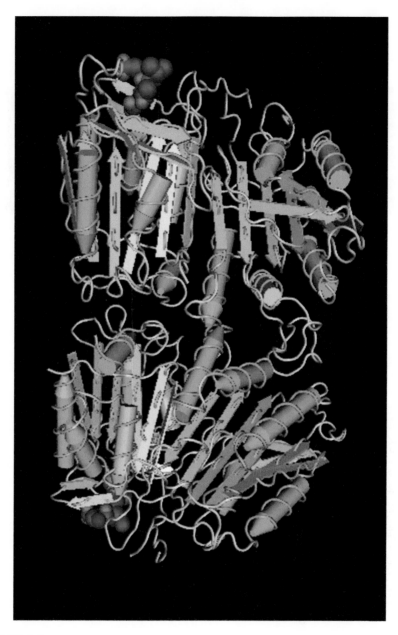

FIGURE 4.3

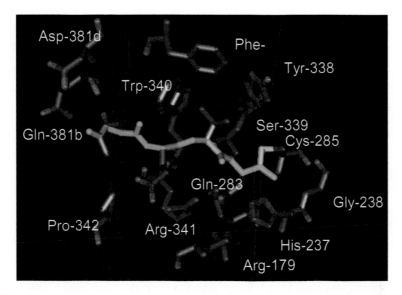

FIGURE 4.5

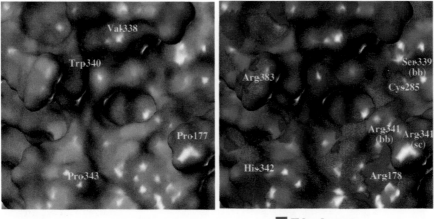

Lipophilicity Map   Hydrogen Bonding Map

■ H-bond acceptor
■ H-bond donor

FIGURE 6.3

FIGURE 6.4

(a)

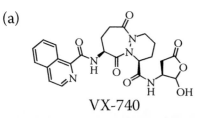 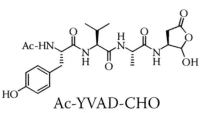

VX-740   Ac-YVAD-CHO

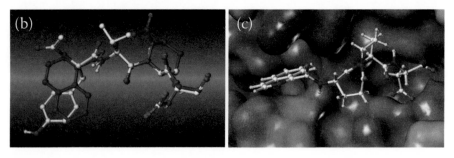

FIGURE 6.8

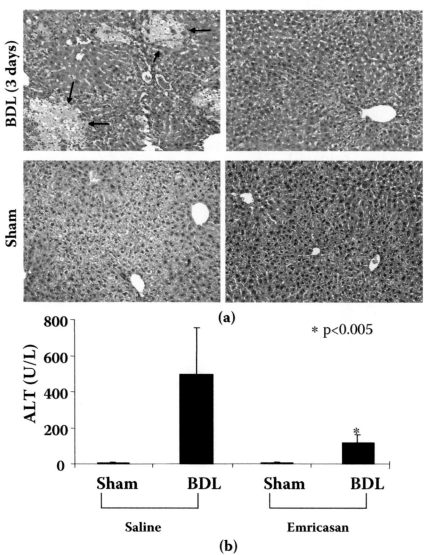

FIGURE 8.2

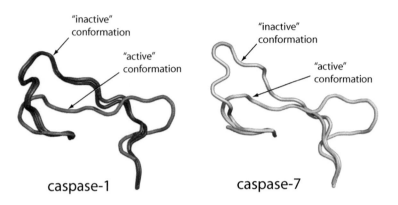

FIGURE 9.5

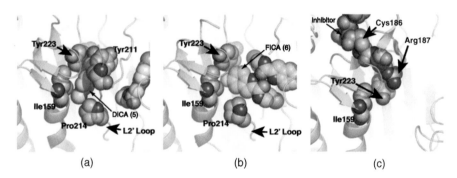

FIGURE 9.6

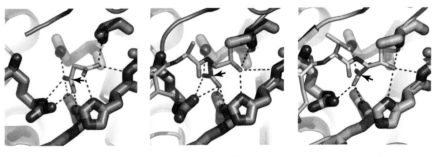

FIGURE 9.8

6 Identification of Inflammatory Caspase Inhibitors

John A. Wos and Thomas P. Demuth Jr.

CONTENTS

6.1 INTRODUCTION

The caspases that modulate inflammatory pathways (caspases-1, -4, -5, -11, and -12) have been intensely investigated by both academia and the pharmaceutical industry over the past several decades. In particular, caspase-1, or interleukin 1β-converting enzyme (ICE), has received considerable scrutiny as a drug target, the result of which has been the recent advancement of several compounds into clinical trials for various indications.[1] In this chapter, we will discuss the design of small molecules that inhibit the inflammatory caspases. The predominance of molecules designed to inhibit caspase-1 will be reflected in the chapter, with latter sections reflecting the emergence of the other, less studied inflammatory caspases. A compilation of scaffolds will be presented at the end of each section, as applicable, to guide readers who would prefer a more global overview of scaffolds for each of the target caspases, with special emphasis given to more recent developments (2002–present). Clinical results will not be discussed at length, as they are covered extensively by other authors within this book. We hope that this overview will provide readers with a fundamental perspective on the principles and challenges of designing small molecule inhibitors for this important family of cysteine proteases.

6.2 CASPASE-1

6.2.1 INTRODUCTION

Since the discovery of caspase-1[2] as a principal mediator of the cytokine pathways for IL-1β and IL-18, it has been the most extensively scrutinized of all of the inflammatory caspases to date, and rivals caspase-3 as the most intensely studied of the family. Caspase-1 is responsible for cleaving an inactive 31 kDa precursor (pro-IL-1β) to release the active 17.5 kDa mature inflammatory cytokine, IL-1β.[3] Therefore, caspase-1 offers an attractive therapeutic target for controlling IL-1β levels, thus potentially providing an effective means of treating a variety of cytokine-mediated diseases, such as rheumatoid arthritis (RA), osteoarthritis (OA), artherosclerosis, septic shock, and inflammatory bowel syndrome.[4] In addition, animal knockout experiments of IL-1β have supported the important role of this cytokine in the inflammatory cascade,[5] lending credence to the inhibitor strategy approach of caspase-1 in mediating IL-1β processing, maturation, and disease modification. As a result, pharmaceutical companies have invested tremendous effort in understanding the biology and chemistry of caspase-1 in order to develop small molecules that would preferentially impact the IL-1β pathway.

A major step forward in understanding the design of inhibitors for caspase-1 was reported in the tertiary protein structure of the enzyme, and subsequent identification of peptide templates that possessed potent and selective inhibitory capability for it.[6–8] These studies allowed for the establishment of the tetrapeptide Ac-YVAD-CHO (**1**) as a selective tetrapeptide scaffold for caspase-1 from which other inhibitor designs have emerged. In addition, the solving of the x-ray crystal structure for caspase-1[9] and the subsequent co-crystallization of various inhibitors into the active site[10,11] afforded a clear premise for the use of structure-based drug design for this target.

Several basic design tenants have emerged that are central to the design of caspase-1-specific inhibitors, and a majority of the compounds currently in the literature have subscribed to these basic features (Figure 6.1). As part of the caspase family, all caspase-1 inhibitors incorporate an (L)-aspartic acid residue in the P1 site to mimic the cleavage site of the pro-IL-1β at the Asp[116]-Ala[117] residues in the cytokine pro-form. This feature is unique to caspases and allows for inhibitor specificity relative to other cysteine proteases (i.e., cathepsins, calpains, etc.). In addition, the Asp residue forms important electrostatic hydrogen bonding interactions with residues' enzyme active sites (Arg-341, Arg-179) to allow for proper orientation of the adjacent electron-deficient atom (trap or warhead) to form a covalent bond with the Cys-285 of the S1 site in the enzyme. This bond could be reversible (**2**) or irreversible (**3**)

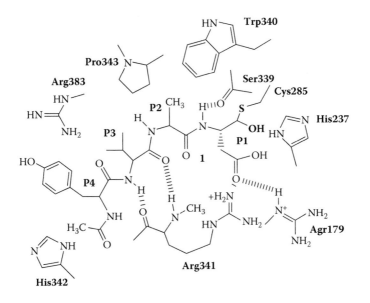

FIGURE 6.1 Key H-bonding interactions of ICE with AcYVAD-CHO.

in nature (Figure 6.2), and many different P1 moieties have been established and reported.[12] Additional stabilization of the tetrahedral intermediate resulting from the Cys-285 attack could be anticipated from the conformational flexible His-237.

There appears to be little hydrogen bonding interaction with the Ala in the P2 pocket of the enzyme (more lipophilic in character); however, important hydrogen bonding elements are incorporated within the P2-P3 region and the Ser-339 and Arg-341 of the pocket. In addition, key lipophilic interactions between the P3 region and nonpolar amino acids (i.e., Pro-343, Trp-340) help to stabilize ligand binding in the

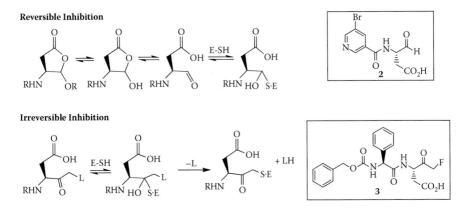

FIGURE 6.2 Basic mechanisms for cysteine protease inhibition.

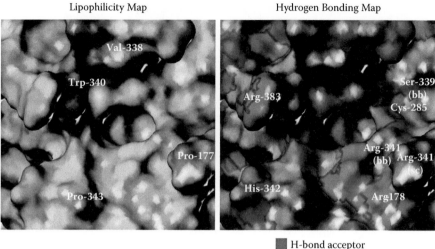

FIGURE 6.3 (See color insert.) Binding site topology for caspase-1.

pocket. Finally, the S4 site of the enzyme pocket is large and highly lipophilic in nature and can accommodate a wide range of aromatic functionality similar to the Ac-Tyr of the tetrapeptide illustrated in Figure 6.1. This observation has driven the design of caspase-1-selective inhibitors relative to the other caspases, which do not possess such a large and lipophilic pocket. For illustrative purposes, the enzyme active site is presented in both hydrogen bonding and lipophilicity models (Figure 6.3).

The docking of the Ac-YVAD-CHO peptide into the homology model illustrates the key components for binding that inhibitor designs have subsequently incorporated into various scaffolds (Figure 6.4).[13] The highly charged pocket (oxy-anion hole) of the S1 site in the enzyme is highly organized for Asp recognition and orientation of the aldehyde trap toward attack of the Cys-285 of the enzyme. The transition zone between S1 and S4 is relatively shallow and hydrophilic in nature, thus exposing a majority of the inhibitor to the solvent face and potential hydrogen bonding interactions, an important consideration in the design of inhibitors. As seen from the modeling, a substantial hydrophobic pocket exists that corresponds to the P4 site of the ligand, and thus explains the majority of highly substituted aryls of biaryls incorporated into typical caspase-1 inhibitors.

6.2.2 HISTORICAL PERSPECTIVE OF CASPASE-1 INHIBITOR DESIGN

The transition from identifying peptide leads as tools for further manipulation into drug-like molecules has a rich historical precedent in medicinal chemistry (i.e., anglo-tensin-converting enzyme (ACE) inhibitors) and was applied to the caspase-1 area in approximately the same timeframe that the x-ray crystal structure of the enzyme was solved. A number of early inhibitors emerged from the establishment of the key hydrogen bond contacts needed for ICE inhibition, and then a subsequent attempt to

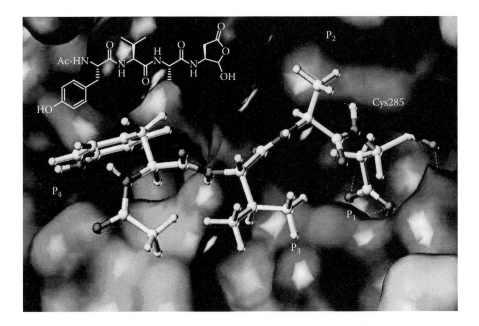

FIGURE 6.4 (See color insert.) Model of Ac-YVAD-CHO bound in caspase-1 active site.

improve upon the absorption, distribution ADME problems associated with peptide drug development of orally active molecules was initiated. The attempted improvements in enhancing orally bioavailability have been mainly focused in two areas, the P1 region (cysteine trap) and the P2-P3 region of the tetrapeptide with cyclic constraints, which are expanded in the following sections.

6.2.2.1 P1 Region (Cysteine Trap)

The aspartic acid group at P1 has been extensively studied with a variety of both reversible and irreversible functional groups in order to optimize cellular permeability and ADME parameters of inhibitors while maintaining the high affinity and selectivity for caspase-1. In particular, ketone replacements have been extensively examined and include benzoyloxymethyl,[14–16] aspartic,[17] phenylalkyl,[18] acyloxymethyl,[19,20] aminomethyl,[21] aryloxymethyl,[22] ketoamides,[23] and alkylsulfonamide[24] for the electrophilic carbonyl center adjacent to the aspartic acid moiety. In general, these inhibitors offer permeability enhancements relative to the parent P1 substituent, Asp-CHO in cellular assays (e.g., THP-1); however, most of the designs result in ejection of a chemically and potentially metabolically reactive leaving group. Additional modifications for the aspartic acid itself have been reported,[10] the potential benefit being the isosteric replacement of the poorly permeable carboxylic acid with a more cell-permeable functionality. As an example, the Yamanouchi group has reported that a sulfonamide analog (**7**) was a potent inhibitor of human caspase-1 and provided significant activity for suppression of IL-1β in a whole cell assay (inhibition of lipopolysaccharide (LPS) induced IL-1β secretion in human monocytes as measured by enzyme-linked immunosorbent assay [ELISA]). The authors suggest that lowering

TABLE 6.1
Yamanouchi Sulfonamide Isosteres at P1 Site[10]

Compound	Csp-1 IC$_{50}$ (μM)	Human Monocytes IC$_{50}$ (μM)
	0.013	0.34
4		
	0.022	0.89
5		
	0.013	0.77
6		
	0.038	0.23
7		
	0.013	0.90
8		

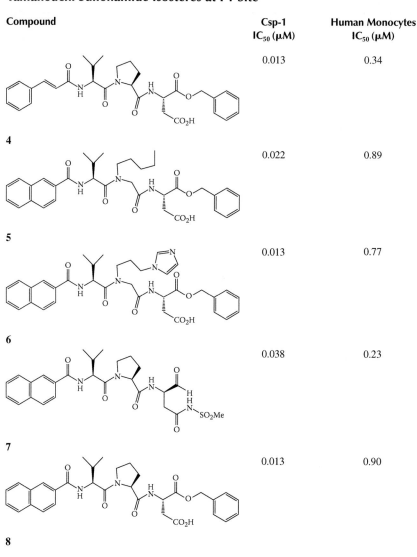

the cLogP value via incorporation of the P1 sulfonamide may enhance cell permeability while reducing protein binding, thus offering a potential design advantage over the more traditional carboxylic acid moiety usually reserved for the P1 site.

It has also been reported that an additional hydrophobic pocket adjacent to the P1 site may be accessible for further optimization, again suggesting that more cell-permeable designs may be attainable by focusing on the P1 site.[25] In this example, the x-ray crystal structure of Ac-Tyr-Val-Ala-Asp-CO-(CH$_2$)$_5$-Ph was used as a design guide to generate truncated sulfonamide analogs (**9**), which proved active at inhibiting caspase-1 (Figure 6.5). The authors noted a major conformational change in which the His-237 catalytic site had rotated from *gauche* to a *trans* orientation, creating a hydrophobic pocket adjacent to the P1 site (Table 6.2).

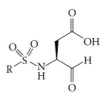

FIGURE 6.5 Sulfonamide core structure for truncated caspase-1 analogs.

By far, the most successful P1 designs to date have been reversible in nature, and based on the hemi-acetal prodrug approach for the parent Asp-CHO. Indeed, two clinical caspase-1 inhibitors that are in clinical trials (VX-740 [pralnacasan] [18] and VX-765 [19]) use various prodrug forms to enhance bioavailability (Figure 6.6). The use of the hemi-acetal approach has been augmented significantly by the work of Merck scientists in establishing a flexible and scalable synthetic route to a variety of different acetal structures in order to probe stability and bioavailability properties.[26] Other reversible inhibitor approaches include the use of α,α-difluoro ketones[27] and semicarbazones[28] as substrates for reversible binding. A comprehensive review on early approaches for P1 trap design has previously been published.[29]

6.2.2.2 P2-P3 Region

A systematic study of the necessary hydrogen bonding elements needed for caspase-1 activity in early tetrapeptide leads via amino acid replacement showed a classical β-sheet hydrogen bonding motif was prevalent within caspase-1, and that the methyl group at P2 (Ala) in the peptide sequence was not necessary for activity.[30] This study was instrumental in identifying key elements for modification that would allow for improved bioavailability and enhanced drug properties. As a result, an approach that utilized constraint of the P2-P3 region with cyclic motifs that would allow for improved pharmacokinetic parameters while maintaining key binding elements was implemented by Dolle et al. (Figure 6.7), and has proven successful in developing a clinically viable caspase-1 candidate.[31]

The pyridazinondiazepine scaffold (**24**) that resulted was subsequently optimized by Vertex scientists and has emerged as the core in pralnacasan (**18**), which was the first caspase inhibitor to enter the clinic (see Chapter 10). A comparison of the conformation of enzyme-bound Ac-YVAD-CHO peptide and a minimized conformation of pralnacasan reveals significant overlap at critical hydrogen bonding regions, and docking experiments of pralnacasan with the enzyme active site reveal significant fit between the inhibitor and the modeled active site (Figure 6.8a and b).[13] In a similar strategy, several other cyclic templates applicable to the P2-P3 region for caspase family inhibitors have been reported, including the oxo-azepino indole,[32] the tetrahydro-1,5-benzodiazepine,[33] and the caprolactam[34] motif.

6.2.2.3 P4 Region

Another area of interest in caspase-1 inhibitor design has been the identification of suitable replacements for the Ac-Tyr residue present in the Ac-YVAD-CHO peptide. The importance for the P4 site's role in selectivity was elegantly confirmed by structure

Design of Caspase Inhibitors as Potential Clinical Agents

TABLE 6.2
BASF Sulfonamide Probes of an Adjacent Hydrophobic Pocket at P1

Compound	R	hCaspase-1 K_i (µm)	hCaspase-1 IC_{50} (µm)
9			
10		9.4	86.8
11		1.6	20.3
12		1.9	34.4
13		6.8	28
14		3.1	37
15		7.2	66.2
16		6.9	11.4
17		2.3	49.8

activity relationship (SAR) studies of Karanewsky et al.,[35] in which the replacement of the carbobenzyloxy unit of caspase-1-selective compound (**26**) with a polar Ac-Asp group resulted in complete selectivity for caspase-3 (**27**). This is an important observation as it allows for potentially altering the selectivity profiles within a specific scaffold readily by adjustments at the P4 site, thus allowing access to both inflammatory (caspase-1) and apoptotic (caspase-3) inhibitors through a common template (Figure 6.9).

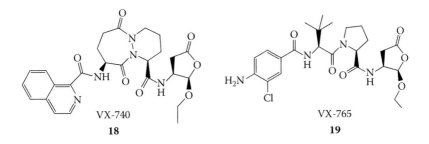

FIGURE 6.6 Caspase-1 inhibitors in clinical trials.

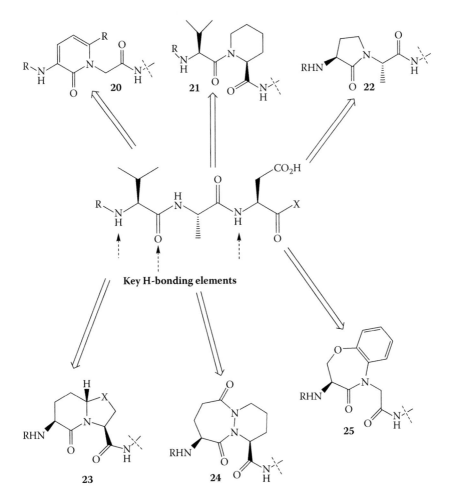

FIGURE 6.7 Cyclic P2-P3 constraints resulting from sequential AA replacements of caspase-1 tetrapeptides.

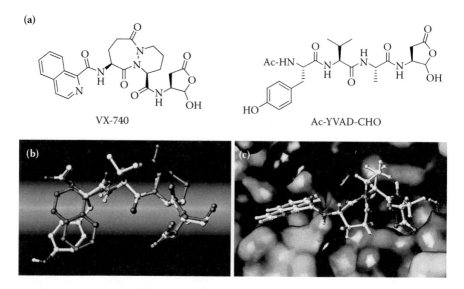

FIGURE 6.8 (See color insert.) (a) Pralnacasan (VX-740) and Ac-YVAD-CHO. (b) Overlap of Ac-YVAD-CHO (gray) with pralnacasan (blue). (c) Pralnacasan docked in caspase-1.

Other SAR studies at P4 have been reported;[36,37] however, there has been less emphasis on this region than the P1-P3 regions. Recently, a report on P4 substituent conformation impacting on inhibitor activity has appeared.[38] The authors report that the S4 pocket of the enzyme can accommodate two different binding motifs for large, hydrophobic moieties (upper or lower groove), and that the lower-groove binding mode may be more energetically favorable, thus producing a more potent inhibitor.

6.2.3 REVIEW OF INHIBITORS: 2002–PRESENT

The emergence of pralnacasan into clinical trials and the acceptance by physicians of the use of anakinra (Kineret®)[39] as a therapeutic approach to modulate IL-1β activity to treat some forms of autoimmune disease have resulted in continued interest in finding orally bioavailable caspase-1 inhibitors with acceptable safety and efficacy profiles to treat diseases. A review of more recent literature suggests that the peptidomimetic strategy is still predominant in caspase-1 design (Table 6.3),

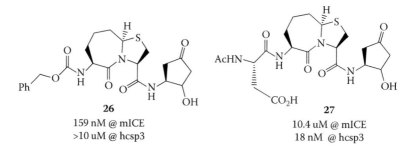

26
159 nM @ mICE
>10 uM @ hcsp3

27
10.4 uM @ mICE
18 nM @ hcsp3

FIGURE 6.9 Influence of P4 substituent on caspase-1 selectivity.

TABLE 6.3
Compilation of Scaffolds with Caspase-1 Inhibitory Activity

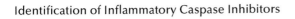

$P_4 \overset{P_3}{\diagup} P_2 \overset{H}{\diagup} N \diagdown Cys\text{-}trap$

Compound	P4	P2-P3	Cys-trap	Comments
1[6]	Ac-Y	Val-Ala	D-CHO	Prototype tetrapeptide for caspase-1 design; IC$_{50}$ 2–5 nM at ICE
31[43]	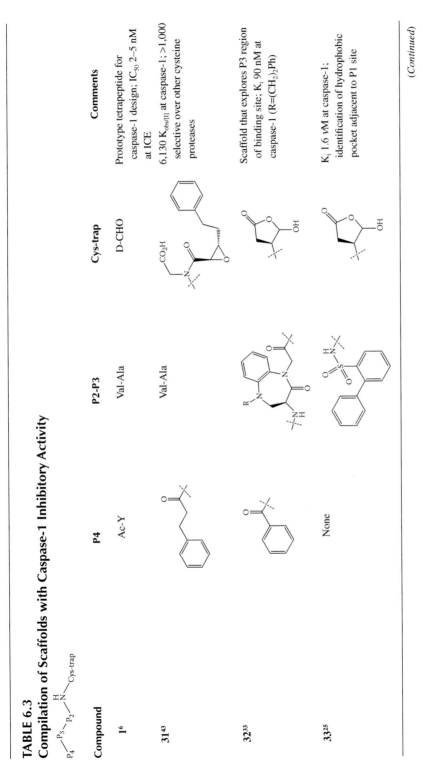	Val-Ala		6,130 K$_{obs/[I]}$ at caspase-1; >1,000 selective over other cysteine proteases
32[33]				Scaffold that explores P3 region of binding site; K$_i$ 90 nM at caspase-1 (R=(CH$_2$)$_2$Ph)
33[25]	None			K$_i$ 1.6 νM at caspase-1; identification of hydrophobic pocket adjacent to P1 site

(Continued)

TABLE 6.3 (Continued)
Compilation of Scaffolds with Caspase-1 Inhibitory Activity

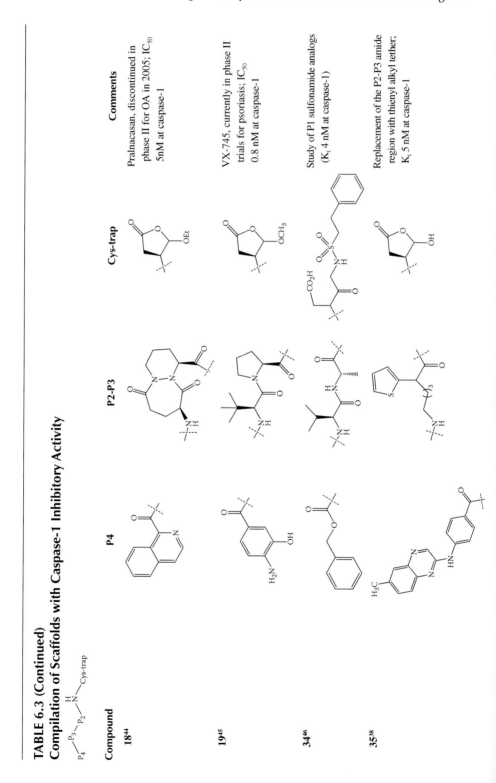

Compound	P4	P2-P3	Cys-trap	Comments
18[44]				Pralnacasan, discontinued in phase II for OA in 2005; IC_{50} 5nM at caspase-1
19[45]				VX-745, currently in phase II trials for psoriasis; IC_{50} 0.8 nM at caspase-1
34[46]				Study of P1 sulfonamide analogs (K_i 4 nM at caspase-1)
35[38]				Replacement of the P2-P3 amide region with thienyl alkyl tether; K_i 5 nM at caspase-1

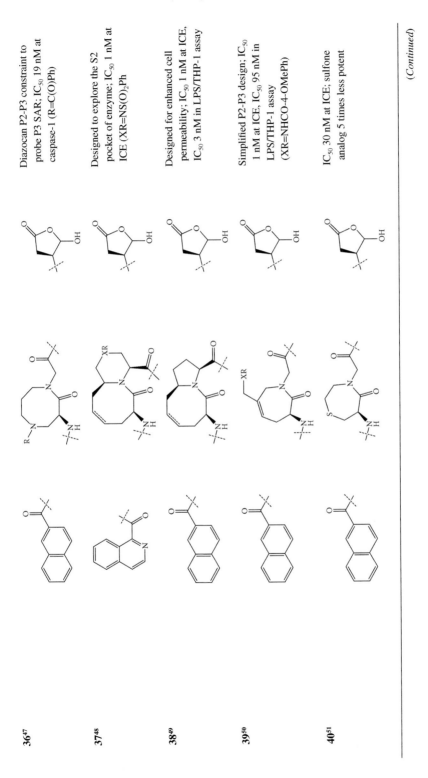

TABLE 6.3 (Continued)
Compilation of Scaffolds with Caspase-1 Inhibitory Activity

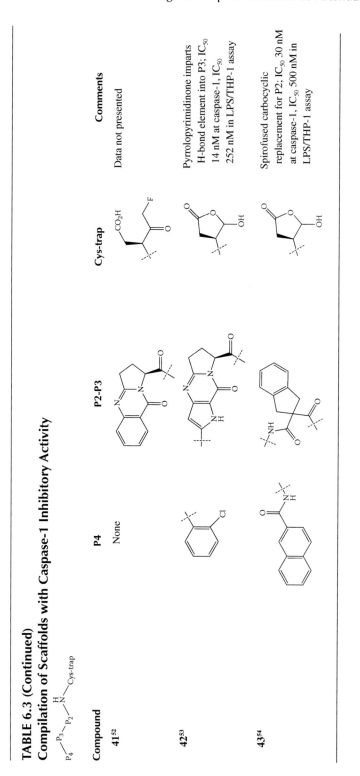

Compound	P4	P2-P3	Cys-trap	Comments
41[52]	None			Data not presented
42[53]				Pyrrolopyrimidinone imparts H-bond element into P3; IC$_{50}$ 14 nM at caspase-1, IC$_{50}$ 252 nM in LPS/THP-1 assay
43[54]				Spirofused carbocyclic replacement for P2; IC$_{50}$ 30 nM at caspase-1, IC$_{50}$ 500 nM in LPS/THP-1 assay

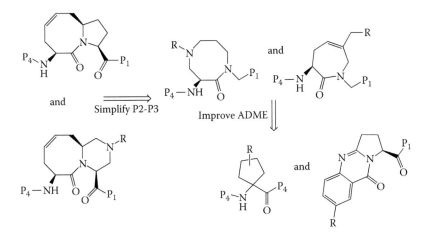

FIGURE 6.10 Efforts to optimize caspase-1 scaffolds in the Procter & Gamble pharmaceuticals program.

although efforts have been directed at simplifying the scaffolds, as exemplified by the attempts to optimize pharmacokinetic (PK) parameters in the Procter & Gamble caspase-1 program (Figure 6.10).[13] The emergence of the backup compound VX-765 (**19**) into the clinic is also indicative of this simplification approach.

There are several reported approaches that have identified nonpeptide caspase-1 inhibitors (Figure 6.11). The early reports of the pyridazine lead (**28**) as a caspase-1 inhibitor[40] have apparently not materialized into small molecule programs that lack a peptide character. Similarly, the natural products EI-1941-1 (**29**) and EI-1941-2 (**30**)[41] have established the potential for natural products as a source of new inhibitor classes, but subsequent reports have not appeared. An in-depth analysis

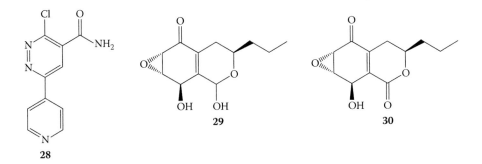

FIGURE 6.11 Nonpeptide caspase-1 inhibitors.

of nonpeptide caspase inhibitors is described in Chapter 5. Finally, a report has appeared using a biotinylation approach to isolate an affinity-labeled caspase-6, but no reports specific to caspase-1 have appeared to date.[42]

6.2.3.1 Tabular Compilation of Caspase-1 Inhibitors (P4/P2-P3/Cysteine Trap)

The creativity of the medicinal chemist in caspase-1 inhibitor design is best appreciated when a partial assessment of the many scaffolds that have appeared in the literature over the last several years is compiled (Table 6.3). As seen in the table, a wide variety of structural manipulations have resulted in a diverse pool of compounds with a range of potency and selectivity profiles for the enzyme. The clinical impact of these modifications remains to be fully evaluated and is worth monitoring as compounds continue to emerge from various programs.

6.3 CASPASE-4 AND CASPASE-5

6.3.1 INTRODUCTION

Caspases-4 (also known as TX, ICH-2, ICE_{REL}-II) and -5 (also known as TY, ICE_{REL}-III) are the other major inflammatory caspases in humans, and can induce apoptosis when overexpressed in several cell lines.[55] Since specific substrates have not been identified for these caspases, their function remains uncertain. Caspase-5 is constitutively expressed at high levels in monocytes, macrophages, and to a lesser extent in neutrophils. It is also highly expressed in the heart and the colon, but is expressed at low levels in most other tissues, including the CNS. Caspase-5 not only cleaves caspase-3[56] and processes the transcription factor Max *in vitro*,[57] but has also been identified as a component of the NALP1 inflammasome, a macromolecular complex involved in the activation of caspase-1.[58]

In contrast, caspase-4 is normally present at high levels in most tissues. Its expression, however, is lower in the brain and colon than caspases-1 and -5, respectively. Caspase-4 may be involved in caspase-1 and caspase-3 activation,[59] but is reported to inefficiently process pro-IL-1β, pro-IL-18, and IL-1F7b.[60] In addition, caspase-4 has been proposed to participate in endoplasmic reticulum (ER) stress-induced apoptosis. Overall, the *in vivo* functions of caspase-4 and caspase-5 are poorly understood.

Structurally, caspase-4 (58%) and caspase-5 (57%) possess high sequence identity when compared to caspase-1, although they have several differing residues at key sites within the substrate binding pocket.[61] Caspase-4 and caspase-5 share 75% identity, and a sequence analysis comparing caspases-4 and -5 with caspases-1 and -3 at the S1 and S2-S4 sites of the enzyme reveals differences in key amino acid residues that may impact the design of ligands for binding specificity (Table 6.4). In the S1 binding site, the QACRG motif (amino acid residues 283–287 in caspase-1, containing the active site cysteine) is conserved in all family members, as are the spatially proximate arginine (R179 in caspase-1, which forms a salt bridge with the P1 aspartate) and histidine (H237 in caspase-1, which comprises the oxyanion hole) within the enzyme catalytic site.

TABLE 6.4
Sequence Alignment between Caspases-1, -3, -4, and -5, about S1 and S2-4 Subsites[62]

Conserved Residues in S1

QACRG

Casp-1	R179	H237	Residues 283–287
Casp-4	R152	H210	Residues 256–260
Casp-5	R193	H251	Residues 297–301
Casp-3	R64	H121	Residues 161–165

Comparison of Key Residues Spanning S2–S4

	S2 Subsite	S2 Subsite	S2 Subsite	S3 Subsite	S4 Subsite
Casp-1	V338	S339	W340	R341	H342
Casp-4	V311	S312	W313	R314	D315
Casp-5	V352	S352	W354	R355	D356
Casp-3	Y204	S205	W206	R207	N208

Of interest to note is that neither caspase-4 nor capsase-5 appears to react with the caspase-1-specific tetrapeptide inhibitor Ac-YVAD-CHO, indicating that these caspases have different binding specificities compared to ICE. Such similarities and differences in substrate and inhibitor specificities between the caspases are suggested by the similarities and differences of the binding sites S2 and S4. For example, the valine residue in the S2 site of the inflammatory caspases-1, -4, and -5 is replaced by a larger aromatic tyrosine residue in caspase-3. In addition, caspases-4 and -5 possess a negatively charged aspartate residue in the S4 site, compared to the positively charged histidine residue in caspase-1 and the neutral asparagine residue in caspase-3. Exploitation of these differences and other changes in the binding site structure could lead to the design of selective, high-affinity caspase-4 or caspase-5 substrates and inhibitors.

The substrate specificities[63] for caspases-1, -4, and -5 have been determined using small peptides *in vitro* and were found to be similar with a preference for the sequence Trp-Glu-His-Asp (WEHD). Overall, caspase-1 was considerably more efficient in processing the peptide substrates. Several peptide substrates have been reported and are included in Table 6.5.[64]

Subsequent to these reports, small peptide inhibitor libraries have been developed that showed some activity at caspase-4 and caspase-5; however, none of these ligands has proven very selective for either caspase[65] (Table 6.6). Of these, the tetrapeptide Ac-WEHD-CHO has proven to be a useful tool in studying inflammatory caspase function as it is subnanomolar at caspase-1 and also active at caspase-4 and caspase-5 (<100 nM).

6.3.2 SMALL MOLECULE INHIBITORS WITH CASPASE-4/5 ACTIVITY

The lack of success in identifying selective peptide inhibitors for caspase-4 and caspase-5, coupled with additional lack of understanding of their respective *in vivo* biology, had resulted in an emphasis on the other caspase family members as targets for inflammation

TABLE 6.5
Specificity for Peptide Substrates

Substrates	Inactivation Constant k_{cat}/K_m (M^{-1}s^{-1})	
	Caspase-1	Caspase-4
Ac-YVAD-AMC	91,000	14,000
Suc-YVAD-pNA	77,000	23,000
PARP	2,000	1,000
Ac-YVAD-pNA	63,000	280
Ac-YEVD-pNA	79,000	2,600
Ac-LEVD-pNA	54,000	3,200
Ac-DEVD-pNA	30,000	1,800
Ac-DQMD-pNA	21,000	270
Ac-VEID-pNA	12,000	750
Ac-VQVD-pNA	2,800	170
Suc-YVAD-pNA	50,000	8,900
Suc-YVAD-AMC	58,000	4,400

Note: DCB = (2,6-dichlorobenzoyl)oxymethyl ketone; AMC = aminomethyl coumarin; pNA = p-nitroanilide; PARP = poly (ADP-ribose) polymerase; FMK = fluoromethyl ketone.

TABLE 6.6
Reversible Inflammatory Caspase Inhibitors

Reversible Inhibitors	Ki (nM)			
	Caspase-1	Caspase-4	Caspase-5	Caspase-3
Ac-WEHD-CHO	0.056	97	43	1960
Ac-YVAD-CHO	0.76	362	163	>10,000
Ac-DEVD-CHO	18	132	205	0.23
Boc-IETD-CHO	<6	400	223	195
Boc-AEVD-CHO	<12	375	438	42

Irreversible Inhibitor	Inactivation Constant k_{cat}/K_m (M^{-1}s^{-1})			
	Caspase-1	Caspase-4	Caspase-1	Caspase-3
Cbz-VAD-FMK	280,000	5,500	130,000	16,000

Note: DCB = (2,6-dichlorobenzoyl)oxymethyl ketone; AMC = aminomethyl coumarin; pNA = p-nitroanilide; PARP = poly(ADP-ribose)polymerase; FMK = fluoromethyl ketone.

and apoptosis. Consequently, small molecule inhibitors of caspase-4/5 are generally broad spectrum in profile, and are only marginally active for these two targets. As an example, the isatin scaffold (Table 6.7) reported by Lee[66,67] is selective for caspase-3, although it possessed some potency for caspase-4 in some examples (i.e., **44**).

TABLE 6.7
Isatin Analogs Possessing Caspase-4 Activity

Compounds	K_i-app (μM)		
	Caspase-1	Caspase-4	Caspase-3
44	12	4	0.5
45	18	5	0.13
46	>25	>25	0.06
47	9	>50	1.4
48	21	30	0.02
49	>50	>50	12
50	17	33	0.02
51	>5	>5	0.001

More recently, workers at Sunesis have reported some caspase-4/5 activity for a series of highly potent caspase-3 inhibitors using an extended tethering technique in a series of salicylic acid derivatives.[68] Several compounds in the series showed marginal activity at caspase-4/5. However, several possessed nanomolar activity against caspase-1 and caspase-3. The co-crystallization of **59** with caspase-3 revealed in high resolution the remarkable similarity in hydrogen bonding motif between this scaffold and the peptide Ac-DVAD-CHO (Table 6.8).

In addition, a novel pyrazinone scaffold with a broad-spectrum caspase inhibitory profile has recently appeared from Merck and has nanomolar inhibitory activity against both caspase-4 and caspase-5.[69] These compounds, based on a modified version of an earlier scaffold reported by Parke-Davis,[70] were designed to be caspase-3 selective; however, they are also potent at inhibiting the inflammatory caspases (Table 6.9).

6.4 PATENT ANALYSIS

Evaluation of the patent landscape for the inflammatory caspases reveals a substantial amount of activity in the caspase-1 area, with much less activity for the other caspases (Table 6.10). Vertex and Idun have been the most active companies with respect to patent filings during this period (2000 to the present), with the former focusing on caspase-1 activity, and the latter on broad-spectrum caspase inhibitors. Several other companies have been active but to a lesser degree (i.e., Novartis, Procter & Gamble, Ono Pharmaceuticals). The filing of specific applications for caspase-4 and caspase-5 has been less obvious, with only examples of claims of activity for these targets embedded in more general claims of broad-spectrum activity. A listing of caspase filings from 2000 to the present reflects the tremendous interest in the area (Table 6.11).

6.5 CONCLUSIONS

The pro-inflammatory caspases continue be a target of high interest for both the pharmaceutical industry and academia. A large volume of knowledge has accumulated surrounding caspase-1 as a viable therapeutic target, as well as numerous clinical candidates for various inflammatory diseases. The validation of the blockade of IL-1β as a viable therapeutic anti-inflammatory strategy (Anakinra®) suggests that inhibition of cytokine production may also be a worthy approach. Consequently, small molecule inhibitors of caspase-1 remain a goal of many research groups, and the fate of the current caspase-1-selective inhibitors in the clinic will have a large influence on future small molecule designs in this area. Although less understood than caspase-1, caspases-4 and -5 continue to draw interest as alternative potential apoptotic pathway mediators in a preclinical setting. Based on the success seen in the identification of selective caspase-1 and caspase-3 inhibitors, one can anticipate that a similar fate waits for these other inflammatory caspases, and that selective nonpeptide small molecule inhibitors are possible in the near future.

TABLE 6.8
Sulfonamide Analogs Possessing Caspase-4/5 Activity

Compounds	K_i-app (mM)			
	Caspase-1	Caspase-4	Caspase-5	Caspase-3
52	0.16	21	12	0.2
53	3.9	92	22	0.05
54	12	96	98	0.09
55	1.5	38	11	0.03
56	25	22	32	0.02
57	8.3	13	23	0.03
58	1.9	38	24	0.12
59	3.2	>100	63	0.02
60	2.7	27	36	0.04

TABLE 6.9
Pyrazinone Scaffold with Caspase 4/5 Activity

Compound	Reversible Inhibitor IC$_{50}$ (nM)			
	Caspase-1	Caspase-4	Caspase-5	Caspase-3
61	50	300	200	5
62	110	500	810	1.4

TABLE 6.10
Caspase-1 Patent Analysis (2000–Present)

Company	Number
Vertex Pharmaceuticals	12
Idun Pharmaceuticals	9
Ono Pharmaceuticals	3
Procter & Gamble	3
LG Sciences	3
Novartis	2
Georgia Tech Research Corp.	2
Mitsubishi Chemicals	2
Others	9

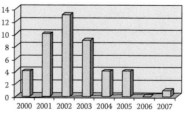

TABLE 6.11
Tabulation of Inflammatory Caspase Patent Filings

Patent Number	Assignee	Representative Structure
WO 2001021600[71]	LG Chemical	
WO 2002070544[72]	Idun Pharmaceuticals	
WO 2007041775[73]	University of Queensland, Australia	
WO2002089749[74]	Abbott GMBH	

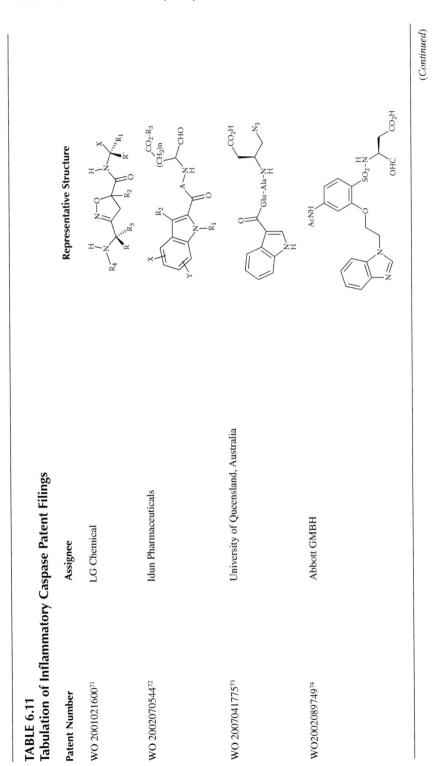

(Continued)

TABLE 6.11 (Continued)
Tabulation of Inflammatory Caspase Patent Filings

Patent Number	Assignee	Representative Structure
US20020523237[75]	Enzyme Systems Products	
US20020183417[76]	Enzyme Systems Products	
WO2004019924[77]	Fundacao de Amparo a Pesquisa do Estado de Sao Paulo	
WO2005080353[78]	Georgia Tech Research Corp.	
WO2004005270[79]	Georgia Tech Research Corp.	

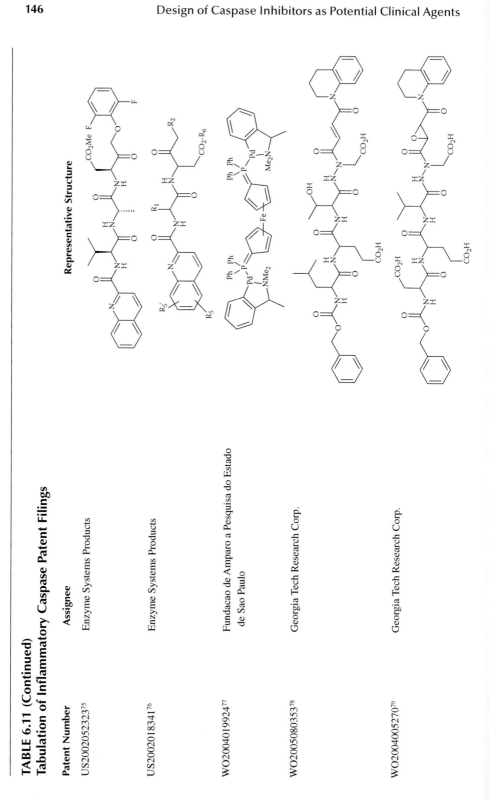

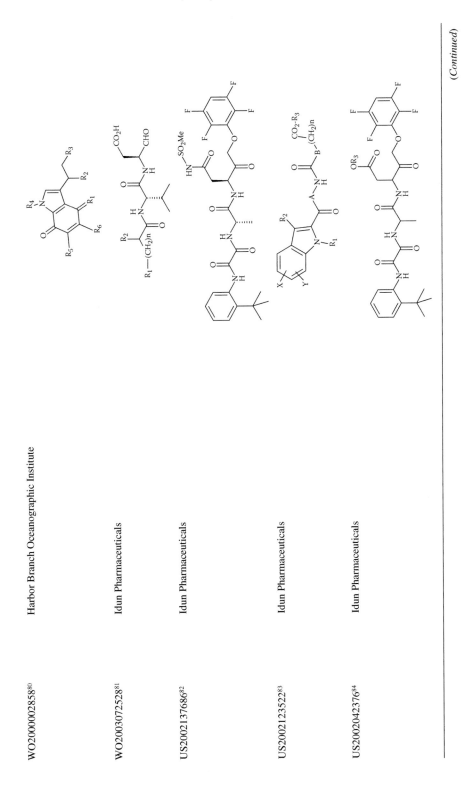

(Continued)

TABLE 6.11 (Continued)
Tabulation of Inflammatory Caspase Patent Filings

Patent Number	Assignee	Representative Structure
WO2001079162[85]	Idun Pharmaceuticals	
WO2001051462[86]	Idun Pharmaceuticals	
US6184244[87]	Idun Pharmaceuticals	
WO2000023421[88]	Idun Pharmaceuticals	

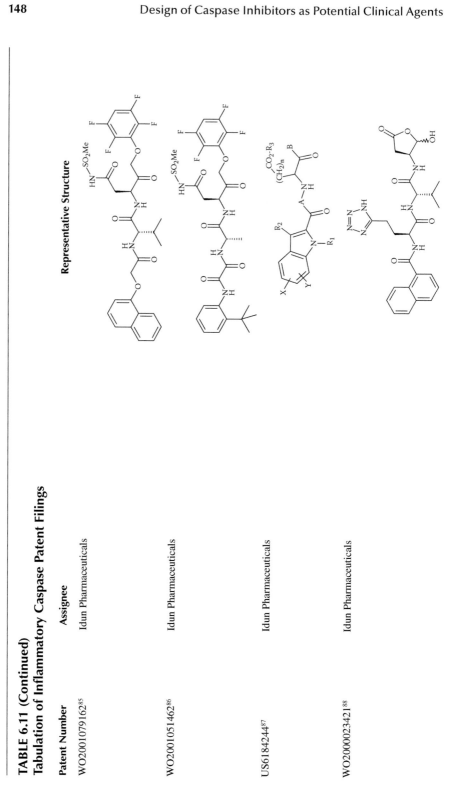

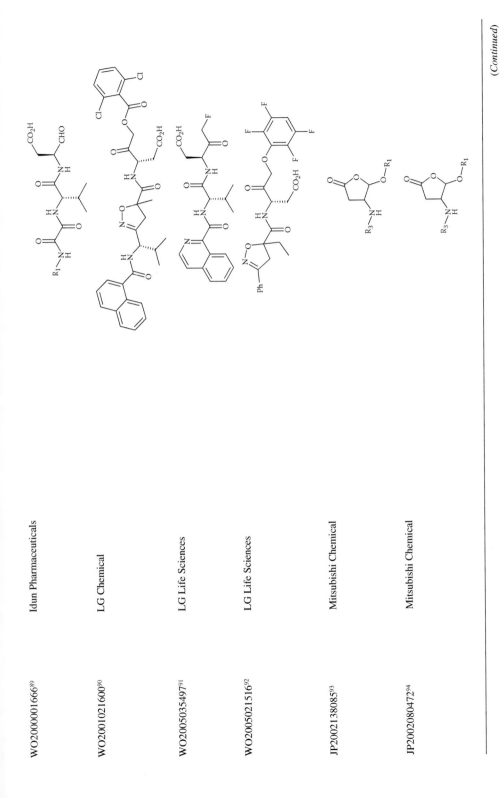

WO200000166689 Idun Pharmaceuticals

WO200102160090 LG Chemical

WO200503549791 LG Life Sciences

WO200502151692 LG Life Sciences

JP200213808593 Mitsubishi Chemical

JP200208047294 Mitsubishi Chemical

(Continued)

TABLE 6.11 (Continued)
Tabulation of Inflammatory Caspase Patent Filings

Patent Number	Assignee	Representative Structure
WO2002076968[95]	Novartis	
WO2002034751[96]	Novartis	
WO2003091202[97]	Ono Pharmaceutical Co.	
WO2001040204[98]	Ono Pharmaceutical Co.	
WO2002096892[99]	Ono Pharmaceutical Co.	

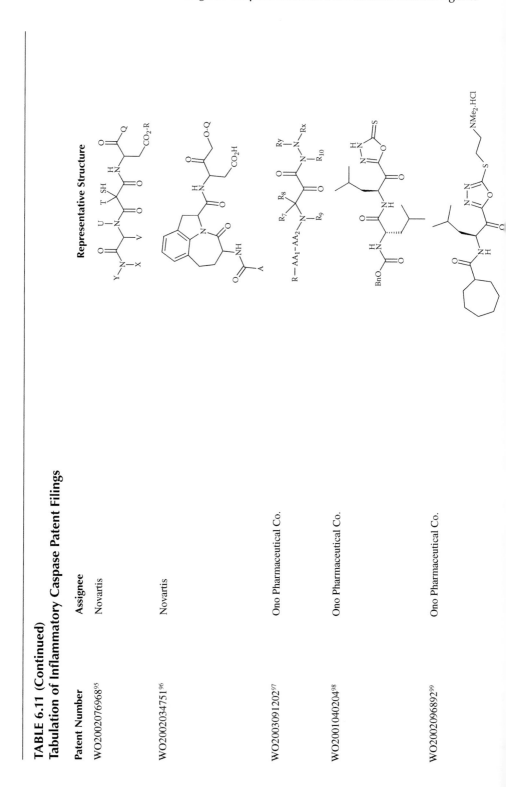

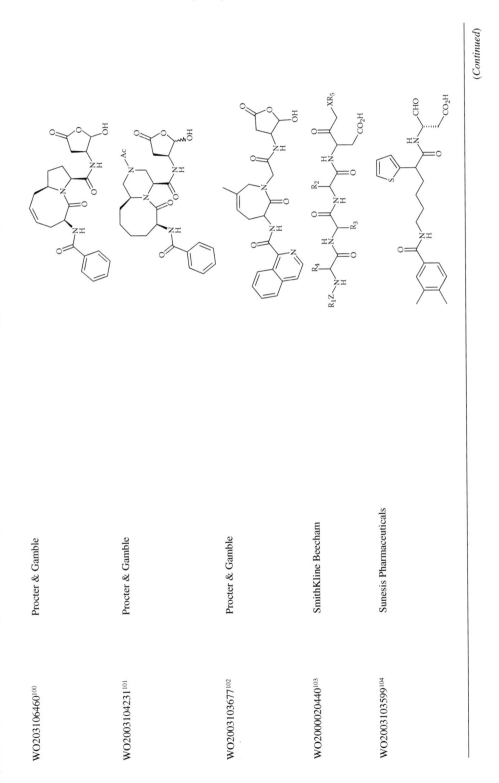

Procter & Gamble

Procter & Gamble

Procter & Gamble

SmithKline Beecham

Sunesis Pharmaceuticals

WO2003106460[100]

WO2003104231[101]

WO2003103677[102]

WO2000020440[103]

WO2003103599[104]

(Continued)

TABLE 6.11 (Continued)
Tabulation of Inflammatory Caspase Patent Filings

Patent Number	Assignee	Representative Structure
WO2007041775[105]	University of Queensland, Australia	
US 6559304[106]	Vertex Pharmaceuticals	
WO2005085236[107]	Vertex Pharmaceuticals	
US2004242494[108]	Vertex Pharmaceuticals	

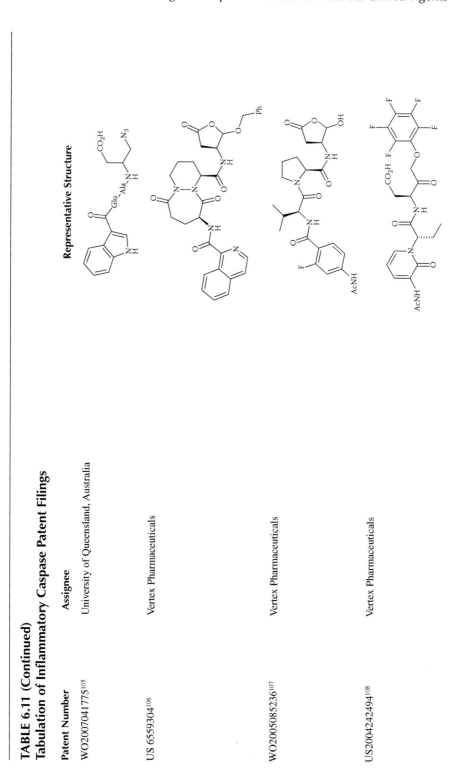

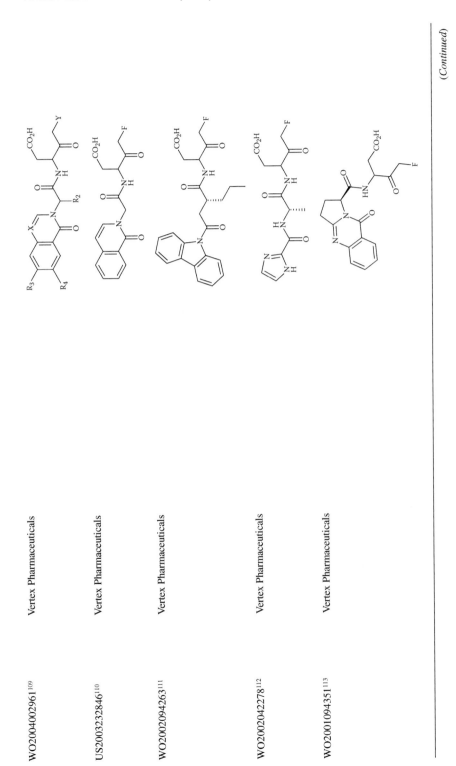

(Continued)

WO2004002961[109]   Vertex Pharmaceuticals

US2003232846[110]   Vertex Pharmaceuticals

WO2002094263[111]   Vertex Pharmaceuticals

WO2002042278[112]   Vertex Pharmaceuticals

WO2001094351[113]   Vertex Pharmaceuticals

TABLE 6.11 (Continued)
Tabulation of Inflammatory Caspase Patent Filings

Patent Number	Assignee	Representative Structure
WO2001090070[114]	Vertex Pharmaceuticals	
WO2001090063[115]	Vertex Pharmaceuticals	
WO2001072707[116]	Vertex Pharmaceuticals	
WO2001042216[117]	Vertex Pharmaceuticals	
WO2003093240[118]	Yungjin Pharmaceutical Co.	

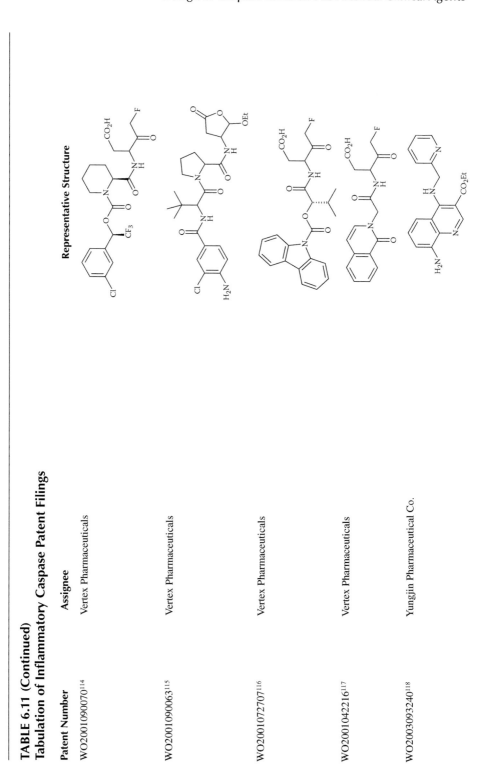

REFERENCES

1. Cornelis, S., et al. 2007. Inflammatory caspases: Targets for novel therapies, P. *Curr. Pharm. Des.* 13:367.
2. Black, R. A., Kronheim, S. A., and Sleath, P. R. 1989. Activation of interleukin-1β by a co-induced protease. *FEBS Lett.* 247:386. Black, R. A., et al. 1989. A pre-aspartate-specific protease from human leukocytes that cleaves pro-interleukin-1β. *J. Biol. Chem.* 264:5323.
3. Dinarello, C. A., and Fantuzzi, G. 2001. Importance of the interleukin-1β-converting enzyme in disease mediated by the proinflammatory cytokines IL-1β and IL-18. In *Cytokine inhibitors,* ed. G. Ciliberto and R. Savino, chap. 1. Basel: Marcel Dekker.
4. Dinarello, C. A., and Wolff, S. M. 1993. The role of interleukin-1 in disease. *N. Engl. J. Med.* 328:106. Goldring, M. B. 2001. Anticytokine therapy for osteoarthritis *Expert Opin. Biol. Ther.* 1:817. Malemud, C. J. 2004. Cytokines as therapeutic targets for osteoarthritis *Biodrugs* 18:23.
5. Kuida, K., et al. 1995. Altered cytokine export and apoptosis in mice deficient in interleukin-1β converting enzyme. *Science* 267:2000. Li, P., et al. 1995. Mice deficient in IL-1β converting enzyme are defective in production of mature IL-1β and resistant to endotoxic shock. *Cell* 80:401.
6. Thornberry, N. A., et al. 1992. A novel heterodimeric cysteine protease is required for interleukin-1β processing in monocytes. *Nature* 356:768.
7. Thornberry, N. A., et al. 1997. A combinatorial approach defines specificities of members of the caspase family and granzyme B. *J. Biol. Chem.* 272:17907.
8. Talanian, R. V., et al. 1997. Substrate specificities of caspace family proteases. *J. Biol. Chem.* 272:9677.
9. Walker, N. P. C., et al. 1994. Crystal structure of the cysteine protease interleukin-1β converting enzyme: A (p20/p10)$_2$ homodimer. *Cell* 78:343. Wilson, K. P., et al. 1994. Structure and mechanism of interleukin-1β converting enzyme. *Nature* 370:270.
10. Okamoto, Y., et al. 1999. Peptide based interleukin-1β converting enzyme (ICE) inhibitors: Synthesis, structure activity relationships and crystallographic study of the ICE-inhibitor complex. *Chem. Pharm. Bull.* 47:11.
11. Brady, K. D., et al. 1999. A catalytic mechanism for caspase-1 and for the bimodal inhibition of caspase-1 by activated aspartic ketones. *Bioorg. Med. Chem.* 7:621.
12. Graczyk, P. P. 2002. Caspase inhibitors as anti-inflammatory and antiapoptotic agents. *Prog. Med. Chem.* 89:1. Talanian, R. V., Brady, K. D., and Cryns, V. L. 2000. Caspases as targets for anti-inflammatory and anti-apoptotic drug discovery. *J. Med. Chem.* 43:3351. Graczyk, P. P. 1999. Caspase inhibitors—A chemist's perspective. *Restorative Neurol. Neurosci.* 14:1.
13. Wos, J. A., et al., 2005. The design and synthesis of medium-sized ring scaffolds as ICE inhibitors for treatment of inflammatory disease. In *230th American Chemical Society National Meeting,* Washington, DC, August 28–September 1, MEDI-269.
14. Thornberry, N. A., et al. 1994. Inactivation of interleukin-1β converting enzyme by peptide (acyloxy)methyl ketones. *Biochemistry* 33:3934.
15. Dolle, R. E., et al. 1994. P1 aspartate-based peptide α-((2,6-dichlorobenzyl)oxy)methyl ketones as potent time-dependent inhibitors of interleukin-1β-converting enzyme. *J. Med. Chem.* 37:563.
16. Revesz, L., et al. 1995. Synthesis of P1 aspartate-based peptide acyloxymethyl and fluoromethyl ketones as inhibitors of interleukin-1β converting enzyme. *Tetrahedron Lett.* 35:9693.
17. Mijalli, A. M. M., et al. 1995. Inhibition of interleukin-1β converting enzyme by N-acyl-aspartic acid ketones. *Bioorg. Med. Chem. Lett.* 5:1405.

18. Mijalli, A. M. M., et al. 1993. Phenylalkyl ketones as potent reversible inhibitors of interleukin-1β converting enzyme. *Bioorg. Med. Chem. Lett.* 3:2689.

19. Brady, K. D. 1998. Bimodal inhibition of caspase-1 by aryloxy and acyloxymethyl ketones. *Biochemistry* 37:8505.

20. Albrecht, H. P. 1998. Aspartate ester inhibitors of interleukin-1β converting enzyme. WO98/16502.

21. Semple, G., et al. 1998. Peptidomimetic aminomethylene ketone inhibitors of interleukin-1β converting enzyme (ICE). *Bioorg. Med. Chem. Lett.* 8:959.

22. Mijalli, A. M. M., et al. 1995. Inhibition of interleukin-1β converting enzyme by N-acyl-aspartyl aryloxymethyl ketones. *Bioorg. Med. Chem. Lett.* 5:1409.

23. Chen, J., et al. 2005. Solid phase synthesis interleukin-1β converting enzyme inhibitors: A case study of molecular modeling-aided design and optimization. In *Pacific-Chem 2005*, Honolulu, HI, December 15–20, MEDI-237, No. 59.

24. Brady, K. D., et al. 1999. A catalytic mechanism for caspase-1 and for bimodal inhibition of caspase-1 by activated aspartic ketones. *Bioorg. Med. Chem. Lett.* 7:621.

25. Shahripour, A. B., et al. 2002. Structure-based design of nonpeptide inhibitors of interleukin-1β converting enzyme (ICE, caspase-1). *Bioorg. Med. Chem.* 10:31.

26. Chapman, K. T. 1992. Synthesis of a potent, reversible inhibitor of interleukin-1β converting enzyme. *Bioorg. Med. Chem. Lett.* 2:613.

27. Majalli, A. M. M., Chapman, K. T., and MacCoss, M. 1993. Synthesis of a peptidyl 2,2-difluoro-4-phenylbutyl ketone and its evaluation as an inhibitor of interleukin-1β converting enzyme. *Bioorg. Med. Chem. Lett.* 3:2693.

28. Graybill, T. L., et al. 1994. Preparation and evaluation of peptidic aspartyl hemiacetals as reversible inhibitors of interleukin-1β converting enzyme. *Int. J. Pep. Pro. Res.* 44:173.

29. Ator, M. A. and Dolle, R. E. 1995. *Curr. Pharm. Des.* 1:191.

30. Dolle, R. A., et al. 1994. Aspartyl α(1-phenyl-3-(trifluoromethyl)-pyrazol-5-yloxy)methyl ketones as interleukin-1β converting enzyme inhibitors. Significance of the P1 and P3 amido nitrogens for enzyme-peptide inhibitor binding. *J. Med. Chem.* 37:3863.

31. Dolle, R. A., et al. 1997. Pyridazinondiazepines as a high-affinity, P2-P3 peptidomimetic class of interleukin-1β converting enzyme inhibitor. *J. Med. Chem.* 40:1941.

32. Deckwerth, T. L., et al. 2001. Long-term protection of brain tissue from cerebral ischemia by peripherally administered peptidomimetic caspase inhibitors. *Drug Dev. Res.* 52:579.

33. Lauffer, D. L., and Mullican, M. D. 2002. A practical synthesis of (S) 3-tert-butoxycarbonylamino-2-oxo-2,3,4,5-tetrahydro-1,5-benzodiazepine-1-acetic acid methyl ester as a conformationally restricted dipeptido-mimetic for caspase-1 (ICE) inhibitors. *Bioorg. Med. Chem. Lett.* 12:1225.

34. Ho, J. Z., Gibson, T. S., and Semple, J. E. 2002. Novel, potent non-covalent thrombin inhibitors incorporating P3-lactam scaffolds. *Bioorg. Med. Chem. Lett.* 12:743.

35. Karanewsky, D. S., et al. 1998. Conformationally constrained inhibitors of caspase-1 (interleukin-1β converting enzyme) and of the human ced-3 homologue caspase-3 (cpp32, apopain). *Bioorg. Med. Chem. Lett.* 8:2757.

36. Rano, T. A., et al. 1997. A combinatorial approach for determining protease specificities: Applications to interleukin-1β converting enzyme (ICE). *Chem. Biol.* 4:149.

37. Karanewsky, D. S., and Bai, X. 1998. C-terminal modified (N-substituted)-2-indoyl dipeptides as inhibitors of the ICE/ced-3 family of cysteine proteases. WO98/11129.

38. Fahr, B. T., et al. 2006. Tethering identifies fragment that yields potent inhibitors of human caspase-1. *Bioorg. Med. Chem. Lett.* 16:559.

39. Anakinra (Kineret) is the property of Amgen and is a recombinant, nonglycosidated form of the human interleukin-1 receptor antagonist (IL-1Ra).

40. Dolle, R. H., et al. 1997. 3-Chloro-4-carboxamido-6-arylpyridazines as a non-peptide class of interleukin-1β converting enzyme inhibitor. *Bioorg. Med. Chem. Lett.* 7:1003.
41. Shoji, M., et al. 2005. Enantio- and diastereoselective total synthesis of EI-1941-1, -2, and -3, inhibitors of interleukin-1β converting enzyme, and biological properties of their derivatives. *J. Org. Chem.* 70:9905.
42. Henzing, A. J., et al. 2006. Synthesis of novel caspase inhibitors for characterization of the active caspase proteome in vitro and in vivo. *J. Med. Chem.* 49:7637.
43. Asgian, J. L., et al. 2002. Aza-epoxides: A new class of inhibitors selective for clan CD cysteine proteases. *J. Med. Chem.* 45:4958. James, K. E., et al. 2004. Design, synthesis and evaluation of aza-peptide expoxides as selective and potent inhibitors of caspases-1, -3, -6, and -8. *J. Med. Chem.* 47:1553.
44. IDdb3 drug alert: VX-740, July 12 2006.
45. IDdb3 drug alert: VX-745, June 1, 2007.
46. Harter, W. G., et al. 2004. The design and synthesis of sulfonamides as caspase-1 inhibitors. *Bioorg. Med. Chem. Lett.* 14:809.
47. Oppong, K. A., et al. 2005. Discovery of novel conformationally restricted diazocan peptidomimetics as caspase-1 (ICE) inhibitors. *Bioorg. Med. Chem. Lett.* 15:4291.
48. O'Neil, S. V., et al. 2005. Synthesis, and evaluation of novel 8,6-fused bicyclic peptidomimetic compounds as interleukin-1β converting enzyme (ICE) inhibitors. *Bioorg. Med. Chem. Lett.* 15:5434.
49. Soper, D. L., et al. 2006. Synthesis and evaluation of novel 8,5-fused bicyclic peptidomimetic compounds as interleukin-1β converting enzyme (ICE) inhibitors. *Bioorg. Med. Chem.* 14:7880.
50. Wang, Y., et al. 2007. Synthesis and evaluation of caprolactams as interleukin-1β converting enzyme (ICE) inhibitors. *Bioorg. Med. Chem.* 15:1311.
51. Ellis, C. D., et al. 2006. Synthesis and evaluation of thiazepines as interleukin-1β converting enzyme (ICE) inhibitors. *Bioorg. Med. Chem. Lett.* 16:4728.
52. Charrier, J.-D., and Brenchley, G. 2001. Caspase inhibitors and uses thereof. WO01/94351.
53. Laufersweiler, M. C., et al. 2005. Synthesis and evaluation of tricyclic pyrrolopyrimidinones as dipeptide mimetics: Inhibition of interleukin-1β-converting enzyme (ICE). *Bioorg. Med. Chem. Lett.* 15:4322.
54. Soper, D. L., et al. 2006. Synthesis and evaluation of novel 1-(2-acylhydrazinocarbonyl)-cycloalkyl carboxamides as interleukin-1β converting enzyme (ICE) inhibitors. *Bioorg. Med. Chem. Lett.* 16:4233.
55. Kamada, S., et al. 1997. Caspase-4 and caspase-5, members of the ICE/CED-3 family of cysteine proteases, are CrmA-inhibitable proteases. *Cell Death Differ.* 4:473.
56. Schwartz, S., Jr., et al. 1999. Frameshift mutations in mononucleotide repeats in caspase-5 and other target genes in endometrial and gastrointestinal cancer of the microsatellite mutator phenotype. *Cancer Res.* 59:2995.
57. Krippner-Heidenreich, A., et al. 2001. Targeting the transcription factor Max during apoptosis: Phosphorylation-regulated cleavage by caspase-5 at an unusual glutamic acid residue in postion P1. *Biochem. J.* 358:705.
58. Martinon, F., Burns, K., and Tschopp, J. 2002. The inflammasome: A molecular platform triggering activation of inflammatory caspases and processing of proIL-beta. *Mol. Cell* 10:417.
59. Kamada, S., et al. 1997. Involvement of caspase-4(-like) protease in Fas-mediated apoptotic pathway. *Oncogene* 15:285.
60. Kumar, S., et al. 2002. Interleukin-1F7B(IL-1H4/IL-1F7) is processed by caspase-1 and mature IL-1F7B binds to the IL-18 receptor but does not induce IFN-γ production. *Cytokine* 18:61.

61. Margolin, N., et al. 1997. Substrate and inhibitor specificity of interleukin-1β converting enzyme and related caspases. *J. Biol. Chem.* 272:7227.

62. Sequence alignment of the caspases: Phylogenetic relationships were determined and polypeptide sequences for the human forms of caspases-1, -3, -4, and -5 were aligned using the PileUp algorithm (http://www.cryst.bbk.ac.uk/pps97/assignments/projects/criekinge/pilup3.html, Gentics Computer Group, Madison, WI). The relationships are based on the full-length proenzymes. The dendrogram is not affected if the same analysis is carried out using caspases without their prodomains or proenzymes.

63. Talanian, R. V., et al. 1997. Substrate specificities of caspase family proteases. *J. Biol. Chem.* 272:9677.

64. Fassy, F., et al. 1998. Enzymatic activity of two caspases related to interleukin-1β-converting enzyme. *Eur. J. Biochem.* 253:76.

65. Garcia-Calvo, M., et al. 1998. Inhibition of human caspases by peptide-based and macromolecular inhibitors. *J. Biol. Chem.* 273:32608.

66. Lee, D., et al. 2000. Potent and selective nonpeptide inhibitors of caspases 3 and 7 inhibit apoptosis and maintain cell function. *J. Biol. Chem.* 275:16007.

67. Lee, D., et al. 2001. Potent and selective nonpeptide inhibitors of caspase 3 and 7. *J. Med. Chem.* 44:2015.

68. Choong, I. C., et al. 2002. Identification of potent and selective small-molecule inhibitors of caspase-3 through the use of extended tethering and structure-based drug design. *J. Med. Chem.* 45:5005.

69. Han, Y., et al. 2005. Novel pyrazinone mono-amides as potent and reversible caspase-3 inhibitors. *Bioorg. Med. Chem. Lett.* 15:1173.

70. Schielke, P. G., and Shahripour, A. 1999. N-[2-(5-Benzyloxycarbonyl-amino-6-oxo-2(4-fluorophenyl)-1,6-dihydro-1-pyrimidinyl)acetoxyl]-L-aspartic acid aldehyde as an in vivo inhibitor of interleukin-1β converting enzyme. WO Patent 99/36426.

71. Kim, E. E.-K., et al. 2001. Isoxazoline derivative caspase inhibitors for pharmaceutical uses. WO Patent 01021600.

72. Aja, T., Ching, B. W., and Gladstone, P. L. 2002. Anti-apoptotic agents or interleukin 1β converting enzyme (ICE/CED-3) inhibitors for preserving antigenicity of markers associated with diseases. WO Patent 02070544.

73. Le, G. T., Abbenante, G., and Fairlie, D. P. 2007. Peptidic compounds possessing an azide functional group as inhibitors of cysteine proteases useful for treatment of diseases. WO Patent 07041775.

74. Knobelsdorf, J., et al. 2002. Preparation of β-amino acid arylsulfonamide ether derivatives as inhibitors of interleukin-1β converting enzyme. WO Patent 02089749.

75. Wang, J. 2002. Preparation of quinolinecarbonyl(multiple amino acids)-leaving group compounds for pharmaceutical compositions and reagents. U.S. Patent 02052323.

76. Wang, J. 2002. Preparation of quinolinecarbonyl(multiple amino acids)-leaving group compounds for pharmaceutical compositions and reagents. WO Patent 02018341.

77. Caires, A. C. F., et al. 2004. Cyclic palladium compounds having coordinated bis(diphenylphosphine)ferrocene ligands which inhibit the activity of proteins and enzymes, and treatment of diseases and disorders associated therewith. WO Patent 04019924.

78. Powers, J. C., et al. 2005. Preparation of peptidyl propenoylhydrazides as protease inhibitors. WO Patent 05080353.

79. Powers, J. C., et al. 2004. Preparation of aza-peptide epoxides as protease inhibitor. WO Patent 04005270.

80. Gunasekera, S. P., et al. 2000. Aminoiminoquinone and aminoquinine alkaloid compounds as caspase inhibitors, isolation from marine sponge, and therapeutic uses. WO Patent 00002858.

81. Karanewsky, D. S., et al. 2003. Preparation of (substituted)acyl dipeptidyl inhibitors of the ICE/ced-3 family of cysteine proteases. WO Patent 03072528.
82. Ternansky, R. J., et al. 2002. Preparation of novel oxamyl dipeptide inhibitors of the ICE/ced-3 family of cysteine proteases. U.S. Patent 02137686.
83. Fritz, L. C., et al. 2002. Preparation of 2-indolecarbonyl amino acid amides for inhibition of inflammation using interleukin-1β-converting enzyme (ICE)/CED-3 family inhibitors. U.S. Patent 02123522.
84. Karanewsky, D. S., et al. 2002. Preparation of C-terminal modified oxamyl dipeptides as inhibitors of the ICE/ced-3 family of cysteine proteases. U.S. Patent 02042376.
85. Ternansky, R. J., et al. 2001. Preparation of sulfonamido dipeptide inhibitors of the ICE/ced-3 family of cysteine proteases. WO Patent 01079162.
86. Ternansky, R. J., et al. 2001. Preparation of novel oxamyl dipeptide inhibitors of the ICE/ced-3 family of cysteine proteases. WO Patent 01051462.
87. Karanewsky, D. S., and Bai, X. 2001. Preparation of (N-substituted)-2-indolyl dipeptides as inhibitors of the ICE/ced-3 family of cysteine proteases. U.S. Patent 6184244.
88. Karanewsky, D. S., et al. 2000. Preparation of (substituted)acyl dipeptidyl inhibitors of the ICE/ced-3 family of cysteine proteases. WO Patent v00023421.
89. Karanewsky, D. S., et al. 2000. C-terminal modified oxamyl dipeptides as inhibitors of the ICE/ced-3 family of cysteine proteases. WO Patent 00001666.
90. Kim, E. E.-K., et al. 2001. Isoxazoline derivative caspase inhibitors for pharmaceutical uses. WO Patent 01021600.
91. Chang, H. K., et al. 2005. Preparation of amino acid derivatives containing a 2-alkyl-4-oxobutanoyl group as caspase inhibitors. WO Patent 05035497.
92. Chang, H. K., et al. 2005. Preparation of carbamoylisoxazoles as caspase inhibitors. WO Patent 05021516.
93. Takuma, Y., and Kasuga, Y. 2002. Preparation of alkyloxyaminofuranones as materials for interleukin 1β converting enzyme inhibitors. JP Patent 02138085.
94. Takuma, Y., et al. 2002. Preparation of alkoxyaminofuranones as intermediates for interleukin-1β converting enzyme inhibitors. JP Patent 02080472.
95. Roggo, S., et al. 2002. Preparation of malonyl amino acid derivatives as inhibitors of the ICE/ced-3 family of cysteine proteases. WO Patent 02076968.
96. Roggo, S., et al. 2002. Preparation of hexahydro-4-oxoazepino[3,2,1-hi]indoles as inhibitors of interleukin-1β converting enzyme and related proteases. WO Patent 02034751.
97. Hatayama, A., et al. 2003. Preparation of diketohydrazine derivatives as cysteine protease inhibitors. WO Patent 03091202.
98. Ohmoto, K., and Itagaki, I. 2001. Preparation of dipeptide analogs contg. 1,3,4-oxadiazoline derivatives as cysteine protease inhibitors and drugs containing these derivatives as the active ingredient. WO Patent 01040204.
99. Ohmoto, K., and Kawabata, K. 2002. Preparation of oxadiazole derivatives as cysteine protease inhibitors and drugs containing these compounds as the active ingredient. WO Patent 02096892.
100. Wos, J. A., et al. 2003. Preparation of pyrrolo[1,2-a]azocinones as interleukin-1β converting enzyme inhibitors. WO Patent 03106460.
101. O'Neill, S. V., et al. 2003. Preparation of condensed azocinone derivatives as interleukin-1β converting enzyme inhibitors. WO Patent 03104231.
102. Wos, J. A., et al. 2003. Preparation of azepine and thiazepine peptide derivatives as interleukin converting enzyme inhibitors. WO Patent 03103677.
103. Lee, D. 2000. Preparation of tetrapeptide thiomethyl-, aminomethyl-, and sulfonamidomethyl-ketone derivs. as caspase inhibitors useful for treatment of apoptosis. WO Patent 00020440.

104. Allen, D., et al. 2003. Preparation of acylamino(formyl)propanoic acids as caspase-1 inhibitors. WO Patent 03103599.

105. Le, G. T., Abbenante, G., and Fairlie, D. P. 2007. Peptidic compounds possessing an azide functional group as inhibitors of cysteine proteases useful for treatment of diseases. WO Patent 07041775.

106. Robidoux, A. L. C., et al. 2003. Synthesis of diazabicycloalkanecarboxamides as caspase inhibitors. U.S. Patent 6559304.

107. Charrier, J.-D., et al. 2005. Preparation of peptides as caspase inhibitors. WO Patent 05085236.

108. Brenchley, G., et al. 2004. Preparation of 3-[2-[(2-oxo-2H-pyridin-1-yl)alkanoyl]amino]pentanoic acid derivatives as caspase inhibitors. U.S. Patent 04242494.

109. Knegtel, R., et al. 2004. Preparation of isoquinolinone and quinazolinone peptide derivatives as caspase inhibitors. WO Patent 04002961.

110. Golec, J. M. C., et al. 2003. Synthesis of peptide heterocyclic derivatives as caspase inhibitors. U.S. Patent 03232846.

111. Mortimore, M., et al. 2002. Caspase inhibitors and therapeutic uses. WO Patent 02094263.

112. Kay, D., and Golec, J. 2002. Preparation of imidazole and benzimidazole as caspase inhibitors and uses thereof. WO Patent 02042278.

113. Charrier, J.-D. and Brenchley, G. 2001. Synthesis of carboxamidoquinazolines as caspase inhibitors. WO Patent 01094351.

114. Golec, J. 2001. Arylmethoxycarbonylpiperidinecarboxamides as caspase inhibitors. WO Patent 01090070.

115. Wannamaker, M., and Davies, R. 2001. Preparation prodrug of an interleukin-1β converting enzyme inhibitor. WO Patent 01090063.

116. Bebbington, D., et al. 2001. Preparation of carbamate caspase inhibitors. WO Patent 01072707.

117. Golec, J., et al. 2001. Synthesis and use of heterocyclic substituted-amido halopentanoate derivatives as caspase inhibitors. WO Patent 01042216.

118. Kim, S.-G., et al. 2003. Quinoline derivatives as selective caspase-3 inhibitors, processes for their preparation, and pharmaceutical compositions comprising them. WO Patent 03093240.

7 Identification of Apoptotic Caspase Inhibitors

Brett R. Ullman

CONTENTS

7.1 INTRODUCTION

The evolution of caspase inhibitors has progressed greatly over the last two decades, from peptide aldehydes and irreversible peptidomimetics to small molecule inhibitors. Herein, research over the last 18 years will be discussed involving the discovery of peptidic and peptidomimetic caspase inhibitors designed to modulate apoptosis. Since the discovery of ICE/caspase-1 in 1989, the study and development of caspase inhibitors has been a very active area of research, evaluating both specific and pan inhibitors.[1]

Researchers at Merck led the way in the identification of peptide inhibitors; Fletcher and coworkers[2] identified Ac-YVAD-CHO (L-709,049) as the optimal peptide recognition sequence for caspase-1. Shortly thereafter, *Chapman and Thornberry*[3] reported that a more specific peptide was Ac-WEHD-CHO (K_i 0.76 nM vs. 56 pM, respectively). Building on this, Thornberry and coworkers[4] optimized the specific tetrapeptide aldehyde sequences for caspases-1 through -9 (Table 7.1). Knowledge of these specificities would guide the development of selective caspase inhibitors. This work showed that caspases fall into three groups as judged by their optimal tetra peptide inhibitor sequence. Group I contains caspases-1, -4, and -5; Group II contains caspases-2, -3, and -7; and Group III contains caspses-6, -8, and -9. There is a tolerance for a range of P-2 amino acids; this flexibility at P-2 lends itself to the creation of pan caspase inhibitors.

There have been many approaches to developing a therapeutic small molecule caspase inhibitor. These varying strategies for the design of caspase inhibitors

TABLE 7.1
Optimal Tetrapeptide Sequence for Each Caspase

Caspase	1	2	3	4	5	6	7	8	9
Optimal peptide	WEHD	DEHD	DEVD	(W/L)EHD	(W/L)EHD	VEHD	DEVD	LETD	LEHD
Group	I	II	II	I	I	III	II	III	III

overlap and interweave. Herein a straight-line approach is attempted although chronological consistency is not maintained. Caspase inhibitors have progressed over the years from reversible peptide aldehydes to more potent reversible and irreversible peptidomimetic inhibitors. This is accomplished with the variation of the warhead on the prime side of the inhibitor. A warhead is defined as an electrophilic moiety that covalently interacts with the active site of the enzyme (Figure 7.1). A warhead is reversible if the enzyme is only temporarily inactivated via the thiohemiketal intermediate (K_i). Irreversible inhibition results in the inactivation of the enzyme through the formation of the thioether adduct (k_3). An inhibitor can also exhibit bimodal inhibition, which results from reversible inhibiton (K_i) followed by slow inactivation (k_3). Brady[5] extensively studied these interactions on caspase-1 with inhibitors with different warheads, and also studied co-crystal structures of inhibitors bound

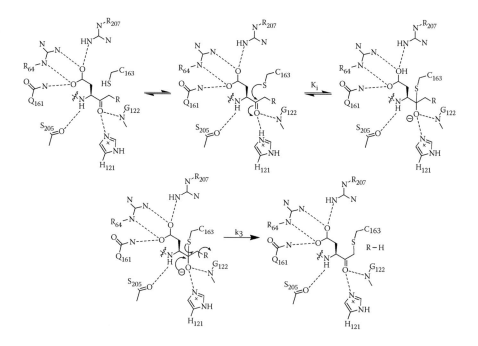

FIGURE 7.1 Stepwise mechanism of bimodal caspase inhbition.

to caspase-1. The pK_a of the leaving group was not necessarily correlated with an inhibitor's inactivation rate, but rather inactivation rates were influenced by initial interactions between the inhibitor and protein, specifically the thiohemiketal complex. The evaluation of crystallographic structures of caspase-1 complexed with the thiohemiketal form of inhibitors from several ketone classes demonstrated the Cys285-C-Ca' leaving group dihedral angle to be near either 60° or 180°. It was proposed that through proper alignment of His237 imidazolium ion with the leaving group, allowing for stabilization of the developing charge, the 180° conformation was the only dihedral angle capable of SN2 displacement of said leaving group. In the context of these modes of inhibition, a range of warhead strategies have been pursued across a wide variety of inhibitor cores, representing the various strategies from both industry and academic research groups.

Initial warhead work focused on inhibitors of caspase-1 as research groups looked to move away from the highly reactive fluoromethyl ketone (FMK) moiety. Mjalli and coworkers reported on various peptide inhibitors with several different warheads: ketones,[6] heterosubstituted methyl ketones,[7] and phenoxymethyl ketones.[8] Building upon data collected on known cysteine protease inhibitors, these warhead moieties were incorporated into a caspase inhibitor backbone. Dolle and coworkers also developed caspase-1 inhibitors[9] (initially developed by Smith, et. al) using known aryloxy and arylacyloxymethyl ketone cysteine inhibitor warheads[10] as well as pyrazoloxymethyl ketones and phosphonomethyl ketones. The introduction of these warheads onto both caspase 3/7–selective and broad-spectrum inhibitors will be discussed in further detail. The more commonly employed warheads are defined (Table 7.2) as aspartylmethyl ketones (Figure 7.2) with their respective abbreviations.

FIGURE 7.2

TABLE 7.2
Common Aspartyl Methyl Ketone Warheads and Their Abbreviations

Abbreviation	FMK	DCB	BTA	PTP	DPP	Ph
Warhead, R	F					

7.2 PEPTIDES

As mentioned previously, Thornberry and coworkers[4] determined the optimal tetra-peptide sequences for caspases-1 through -9. That work opened the door for others to pursue caspase inhibitors as potential therapeutic candidates. Due to pharmacological constrains, a tetrapeptide aldehyde does not readily lend itself to therapeutic utility. Apart from potency, solubility, stability, and selectivity are the main issues faced when designing a caspase inhibitor. Tetrapeptide caspase inhibitors such as these require modification in order to develop them into potential candidates for clinical evaluation.

Grimm and coworkers reported on the warhead optimization[11] of the Ac-DEVD core (Figure 7.3) for the development of caspase-3-specific inhibitors. The simple use of a methyl or propyl ketone (**2** and **3**, respectively) led to a 20- to 30-fold increase in inhibitor activity (Table 7.3). They found that a phenyl ketone (**4**) was less effective than the aldehyde parent. However, the use of alkylaryl ketones provided a slight increase in activity, with the naphthyl derivative (**7**) having over a 10-fold increase

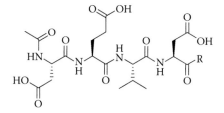

FIGURE 7.3

TABLE 7.3
Ac-DEVD Warhead Analogues

Cmpd	R	Csp-1	Csp-3	Csp-6	Csp-8	NT2 Cells IC$_{50}$ (μM)
		Enzyme Assays IC$_{50}$ (nM)				
1	H	292	44	716	286	> 100
2	CH$_3$	3775	1.3	12	185	—
3	(CH$_2$)$_2$ CH$_3$	4575	1.9	14	140	—
4	Ph	1395	46	363	656	—
5	(CH$_2$)$_3$Ph	307	0.8	7.7	16	30
6	(CH$_2$)$_5$Ph	24	0.8	7.2	9.6	—
7	(CH$_2$)$_3$-1-Np	14	0.1	1.2	1.9	26

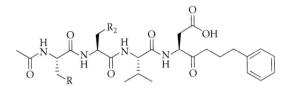

FIGURE 7.4

TABLE 7.4
Modification of the Ac-DEVD Backbone

Cmpd	R	R$_2$	IC$_{50}$ (nM) Csp-3	IC$_{50}$ (μM) NT2 Cells
5	CO$_2$H	CH$_2$CO$_2$H	0.8	30
8	CO$_2$H	CH$_2$SO$_2$CH$_3$	1.7	10
9	CO$_2$H	H	1.4	5
10	CO$_2$CH$_3$	H	18	10
11	SO$_2$CH$_3$	H	1.9	12

---

in caspase-3 activity over the alkyl ketones. Cellular data were also reported for the two propylaryl derivatives (**5** and **7**).

Working from the premise that cellular penetration can be improved by increasing the lipophilicity of a small molecule, it was shown that removal of one of the two carboxylate groups from the aspartic acid or glutamic acid residues (Table 7.4), by conversion to an ester, sulfone, or simple methyl group (Figure 7.4), could increase the cellular activity of the inhibitors with only a slight decrease in the enzymatic activity. This work by Grimm and coworkers also demonstrated that the warhead pocket can accommodate various branched and aryl hydrophobic moieties. This in turn showed that simple modifications could lead to *in vitro* activity. Although modification of side chain carboxylic acids resulted in a reduction in caspase-3 activity, an improvement in cellular activity was observed.

Garcia-Calvo and coworkers reported the use of the tripeptide Z-VAD-FMK[12] (Figure 7.5) as a broad-spectrum caspase inhibitor (Table 7.5). Z-VAD-FMK has become a preferred positive control as an irreversible caspase inhibitor and has been used extensively in proof-of-concept experiments (see Chapter 3). For example, in rat models of ischemia-reperfusion injury, z-VAD-FMK was used to demonstrate significant reductions in both DNA fragmentation and infarct size at 3 and 24 hours after transient ischemia (**61**).[13] The fluoromethyl ketone[14] warhead has been reported on extensively and is a preferred warhead utilized by many research groups.

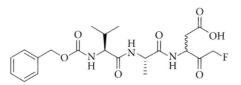

FIGURE 7.5

TABLE 7.5
Inactivation Rates of Z-VAD-FMK

Caspase	1	2	3	4	5	6	7	8	9
k_3/K_i (M^{-1}, s^{-1})	280,000	290	16,000	5,500	130,000	7,100	18,000	280,000	180,000

7.3 PEPTIDOMIMETICS

A peptidomimetic can be defined as a molecule that mimics a peptide sequence while eliminating or reducing the enzymatically scissile peptide bonds. Most research groups took the approach of truncating a terapeptide down to a dipeptide and then modifying the N-capping group. This approach proved to be quite successful, result-ing in stable compounds with *in vivo* activity.

Karanewsky and Bai[15] reported the use of indoyl dipeptides (Figure 7.6) as cas-pase inhibitors. The initial focus was on amino acid substitutions at the P-2 posi-tion (Table 7.6) utilizing reversible aspartyl aldehyde inhibitors (typically caspase-1 selective). Alanine and valine provided the most activity (**12** and **14**, respectively), while the use of proline or glycine (**13** and **17**, respectively) resulted in a large decrease in activity, with **17** being inactive. The leucine analogue (**15**) was not selective and resulted in a decrease in activity. Incorporation of a carboxylic acid side chain attached to the indole nitrogen increased selectivity for caspase-1 (**20** and **21**), and substitution of the indole nitrogen with a benzyl group (**18**) also increased selectivity for caspase-1. Substitution of the indole 3-position with a methyl group and incorporation of a 2,6-dichlorobenzoyloxymethyl ketone (DCB) warhead (**22**) provided a caspase-3-selective inhibitor. The use of an FMK warhead (**23**) resulted in an increase in caspase inhibition, albeit with a loss of selectivity.

The effect of indole ring substitutions on caspases-1, -3, and -8 activity was then evaluated using a VD-FMK backbone (Figure 7.7). Substitution of the indole nitrogen with a methyl group generated inhibitors that were selective for caspase-3 and -8 over caspase-1 (Table 7.7). The phenyl analogue (**27**) maintained selectivity for caspase-3 over caspase-1, whereas the phenpropyl analogue (**26**) provided an

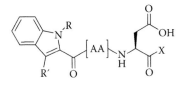

FIGURE 7.6

TABLE 7.6
Indolyl Dipeptide Caspase Inhibitors

					IC$_{50}$ (µM)	
Cmpd	R	R'	AA	X	Csp-1	Csp-3
12	CH$_3$	H	Ala	H	0.177	>10
13	CH$_3$	H	Pro	H	11.7	>50
14	CH$_3$	H	Val	H	0.531	2.48
15	CH$_3$	H	Leu	H	5.52	5.62
16	CH$_3$	H	Phe	H	3.34	49.8
17	CH$_3$	H	Gly	H	34.7	>50
18	Bn	H	Ala	H	0.393	>50
19	(CH$_2$)$_2$CH:CH	H	Val	H	0.313	1.45
20	CH$_2$CO$_2$H	H	Ala	H	1.63	>50
21	(CH$_2$)$_2$CO$_2$H	H	Ala	H	0.198	>50
22	CH$_3$	CH$_3$	Val	CH$_2$DCB	>50	17.3
23	CH$_3$	CH$_3$	Val	CH$_2$F[a]	1.40	0.96

[a] Racemic at Asp.

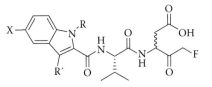

FIGURE 7.7

TABLE 7.7
Substituted Indolyl Dipeptide Fluoromethyl Ketones

				k$_3$/K$_i$ (M^{-1}, s^{-1})		
Cmpd	R	R'	X	Csp-1	Csp-3	Csp-8
23	CH$_3$	CH$_3$	H	2,860	13,400	21,500
24	CH$_3$	Cl	H	6,150	25,900	37,000
25	CH$_3$	Cl	F	7,120	72,700	52,500
26	(CH$_2$)$_3$Ph	H	H	45,100	33,700	32,500
27	Ph	H	H	8,900	74,200	35,500
28	CH$_2$CO$_2$H	H	H	16,800	58,700	127,000

increase in activity for caspase-1 resulting in a pan caspase inhibitor. Replacement of the 3-methyl group with chlorine (compounds **24** and **25**) resulted in an increase in inhibitor activity, and incorporation of an additional halogen in the 5-position (**25**) increased activity and selectivity for caspase-3 and -8. Caspase-8 activity was further enhanced by the addition of a carboxylic acid side chain to the indole nitrogen (**28**).

This work led directly to the identification of IDN-1965 (**23**), which is an irreversible inhibitor that has been utilized as a benchmark pan-caspase inhibitor for *in vivo* investigations. Indoyl dipeptides were a big step away from tetrapeptides; use of a heterocycle and truncation to a dipeptide resulted in a less peptidic, more drug-like molecule. The indole moiety also allows for ring modification and probing of the P-3 pocket. The use of an acid side chain off the indole nitrogen increases activity and in turn leads to pan-caspase inhibitors.

Gores and coworkers[16] reported the use of IDN-1965 for the prevention of apoptosis associated with orthoptic liver transplant (OLT). Utilizing this pan-caspase inhibitor, they sought to attenuate apoptosis resulting from cold ischemia/warm reperfusion injury (CI/WR) and demonstrated that intravenous administration of IDN-1965 to donor animals, the University of Wisconsin solution (organ storage media, UW solution), and the reperfusion media reduced caspase-3 activation and prolonged animal survival during OLT. They reported that for IDN-1965 to be efficacious it must be added to the UW solution. These results show that a caspase inhibitor can attenuate liver damage associated with CI/WR. The use of a caspase inhibitor in conjunction with UW solution may also be a means to preserve other organs during transplant.

Hoglen and coworkers[17] reported the use of IDN-1965 in an α-Fas-mediated model of liver disease. IDN-1965 was shown to reduce alanine aminotransferase (ALT) levels in a dose-dependent manner when administered (i.p.) simultaneously with α-Fas. Complete inhibition was seen when 1.0 mg/kg of IDN-1965 was administered; a single dose of IDN-1965 (1 mg/kg, i.p.) blocked apoptosis after 6 hours, attenuated liver damage after 24 hours, and eliminated lethality for up to 7 days. IDN-1965 (3 mg/kg, i.p.) was also shown to be efficacious up to 3 hours post α-Fas inoculation by reducing liver damage by 61%. However, treatment with IDN-1965 4 hours post α-Fas inoculation did not have a statistically significant effect on ALT levels. Pretreatment of IDN-1965 was also evaluated prior to α-Fas administration. Pretreatment (3 mg/kg) 1 hour prior to α-Fas infusion was efficacious in preventing liver damage, but pretreatment at 3 or 5 hours prior to insult resulted in no protection. This result is consistent with pharmacokinetic data showing that IDN-1965 is cleared quickly, having a $t_{1/2}$ of ~40 minutes (i.v.).

Cai and coworkers[18] developed dipeptide caspase inhibitors by truncating the caspase-3-selective tetrapeptide, DEVD, to the dipeptide Z-VD-FMK (**29**) and then focused on optimization of the P-2 amino acid for caspase-3 activity (Figure 7.8). In summary, the P-2 SAR study in the context of this backbone was extensive (Table 7.8) but largely unsuccessful. Of the many amino acids incorporated, isoleucine (**30**) was the only substitution that resulted in a less than 2-fold decrease in activity. The use of a charged side group greatly decreased activity, i.e., lysine or glutamine (**32** and **33**, respectively), and the incorporation of α-methylamino acids (**35**) also resulted in decreased caspase-3 activity (~76-fold). It was also noted that

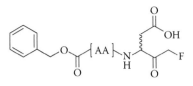

FIGURE 7.8

TABLE 7.8
P-2 SAR Carbobenzoyloxy-AA-
Aspartylfluoromethyl Ketones

Cmpd	Amino Acid	Caspase-3 IC$_{50}$ (nM)
29	Valine	30
30	IsoLeusine	70
31	Leusine	200
32	Lysine	1,600
33	Glutamine	14,000
34	Phenyl glycine	100
35	α-CH$_3$-Valine	2,300

derivatizing the β-acid of compound **29** as a methyl ester (**49**) and altering its ability to interact with arginines and other residues in the active site reduced inhibitor activity greatly (1.1 µM, caspase-3) and eliminated the selectivity against other cysteine proteases, mainly cathepsin-B and calpain-1 (Table 7.11). It should also be noted that compound **29** (MX1013) is a broad-spectrum caspase inhibitor, with IC$_{50}$ values ranging from 5 to 30 nM.

The *in vivo* activity of MX1013 was reported in models of stroke, liver disease, and myocardial infarction.[19] In a rat transient middle cerebral artery occlusion (MCAO) model, MX1013 was administered as an i.v. bolus (20 mg/kg) 10 minutes postischemia followed by infusion (5 mg/kg/h) for 6 or 12 hours and was shown to reduce infarct size by 46 and 57%, respectively. These results were similar to those found for Z-VAD-FMK (i.c.v) in a similar MCAO model, which reduced infarct size by 40–50%.[20] In an anti-Fas (α-Fas)-induced murine model of liver failure, MX1013 administered i.v. (0.25 mg/kg) reduced lethality by 66% (n = 6) after 3 hours. Administration of MX1013 (1 and 10 mg/kg) resulted in complete survival after 3 hours, whereas the vehicle treated animals resulted in complete mortality. These results were similar to those of IDN-1965 (1 mg/kg, i.p.) in a similar α-Fas liver disease model.[17] In a rat model of myocardial infarction (MCI), MX1013 was administered i.v. bolus (20 mg/kg) at t = 0 followed by continuous infusion (5 mg/kg/h) for 12 hours and was shown to reduce infarct volume by 52%. This was a much improved affect over that seen with i.v. administration of Z-VAD-FMK in a similar MCI model,[21] which resulted in a 21% reduction in infarct size.

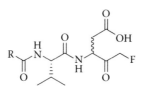

FIGURE 7.9

TABLE 7.9
P-3 SAR of Dipeptide Fluoromethyl Ketones

Cmpd	R	Caspase-3 IC_{50} (nM)
29	$PhCH_2O$	30
36	CH_3	250
37	$PhCH_2$	61
38	$PhCH_2CH_2$	98
39	CH_3O	37
40	$Ph(CH_2)_3O$	46
41	$PhCH_2O_2C(CH_2)_3$	33
42	$2,4-Cl_2-PhCH_2O$	25
43	$HO_2C(CH_2)_3$	50

The dipeptide MX1013 was further optimized by exploring the P-3 SAR (Table 7.9) on the VD-FMK backbone[22] (Figure 7.9). Selected compounds were tested in a HeLa cell apoptosis protection assay that measures the protecting effects of caspase inhibitors against TNF-α-induced apoptosis. The viability of the cells was quantified by calcein AM uptake.[19] Amides, carbamates, and glutaric amides were evaluated. Starting with MX1013, they probed the P-3 site by varying the side chain. Amides were first investigated, resulting in a direct correlation observed between chain length and potency. The benzyl amide (**37**) proved to be more active than the phenethyl analogue (**38**), which is unexpected since the phenethyl amide is more analogous to the carbobenzyloxy carbamate (Cbz, Z) group of MX1013. A simple methyl carbamate (**39**) was more potent than any of the amides reported. In general, inhibitors employing a carbamate at P-3 were more potent than amides.

Two glutaric amides were also tested, with the benzyl ester being more potent than the parent acid (**41** and **43**, respectively). In a HeLa cell protection assay, it was noted that the unprotected **43** was inactive, whereas compounds with halogen substitution on the carbamate phenyl ring (**42**) were more active (100–150 nM vs. 250 nm) than the parent MX1013. Additional caspase data were provided for compound **42** (MX1122), which showed broad-spectrum caspase inhibition with selectivity over other cysteine proteases (Table 7.11). In a time-dependent murine α-Fas liver model, MX1122 provided dose-dependent protection against mortality for up to 5 days at 10 mg/kg.

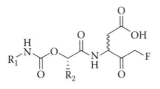

FIGURE 7.10

TABLE 7.10

α-Carbamoyl-Alkylcarbonyl-Aspartyl Fluoromethyl Ketones

Cmpd	R₁	R₂	Caspase-3 IC₅₀ (nM)
44	Ph	CH₃	66
45	Ph	2-Pr	17
46	3-F-Ph	2-Pr	6
47	2,4-Cl₂-Ph	2-Pr	14
48	Bn	2-Pr	20

Building on their earlier work with dipeptides, Cai and coworkers developed a P-2 peptidomimetic replacement using α-hydroxy acid, creating a reversed carbamate[23] (Figure 7.10). First they varied the P-2 alkyl chain and demonstrated that the isopropyl side chain was preferred (Table 7.10). Second, they modified the P-3 aryl group to show a graded increase in potency going from benzyl to phenyl to substituted phenyl. In a murine α-Fas liver model, compound **45** (MX1153) provided dose-dependent 83% reduction in mortality for up to 3 days at the high dose of 10 mg/kg.

A summary of the Maxim lead compounds from the previously mentioned three series is shown in Table 7.11. Z-VD(OCH₃)-FMK (**49**) was also tested in the caspase panel. The conversion of the aspartyl acid to an ester resulted in a loss of activity and selectivity over other cysteine proteases. The Maxim strategy is elegant in its simplicity: truncation of a tetrapeptide down to a dipeptide resulting in potent caspase inhibitors with *in vivo* activity and potentially improved blood-brain barrier (BBB) penetration. The approach taken to understand the SAR of these series was very systematic with respect to the investigation of the P-2 amino acid moiety and the P-3 end group. The leads from these series proved to be potent in a α-Fas-induced liver injury model, and MX1013 was also effective in stroke and MI models. Also, the three leads were very selective for caspases versus other proteases (e.g., >1,000-fold selective against calpain-1, cathepsin B, and factor Xa).

Researchers at Idun also reported on the truncation of tetrapeptide inhibitors to dipeptides (Table 7.12).[24] Starting with Ac-DEVD-H (**1**), they studied tripeptides (**50**, **51**) and dipeptides (**52**, **53**). The potency dropped off precipitously with the use of dipeptides. A strategy was pursued making use of an N-terminal substitution in an effort to improve upon the binding interactions observed with a Cbz group. This

TABLE 7.11
Summary of Maxim Caspase Inhibitors

Cmpd	IC$_{50}$ (nM)			
	MX1013	MX1122	MX1153	49
Csp-1	20	26	12	900
Csp-3	30	25	17	1,100
Csp-6	18	ND	50	—
Csp-7	7	10	11	—
Csp-8	7	6	17	—
Csp-9	5	9	30	—
Cathepsin B	>10,000	10,000	>100,000	300
Calpain-1	>10,000	5000	>100,000	120
Thrombin	>100,000	—	—	>100,000
Factor Xa	>100,000	>100,000	>100,000	>100,000

yielded analogues such as (**54**) employing a 2-naphthyloxy-acetyl group in place of the Cbz, group, resulting in an improvement in inhibition of caspase-3 activity, and thus identifying an important moiety for exploring structure-activity relationships. An extensive SAR study was designed to discover a replacement for the Cbz group on a dipeptide backbone. However, a simple analogue of the initial lead, the 1-naphphthyloxy-acetyl analogue (**55**), demonstrated good broad-spectrum activity and provided the basis for a new series of acyloxy dipeptide caspase inhibitors.

With this knowledge in hand, Ullman and coworkers reported the use of various warheads on the 1-naphthyloxyacetyl-Val-Asp backbone[25] (Figure 7.11). Initially the more common[6-9] warheads were investigated (Table 7.13). The fluoromethyl ketone (**56**) exhibited the best overall activity; the α-(diphenylphosphinyl)oxymethyl ketone (DPP) (**60**) exhibited activity similar to that of the FMK, but unfortunately had much lower cellular activity.

TABLE 7.12
Truncation SAR: from Tetrapeptide to Dipeptide

Cmpd	Structure	Enzyme Assays IC$_{50}$ (μM)			
		mCsp-1[a]	Csp-3	Csp-6	Csp-8
1	Ac-DEVD-H	0.05	0.0035	0.01	0.08
50	Z-ELD-H	0.0065	0.0023	—	0.03
51	Z-FLD-H	0.043	0.137	2.53	8.70
52	Z-VD-H	5.85	1.75	10	11.96
53	Z-LD-H	18.82	3.47	14.03	50
54	2-NpOAc-LD-H	10	0.94	18.56	10
55	1-NpOAc-LD-H	0.57	0.135	0.94	0.77

[a] Murine caspase-1.

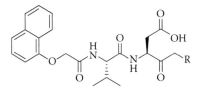

FIGURE 7.11

TABLE 7.13
1-Naphthyloxy-Acetyl Dipepetide Methyl Ketones

Cmpd	R	Enzyme Assays k_3/K_i (M⁻¹ s⁻¹)				Cellular Assays IC$_{50}$ (μM)		
		mCsp-1[a]	Csp-3	Csp-6	Csp-8	Con A[b]	Monocyte[c]	Jurkat[d]
56	F	129,000	207,000	36,842	71,667	3.05	0.95	0.24
57	DCB	52,109	725,222	56,883	17,010	2.15	11.50	10.61
58	OPh	3,630	19,152	2,399	4,258	38.00	11.17	13.95
59	PTP	7,572	71,138	23,909	2,266	—	—	28
60	DPP	142,948	350,785	39,714	59,312	ND	3.48	18.42

[a] Murine caspase-1.

[b] Concanavalin A–induced apoptosis in rat embryonic cortical neurons.

[c] Staph A–induced IL-1β secretion from isolated human monocytes.

[d] Anti-Fas-induced apoptosis in Jurkat clone E6-1 cells.

The use of α-phenolic ketones as warheads was one of many areas of focus in an extensive SAR study (Table 7.14) using the 1-naphthyloxy acetyl-Va Asp (1-N$_p$OAc-VD) backbone (Figure 7.11). The P-1' pocket has limited space with respect to linear warheads; 1-naphthyl (**62**) is tolerated, wherein the 2-naphthyl analogue (**63**) shows a large decrease in activity. This was confirmed with the use of a 4-biphenyl warhead (**64**). The incorporation of electron-withdrawing groups, especially halogens, increased activity. Ortho substitution of the warhead with fluorine resulted in an increase in activity, with the 2,6-difluoro substitution being favored. The tetrafluorophenoxymethyl ketone (TFP) analogue (**71**) exhibited the best pan-caspase activity with good cellular activity, and as a result became the warhead of choice within the Idun group. It was shown that the size, lipophilicity, and binding affinity of the prime side of the inhibitor (as indicated by K$_i$), as well as the conformation of the warhead[8] and resulting leaving group ability (as indicated by K$_3$), all play a role in inhibitor activity.

Further analysis of the warhead moiety on the 1-N$_p$OA$_c$-VD backbone (Figure 7.11) involved the use of heterocyclic warheads[26] (Table 7.15). Starting with 5-membered heterocycles (**73–77**), select compounds were found to exhibit activity against the caspases studied and moderate activity in a Jurkat cellular

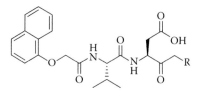

FIGURE 7.11

TABLE 7.14
Warhead SAR,1-Naphthyloxy-Acetyl Dipepetide Methyl Ketones

Cmpd	R	Enzyme Assays k_3/K_i (M^{-1} s^{-1})				Cellular Assays IC_{50} (μM)		
		mCsp-1[a]	Csp-3	Csp-6	Csp-8	Con A	Monocyte	Jurkat
61	OPh	3,630	19,152	2,399	4,258	38	11.17	13.95
62	O(1-Naphthyl)	354,241	1,958,905	0[b]	0[b]	2.65	19	>100
63	O(2-Naphthyl)	263	7,521	0[b]	0[b]	>50	15.90	>100
64	O(4-Ph-Ph)	1,554	9,993	270	1311	>50	38.70	87.85
65	O(2-F-Ph)	6,019	26,835	0[b]	233	15	15.73	>100
66	O(3-F-Ph)	2,732	17,813	48	461	>50	11.96	>100
67	O(4-F-Ph)	9,119	8,561	0[b]	699	>50	14.26	>100
68	O(2,6-F$_2$-Ph)	53,875	200,052	9,979	9,075	2.40	>100	2.04
69	O(2,3,6-F$_3$-Ph)	47,081	187,700	81,749	16,972	2.433	—	1.195
70	O(2,4,6-F$_3$-Ph)	20,681	114,035	5,144	9,076	2.35	14.87	6.84
71	O(2,3,5,6-F$_4$-Ph)	77,528	267,953	78,887	30,173	2.17	6.25	4.50
72	O(2,3,4,5,6-F$_5$-Ph)	14,675	56,615	13,576	8,040	—	8.89	3.05

[a] Murine caspase-1.

assay. As proposed, this did not necessarily correspond to the pKa of the potential leaving group, as exemplified by **75** and **77**. The use of 6-membered heterocycles (**78–82**) provided interesting results with respect to substitution patterns, with the 2-trifluoromethyl-4-pyrimidinol (**79**) being the most active. It is presumed that the warhead moiety of **79** adopts a dihedral angle near 180°.[5] The use of more bulky warheads once again showed that 1-substitution was favored over 2-substitution, with the lone exception being the benzotriazone (**82**). This warhead SAR provided the groundwork for further caspase inhibitor development within the Idun group.

Returning to optimization of the 1-naphthyloxy dipeptide backbone (Figure 7.12), Linton and coworkers investigated the addition of an acidic moiety in an effort to pick up interactions made by the P-3 glutamic acid in AcDEVD caspase-3-specific inhibitors, thus increasing caspase-3 activity[27] (Table 7.16). The incorporation of an acidic moiety onto the naphthyl ring increased the specificity for caspase-3 in one example (**87**), and the use of an acidic side chain attached to the acyl moiety as a glutamic acid mimic greatly enhanced caspase-3 specificity (**85, 86**). In contrast, the incorporation of a triflamide to mimic an acid moiety (**89**) only resulted in a slight increase in inhibitor activity.

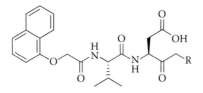

FIGURE 7.11

TABLE 7.15
Heterocyclic Warhead SAR, 1-Naphthyloxy-Acetyl Dipepetide Methyl Ketones

| Cmpd | R | Enzyme Assays k_3/K_i (M^{-1} s^{-1}) | | | | IC$_{50}$ (µM) |
		mCsp-1[a]	Csp-3	Csp-6	Csp-8	Jurkat[c]
73	PTP	7,572	71,138	23,909	2,266	28
74		9,396	35,478	509	295	46
75		1,445	5,674	0[b]	1,141	>200
76		3,671	21,844	220	279	>200
77		42,463	143,850	5,804	6,344	71
78		138	733	5	8	>200
79		24,834	113,766	5,147	3,199	37
80		479	1,752	156	0[b]	>200
81		0[b]	0[b]	0[b]	0[b]	>200
82		45,379	168,736	44,007	18,230	137

[a] Murine caspase-1.

[b] Denotes reversible inhibition.

[c] Anti-Fas-induced apoptosis in Jurkat clone E6-1 cells.

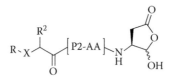

FIGURE 7.12

TABLE 7.16
SAR of Acyl Dipeptide Caspase Inhibitors

| Cmpd | R | X | R² | P-2 | Caspase Activity IC$_{50}$ (µM) | | | |
					mCsp-1[a]	Csp-3	Csp-6	Csp-8
83	1-Np	O	H	Leu	0.570	0.135	0.940	0.770
84	1-Np	O	H	Val	0.336	0.335	>10	1.20
85	1-Np	O	R-(CH$_2$)$_2$CO$_2$H	Leu	2.55	0.021	0.015	0.012
86	1-Np	O	S-(CH$_2$)$_2$CO$_2$H	Leu	4.86	0.0038	0.0035	0.031
87	1-Np-2-CO$_2$H	O	H	Val	0.385	0.054	1.43	0.048
88	1-Np-3-CO$_2$H	O	H	Val	1.89	0.731	1.90	0.200
89	1-Np-5-NHSO$_2$CF$_3$	O	H	Val	0.127	0.207	1.01	0.615
90	1-Np	NH	H	Leu	0.033	0.013	0.037	0.0076
91	1-Np	NH	CH$_3$	Leu	0.087	0.512	0.310	0.017
92	PhCH$_2$CH$_2$(Ph)	NH	H	Leu	0.790	0.320	0.589	0.990
93	1-Np	NH	=O	Leu	0.027	0.010	1.50	0.179

[a] Murine caspase-1.

It was noted that the P-3 ether (**83**) could be replaced with an amine (**90**) (Figure 7.13) while maintaining activity (Table 7.17). The use of a 1-naphthyl group was very effective in increasing inhibitor activity, but in contrast, the addition of a methyl group onto the acyl side chain was detrimental to activity (**91**). Unfortunately, the aryl amines proved to not be particularly stable. In order to improve upon this, the aryl glycinate was replaced with an oxamide (**93**).[28] The conversion to an oxamide functionality not only resulted in an increase in enzyme activity, but also enhanced cellular activity (Table 7.17).

Modeling of some of these oxamide analogues in caspase-1, as well as in crystal structures of various inhibitors bound to caspase-3 (unpublished results), suggests that the key interactions in the P-3/4 region are primarily between Arg 207 and the oxamide functionality itself. More specifically, in a crystal structure of an inhibitor bound to caspase-3, a 2.8 A interaction between the carbonyl of Arg 207 and the distal oxamyl NH was noted, as well as a 2.7 A interaction between the amide hydrogen of Arg 207 and the proximal oxamyl carbonyl.

An extensive SAR investigation was pursued on the oxamide series. Several P-2 amino acids were investigated (Table 7.18) on the 1-Np-NHCOCO-[AA]-DCH$_2$TFP backbone (Figure 7.14). The P-2 position favored hydrophobic residues, with valine (**94**) and t-butyl glycine (**97**) being among the more potent. The use of cyclic amino

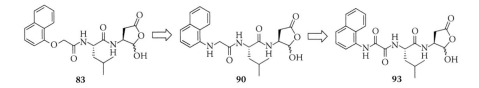

FIGURE 7.13

TABLE 7.17
Transition from Acyl to
Oxamyl Dipeptide

Cmpd	Monocyte Assay % Inhib at 10 μM
83	17
90	18
93	79[a] [c]

[a] IC_{50} = 3.08 μM.

---

acids (**96, 98**) was not tolerated; however, the spiro derivative (**100**) maintained activity in both enzyme and cellular assays. It was speculated that cyclic amino acids in the P-2 position might induce sufficient conformational change in the peptide backbone such that the interactions between the oxamide functionality and the protein were adversely affected, causing a loss in overall binding affinity of the inhibitor.

A wide variety of warheads were also investigated on the optimized 1-naphthyl-oxamyl-Val-Asp backbone (Figure 7.15), resulting in several broad-spectrum inhibitors with good cellular activity (Table 7.19). Results were similar to that those observed with the 1-naphthyloxy-acetyl backbone (Table 7.14), with the additional use of phosphonate warheads (**105, 106**). The DPP warhead (**106**) was among the analogues exhibiting the most potent binding and cellular activity. The naphthyl analogues exhibited the same results as with the 1-naphthyloxyl backbone warhead analysis, wherein the 1-naphthyloxy warhead (**111**) was more potent than the 2-naphthyloxy warhead (**112**). The 2,6-difluorophenoxy warheads (**103, 104**, and **107**) exhibited good cellular and enzyme activity.

Finally, the bulk of the reported data focused on the P-4 SAR (Table 7.20) and optimization of the distal amide of the oxamide functionality (Figure 7.16). A wide range of groups were investigated, with polycyclic aromatic groups being preferred (**115, 118**). The 1-naphthyl(CH₃) disubstituted amine (**117**) proved to be the least active. The t-octyl (**124**) exhibited potent broad-spectrum caspase activity as well as good cellular activity. The use of hetoraryl oximdes (**123**) drastically reduced activity. The use of hydrazines provided interesting results; the phenyl hydrazine (**129**) exhibited a drop in enzyme and cellular data, whereas the disubstituted hydrazines (**127, 128**) exhibited a large increase in caspase-1 binding and modest Jurkat activity.

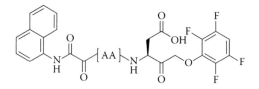

FIGURE 7.14

TABLE 7.18
P-2 SAR of 1-Naphthyloxamyl Dipeptide Tetrafluorophenoxymethyl Ketones

Cmpd	AA	Enzyme Assays k_3/K_i (M^{-1} s^{-1})				Cellular Assays IC$_{50}$ (μM)	
		mCsp-1[a]	Csp-3	Csp-6	Csp-8	THP-1	Con A
94	Val	892,596	288,748	142,903	447,761	0.77	0.55
95	Leu	253,514	180,453	83,651	177,519	2.13	2.58
96	hPro	35,955	9,849	11,557	19,481	NT	16
97	t-BuGly	2,323,255	98,551	160,076	456,491	0.27	2.9
98	4,5-BnPro	912	511	5,583	1,136	22.14	50
99	PhGly	1,075,916	275,734	229,873	1,305,449	2.77	0.5
100	HNcC$_5$H$_{10}$CO	767,316	70,631	8,227	51,317	0.93	2.5

Note: NT = not tested.

[a] Murine caspase-1.

[b] Human monocytic cell line primed with interferon-γ and then stimulated with LPS to induce secretion of IL-1β.

Building off of the 1-naphthyl oxamide, 2-sustituted aryl oxamides (Figure 7.17) were investigated (Table 7.21). These compounds tended to be more potent inhibitors when they contained an electron-withdrawing group. The larger groups tended to be more active, with the benzyl and phenoxy analogues (**138** and **139**, respectively) exhibiting good potency.

While exploring the use of caspase inhibitors in the context of liver disease, additional optimization of the oxamide series was investigated. This particular P-4 SAR study was focused on a similar set of tetrafluorophenoxymethyl ketones with a P-2 replacement of valine with alanine.[29] Previous SAR[28] studies on the oxamide series revealed that 2-substituted phenyl oxamides generally exhibited better activity (Table 7.21). Some of the more potent oxamides were then compared using three different warheads (Table 7.22) on the Ala-Asp backbone (Figure 7.18), which led directly to the discovery of IDN-6556/PF-03491390 (emricasan) (**140**), a first-in-class pan-caspase inhibitor (PCI). Emricasan is currently being clinically evaluated for the treatment of liver disease as an antifibrotic agent. *In vivo* profiling and clinical studies with emricasan will be discussed in further detail in Chapters 8 and 10.

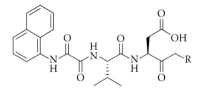

FIGURE 7.15

TABLE 7.19
Warhead SAR, 1-Naphthyloxamyl Dipeptide Methyl Ketones (P-2 = Valine)

| Cmpd | R | Enzyme Assays k_3/K_i (M^{-1} s^{-1}) | | | | Cellular Assays IC$_{50}$ (μM) | |
		mCsp-1[a]	Csp-3	Csp-6	Csp-8	Jurkat	Con A
101	F	2,170,000	413,000	199,000	1,157,000	0.12	1.65
102	DCB	1,052,775	740,466	234,318	119,197	0.004	0.35
103	O(2,3,5,6-F$_4$-Ph)	893,000	289,000	143,000	448,000	0.006	0.5
104	O(2,4,6-F$_3$-Ph)	879,183	340,398	37,438	189,358	0.54	0.30
105	DPP	2,950,000	876,478	222,420	2,447,090	0.005	0.65
106	OPO(CH$_3$)Ph	1,404,853	529,376	165,406	847,197	0.09	2.00
107	O(2,6-F$_2$-Ph)	1,295,045	492,401	49,970	311,465	0.06	0.45
108	PTP	1,069,270	293,818	13,349	60,347	0.13	2.00
109	O(5-CO$_2$CH$_3$-3-isoxazyl)	532,101	264,194	12,805	88,785	2.8	9.00
110	OSOPh	2969	342	459	1	66	>50
111	O(2-Naphthyl)	155,894	9,455	31	20	25	7.30
112	O(1-Naphthyl)	378,200	11,653	268	12	>50	2.70

[a] Murine caspase-1.

The discovery of the oxamyl dipeptides greatly increased binding affinity for caspases and represented a move away from tetrapeptides. The Idun group found that a large variety of oxamides could be employed while maintaining inhibitor activity; even the methyl analogue (**113**) exhibits modest activity (Table 7.20). The binding interactions of the oxamide moiety were discussed, but no crystal structures were provided. The use of disubstituted hydrazines was also reported to provide affinity for caspase-1.

Han and coworkers[30] reported on a series of reversible dipeptide caspase inhibitors, initiated through the investigation of P-3 amide modifications (Table 7.23) on the ValAsp-CH$_2$SBn backbone (Figure 7.19). The phenylacetamide analogue (**152**) was quickly identified as an interesting lead compound. Modification of the distance between the amide and the aryl moiety was explored but with no improvement in activity. They then examined substituted phenylacetamide derivatives, which

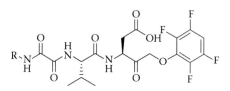

FIGURE 7.16

TABLE 7.20
P-4 SAR Oxamyl Dipeptide Tetrafluorophenoxymethyl Ketones

Cmpd	R	Enzyme Assays k_3/K_i (M^{-1} s^{-1})				Cellular Assays IC$_{50}$ (μM)	
		mCsp-1[a]	Csp-3	Csp-6	Csp-8	Jurkat	Con A
113	CH$_3$	17,000	149,000	126,000	21,000	0.11	7.0
114	Ph	159,000	327,000	275,000	282,000	0.43	1.5
115	1-Np	893,000	289,000	143,000	448,000	0.0061	0.5
116	2-Np	625,000	395,000	484,000	589,000	0.0007	3.3
117	1-Np(CH$_3$)	236	2	78	0	>100	>50
118	1-Anthryl	17,600,000	260,000	264,000	1,920,000	0.01	0.85
119	2-Anthryl	43,300	142,000	191,000	39,000	0.15	0.55
120	1,2,3,4-H$_4$-1-Np	46,000	20,500	21,000	18,700	0.37	2.4
121	5,6,7,8-H$_4$-1-Np	905,000	139,000	132,000	1,150,000	0.18	2.0
122	4-F-Ph	106,345	161,648	247,449	179,386	0.01	2.5
123	2-Pyridyl	22,030	117,608	64,055	42,243	0.14	9.4
124	t-Octyl	241,885	243,190	444,855	643,095	0.013	2
125	Bn	259,386	388,350	218,579	12,416	0.077	1.5
126	PhCH$_2$CH$_2$	345,243	243,319	317,122	206,906	4.7	3.5
127	Ph(Bn)N	5,594,712	104,026	123,918	526,289	0.07	2.3
128	Ph$_2$N	1,234,278	168,506	130,873	666,672	0.19	4.87
129	PhNH	59,480	113,104	143,499	70,373	2.35	20

[a] Murine caspase-1.

resulted in the 2,5-dimethoxy phenyl acetamide (**157**), which exhibited increased selectivity toward caspase-3.

Over 100 thiol warhead derivatives were investigated, which led to the identification of the 2-chloro-6-fluoro-benzylthio ether warhead (compound **158**), which exhibited a slight increase in cellular activity. Modification of the P-3 aryl ring (Figure 7.20), at the 5-position, led to the discovery of compound **159**, an acyloxy ester (Table 7.24). Computational docking of the compound suggested that the acid derivative bound in the active site of caspase-3 much like Ac-DEVD-H, which showed that the ester was acting as a prodrug. These findings led to the acrylate (**161**) and sulfone (**162**) derivatives, which proved to be potent, selective, reversible caspase-3 inhibitors, and also exhibited good cellular activity.

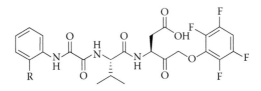

FIGURE 7.17

TABLE 7.21
2-Substituted Phenyloxamyl Dipeptide Tetrafluorophenoxymethyl Ketones

Cmpd	R	Enzyme Assays k_3/K_i (M^{-1} s^{-1})				Cellular Assays IC$_{50}$ (μM)	
		mCsp-1[a]	Csp-3	Csp-6	Csp-8	Jurkat	Con A
130	F	195,000	221,000	192,000	248,000	0.097	0.97
131	Cl	2,019, 000	1,170,000	655,000	2,140,000	0.016	0.3
132	Br	7,160,000	1,270,000	504,000	2,350,000	0.36	0.4
133	I	1,980,000	564,000	373,000	2,330,000	0.081	1
134	t-Bu	956,000	268,000	424,000	2,302,000	0.0022	5.7
135	CF$_3$	1,940,000	842,000	746,000	1,630,000	0.0017	0.65
136	CH$_3$O	369,000	100,000	98,500	442,000	0.31	2.5
137	Ph	2,593,143	158,011	335,251	1,269,323	0.2	2.5
138	Bn	1,156,360	187,179	213,969	935,882	0.0090	2
139	PhO	1,502,281	78,594	202,138	933,186	0.66	2.5

[a] Murine caspase-1.

The inhibitors developed by Han and coworkers[36] demonstrated that the tetrapeptide Ac-DEVD-H could be truncated to a dipeptide, accomplished in this case via modification of the phenylacetamide moiety of the dipeptide terminal capping group. Han and coworkers also examined warhead variants utilizing benzylthio ethers, which resulted in reversible inhibitors. Finally, incorporation of an acidic side chain off of the phenyl acetamide moiety was shown to increase caspase-3 activity and selectivity.

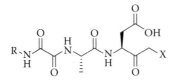

FIGURE 7.18

TABLE 7.22

P-4/Warhead SAR Oxamyl Dipeptide Methyl Ketones (P-2 = Alanine)

Cmpd	R	X	Enzyme Assays k_3/K_i ($M^{-1} s^{-1}$)				Cellular Assays IC_{50} (μM)		α-Fas Liver, IP ED_{50} (mg/kg)
			mCsp-1[a]	Csp-3	Csp-6	Csp-8	Jurkat	THP-1	
140	2-t-Bu Ph	TFP	689,000	75,700	58,700	2,940,000	0.025	0.6	0.08 (0.06–0.12)
141	2-t-Bu Ph	DPP	2,158,000	99,100	52,100	2,010,000	0.081	9.09	0.14
142	2-t-Bu Ph	DCB	8,330,000	236,000	105,000	4,300	0.143	11.8	45.8
143	2,5-(t-Bu)$_2$ Ph	TFP	16,300,000	38,200	15,400	365,000	0.062	1.84	0.05 (0.02–0.11)
144	2,5-(t-Bu)$_2$ Ph	DPP	2,710,000	18,300	7,640	970,000	0.109	5.87	0.5
145	2,5-(t-Bu)$_2$ Ph	DCB	4,025,000	30,000	6,400	68	1.38	3.13	ND
146	2-Cl Ph	TFP	13,400,000	219,000	26,000	228,000	0.064	0.54	0.05 (0.02– 0.16)
147	2-Cl Ph	DPP	3,960,000	211,000	37,100	1,410,000	0.068	1.04	0.04 (0.03–0.05)
148	2-Cl Ph	DCB	8,312,000	256,000	44,000	109	1.17	0.32	ND
149	2-Ph Ph	TFP	323,000	3,400	5,900	154,000	0.13	0.5	0.14 (0.08–0.23)
150	2-Ph Ph	DPP	5,000,000	28,400	22,700	1,570,000	3.94	4.12	0.44
151	2-Ph Ph	DCB	7,280,000	29,000	61,000	342,000	0.39	4.36	ND

Source: Modified from Linton, S. D., et al., *J. Med. Chem.*, 48, 6779, 2005.

[a] Murine caspase-1.

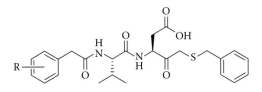

FIGURE 7.19

TABLE 7.23
P-3 SAR Dipeptide Benzylthiomethyl Ketones

Cmpd	R	IC$_{50}$ (μM)				
		Csp-1	Csp-3	Csp-7	Csp-8	NT2[a]
152	H	22	0.75	8.4	22.5	—
153	2-CH$_3$O	28	0.31	2.7	15.5	—
154	3-CH$_3$O	14	0.21	3.1	14	60
155	2-CH$_3$O, 5-Br	3.1	0.06	3.1	2.3	9.0
156	2-CH$_3$O, 5-Ac	5.9	0.01	0.13	2.9	2.0
157	2,5-(CH$_3$O)$_2$	6.0	0.048	3.2	6.6	10

[a] Camptotherin-induced DNA fragmentation in neuronal precursor (NT2) cells.

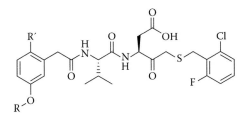

FIGURE 7.20

TABLE 7.24
2,5-Disubstituted Phenacyl Dipeptide 2-Chloro-6-Fluorobenzylthiomethyl Ketones

Cmpd	R	R'	IC$_{50}$ (μM)				
			Csp-1	Csp-3	Csp-7	Csp-8	NT2
158	CH$_3$	OCH$_3$	5.0	0.073	2.5	2.4	9.0
159	CH$_2$CH$_3$	OCH$_2$CO$_2$CH$_3$	6.5	0.053	1.7	9.3	0.7
160	CH$_2$CH$_3$	OCH$_2$CO$_2$H	3.8	0.005	0.36	0.81	5.0
161	CH$_2$CH$_3$	CH=CHCO$_2$CH$_3$	1.8	0.008	0.024	3.85	0.6
162	CH$_3$	SO$_2$CH$_3$	1.6	0.005	0.12	0.49	1.0

7.4 CONFORMATIONAL CONSTRAINTS IN PEPTIDE BACKBONES

The use of conformationally constrained peptides has been successfully incorporated into peptide scaffolds. Dolle and coworkers[31] first reported the use of constrained peptides as inhibitors of caspase-1[32] with the development of pyrimidone peptidomimetics (Figure 7.21). Starting from Ac-YVAD-H, the inhibitor backbone was modified to maintain the key hydrogen bond interactions with the P-3 carbonyl and P-3 N-H. The binding interactions of Ac-DEVD-H in the caspase-3 active site[33] are shown in Figure 7.22. The P-2 N-H is not necessary for binding affinity, which allows for the cyclization of the P-3 portion onto the P-2 N-H. This was followed by the work of Semple and coworkers,[34] who employed a similar strategy using a pyridone and pyrimidone backbones (Figure 7.23). This work was later resurrected with the development of a pyrazone (Figure 7.24) core, which added stability to the inhibitors over the pyridone core.[35] Constrained peptide backbones provide a lower free energy of binding, and therefore can improve the potency of one's inhibitor. New constrained scaffolds can be planned and evaluated using molecular modeling; however, given the uncertainties associated with computation modeling, it is always prudent to generate and assay a select number of specific scaffolds.

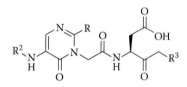

FIGURE 7.21 Pyrimidone peptidomimetic aspartylmethyl ketones.

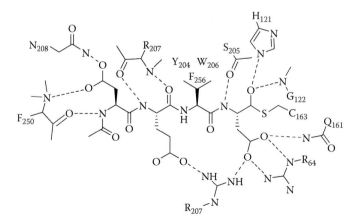

FIGURE 7.22 Caspase-3 active site with AcDEVD-H bound; no direct interaction with P-2 NH.

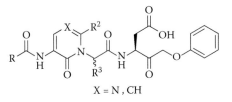

FIGURE 7.23 Pyridone peptidomimetic aspartyl phenoxymethyl ketones.

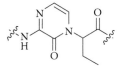

FIGURE 7.24 Pyrazone peptidomimetics.

7.4.1 CYCLIC PEPTIDES

Karanewsky and coworkers[36] reported the use of conformationally constrained pep-tides[37] as inhibitors of caspases-1 and -3. Their approach was different as the starting point for these modifications was the tripeptide Z-VAD-H. Constraining the pep-tide involved the P-2 amide nitrogen being "tied back" to the P-3 or P-2 side chain (Figure 7.25). The tying back of the P-2 NH to the P-3 side chain to give pyrrolidone-type analogues (Figure 7.25, structural analogue I) provided compounds with little or

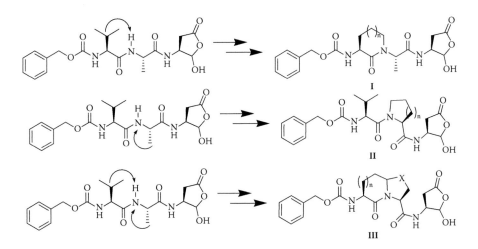

FIGURE 7.25

TABLE 7.25
Constrained Dipeptides Based on Z-VAD-H:
P-3 vs. P-2 Constraints

				IC$_{50}$ (μM)	
Cmpd	Type	n	x^a	Csp-1	Csp-3
Z-VAD-H	NA	NA	NA	0.064	47
163	I	0	NA	3.7	>100
164	I	2	NA	0.186	>10
165	II	1	NA	0.087	14
166	II	2	NA	0.095	3.49
167	III	1	S,β	4.8	86
168	III	2	S, α	0.159	>10
169	III	2	(CH$_2$)$_2$, α	0.036	>10

Note: NA = not applicable.

^{a} Bridge chirality.

no activity. However, activity began to emerge as the constraint was relieved with the use of 6- and 7-membered rings (n = 1, 2), allowing more conformational flexibility.

The use of 5- and 6-membered rings to tie back the P-2 NH to its own side chain provided proline/pyrolidine-type analogues (**165, 166**), which increased their inhibitory activity for caspases-1 and -3 (Table 7.25). The combination of these two types of backbones resulted in a bicyclic core (**167–169**), wherein the 7,5- and 7,6-bicyclic inhibitors proved to be potent and selective for caspase-1. As the contraint of the bicycle was relieved Caspase-1 activity increased, as was observed with the pyrrolidinone-type analosues.

The 7/5 bicylic core was then incorporated into the Ac-DEVD backbone, providing the most interesting inhibitor discussed in this study (Figure 7.26). A pivotal observation of this paper revealed that although an increase in caspase-3 potency (IC$_{50}$ = 0.018 μM) was expected upon returning to the caspase-3-specific backbone, the observed loss in

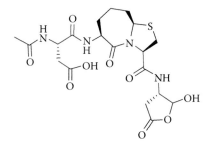

FIGURE 7.26 P-3/2 conformationally constrained 7/5 bicyclic aspartyl aldehyde based on Ac-DEVD-H.

caspase-1 activity (10.4 µM) was not expected. It was proposed that this was possibly due to poor interactions of the bicyclic ring structure, not seen in an unconstrained ana-logue with an acetyl aspartic acid residue in the same P-3/4 region. Observations from these studies, combined with work from Dolle and coworkers described earlier in this chapter, paved the way for other constrained caspase inhibitors, including caspase-1 inhibitors from Vertex and Procter & Gamble, described in Chapter 6.

Karanewsky and Linton[38] first reported the use of oxo azepino indoles (OAIs) (Figure 7.27, Table 7.26) as a caspase inhibitor scaffold employing constrained replacements for P-2/3. This work continued in conjunction with the Idun/Novartis collaboration and was further reported by Roggo.[39] This was initially investigated as a replacement for tri- and tetrapeptide aldehyde inhibitors (Table 7.26). In the case of the latter, the P-4 carbobenzyloxy group contributes to activity against caspase-1 and -8 activities (**170** vs. **171**). DEVD analogues based on an OAI scaffold demonstrated that incorporation of an aspartate at P-4 (**172**) increased activity as well as selectivity for caspase-3.

A variety of warheads were investigated using the Z-OAI-D backbone (Table 7.27), and the FMK (**173**) exhibited the best overall activity, with the TFP warhead displaying comparable results (**175**). Lastly, they optimized the P-4 cap-ping group (Figure 7.28), investigating amides, carbamates, ureas, and sulfonamides (Figure 7.28, Table 7.28) on the OAI-DCH$_2$TFP backbone; it was determined that the phenylcarbamate (**176**) had the best overall activity.

Deckworth and coworkers[42] reported the use of two caspase inhibitors utiliz-ing the OAI core, **173** (IDN5370) and **176** (IDN7866), in an MCAO stroke model. IDN5370 was administered 60 minutes preocclusion (300 ng, intracerbroventricular [i.c.v.]), followed by infusion (40 ng/h, i.c.v.) for 24 hours in a rat permanent MCAO stroke

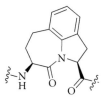

FIGURE 7.27

TABLE 7.26
Oxo Azepino Indole (OAI) Aspartyl Aldehydes

Cmpd	Structure	Enzyme Assays IC$_{50}$ (µM)			
		mCsp-1	Csp-3	Csp-6	Csp-8
170	Z-OAID-H	0.046	1.8	20	7.6
171	Ac-OAID-H	3.07	3.9	10	50
172	Z-D-OAID-H	0.096	0.0052	—	1.19

TABLE 7.27
Warhead SAR of Z-OAID Methyl Ketones

Cmpd	Structure	Enzyme Assays K_i (nM)				Cellular Assays IC_{50} (µM)	
		mCsp-1[a]	Csp-3	Csp-6	Csp-8	Jurkat	THP-1
173	Z-OAID-FMK	0.5	12	33	22	0.811	0.1
174	Z-OAID-CH$_2$DPP	14	7.7	5.6	12	3.5	1.6
175	Z-OAID-CH$_2$TFP	14	11	25	64	4	0.55

[a] Murine caspase-1.

model. Infarct volume was reduced 38% after 48 hours compared to vehicle-treated animals, and brain atrophy was also reduced by 43% after 4 weeks, suggesting long-term protection resulting from administration of IDN5370.

In a murine kainic acid–induced seizure model, IDN5370 was reported in the same publication to be ineffective in reducing hippocampal caspase-3 activity when administered i.p., thus suggesting that IDN5370 was unable to effectively

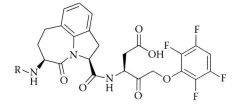

FIGURE 7.28

TABLE 7.28
P-4 SAR of OAI Aspartyl Tetrafluorophenoxymethyl Ketones

Cmpd	R	Enzyme Assays K_i (nM)				Cellular Assays IC_{50} (µM)	
		mCsp-1[a]	Csp-3	Csp-6	Csp-8	Jurkat	THP-1
176	4-CH$_3$O-PhOCO	1	3	26	11	2	0.2
177	4-CH$_3$O-PhCH$_2$NCO	6	24	15	171	ND	3
178	F3CO-PhSO$_2$	1	7	186	152	12	5.2
179	(CH$_3$)$_2$NSO$_2$	8	7	14	41	0.33	0.73
180	1-NpCH$_2$CO	1	3	9	51	5.9	4.2
181	BnCO	1	16	38	36	0.52	1.4

[a] Murine caspase-1.

penetrate the blood-brain barrier (BBB). Deckworth and coworkers then studied IDN7866, another caspase inhibitor of the same structural class. Administration of IDN7866 (30 mg/kg, i.p.) returned hippocampal caspase-3 activity to baseline levels in the murine kainate model. This suggested that IDN7866 could penetrate the BBB, and could therefore be administered in a more therapeutically acceptable manner. Intravenous administration of IDN7866 was subsequently examined in a transient MCAO model, demonstrating a reduction in infarct size of 24% when administered at three time points (5 minutes preocclusion and 2 and 4 hours postocclusion). In the permanent MCAO model, IDN7866 reduced infarct size by 17% when administered in a similar manner at doses of 10 and 30 mg/kg.

The tricyclic core incorporated into caspase inhibitor scaffolds first reported by Karanewsky and Linton has led to potent pan-caspase inhibitors. These initial compounds, such as IDN5370, were promising and quite potent with respect to their enzyme and cellular data (utilized as a positive control by the Idun group in several and varied *in vitro* assays), but proved ineffectual in a stroke model. Modification of the terminal group and optimization of the warhead by *Roggo and coworkers*[42] resulted in improved inhibitors that could penetrate the BBB, providing promising results in accepted animal models of stroke (IDN7866). This is somewhat surprising wherein the high molecular weight of IDN7866 (>650) at first glance would lead one to believe this might limit its ability to penetrate the brain.

In addition, Becker and coworkers[40] reported the use of the tri-cycle core in their structural analysis of caspase inhibitors. Becker and coworkers reported caspase-3 activity for Ac-D-OAI-D-CH$_2$SBn (Figure 7.29) as 3 nM. Willoughby and coworkers[41] even reported the incorporation of the OAI core into granzyme B inhibitors (Figure 7.30). It should be noted that the OAI core is similar to the 7/5 bi-cycle core (Figure 7.26) with the added benefit of containing one less chiral center.

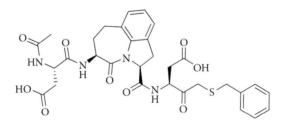

FIGURE 7.29 OAI aspartyl benzylthiomethyl ketone caspase-3 inhibitor.

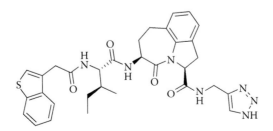

FIGURE 7.30 OAI-based granzyme B inhibitor.

7.4.2 HETEROCYCLIC P-2/3

Han and coworkers performed an extensive SAR analysis of pyrazinone monoamides as caspase-3 inhibitors[43] based on aspartyl benzylthiomethyl ketones as reversible inhibitors. These mimic a tetrapeptide with the cyclization of the P-3 side chain onto the P-2 nitrogen, shown earlier to be an effective modification to generate constrained peptides.[36,37] Analysis of the pyrazinone 3-position amine SAR (Figure 7.31) shows that the isoxazole moiety (**188**) provides the best selectivity for caspase-3 (Table 7.29).

The warhead portion of the inhibitors was also investigated by comparing a simple aldehyde to a series of ketones, ultimately focusing on substituted aminomethyl ketones (**190–195**) (Table 7.30, Figure 7.32). The methyl-alkyl amines showed the best activity, tuning out caspase-8 activity and increasing cellular activity.

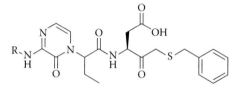

FIGURE 7.31

TABLE 7.29
Substituted 3-Amino-Pyrazinone Aspartyl Benzylthiomethyl Ketones

Cmpd	R	IC$_{50}$ (μM)				
		Csp-1	Csp-3	Csp-7	Csp-8	NT2
182		1.70	0.053	2.60	>10	7.10
183		0.67	0.048	4.68	6.42	5.33
184		0.22	0.088	2.42	2.73	37.9
185		0.92	0.071	9.73	7.65	>10
186		0.13	0.026	1.13	1.75	3.08
187		0.23	0.051	2.54	3.30	6.41
188		0.22	0.0079	0.73	2.46	1.39

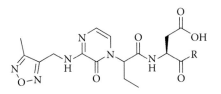

FIGURE 7.32

TABLE 7.30
3(4-Methyl-1,2,5-oxadiazol-3-yl)Methylaminopyrazinone Caspase Inhibitors

Cmpd	R	IC₅₀ (µM)				
		Csp-1	Csp-3	Csp-7	Csp-8	NT2
189	H	0.15	0.078	1.00	0.99	3.56
190		0.0087	0.014	0.13	4.29	1.52
191		1.75	0.10	1.01	>10	3.99
192		0.62	0.021	0.13	>10	0.29
193		0.031	0.015	0.13	>10	0.30
194		0.028	0.011	0.14	>10	0.13
195		0.025	0.009	0.10	>10	0.15

Lastly, Han et al.[43] continued to optimize this series of inhibitors by evaluating substitutions at the pyrazinone 5-position (Figure 7.33). Incorporation of a t-butyl group led to compounds **196** (M826) and **197** (M867), which exhibited greatly enhanced cellular potency (Table 7.31), and were subsequently used as tool compounds to evaluate the *in vivo* efficacy of this series.

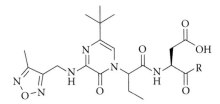

FIGURE 7.33

TABLE 7.31
3(4-Methyl-1,2,5-oxadiazol-3-yl)-Methylamino-5-(t-Butyl)-pyrazinone Caspase Inhibitors M826 and M867

Cmpd	R	IC$_{50}$ (μM)					Whole Cell IC$_{50}$ (μM)			
		Csp-1	Csp-3	Csp-6	Csp-8	NT2	mCGN[a]	rCort[b]	hWBC(−)[c]	hWBC(+)[d]
196 (M867)		0.05	0.005	1.7	2.1	0.021	0.05	0.09	0.05	1.2
197 (M26)		0.11	0.0014	3.44	4.03	0.027	0.06	0.06	0.04	0.7

[a] Etoposide-induced apoptosis in mice cerebellar granule neurons.
[b] Etoposide-induced apoptosis in rat cortical neurons.
[c] Cyclohexamide-induced cell apoptosis in human white blood cells in the absence of plasma proteins.
[d] Cyclohexamide-induced cell apoptosis in human white blood cells in the presence of plasma proteins.

Toulmond and coworkers[44] reported the use of M826 in a rat malonate-induced model of Huntington's disease. Huntington's disease (HD) is a genetic disorder that results in the degeneration of the central nervous system, resulting in chorea, gliosis, neuronal necrosis, and dementia. In the model of HD studied, mice were treated with malonate via intrastratial infusion (i.str.) and M826 was administered i.str. 30 minutes prior (3 mg/kg), 90 minutes post (1.5 mg/kg), and 180 minutes post malonate treatment (1.5 mg/kg). M826 was shown to reduce caspase-3 activity by 66%, cell death by 24% (DNA fragmentation), and lesion volume by 39%.

Holtzmann and coworkers[45] tested M826 in a rat neonatal hypoxia-ischemia (H-I) brain injury model. Reduction of oxygen to the prenatal brain (hypoxia) results in brain injury that can lead to impairment, cerebral palsy,[46] and mortality. Administration of M826 (30 nmol and 3 nmol) by i.c.v. post-insult resulted in a significant reduction in caspase-3 activity at 24 and 48 hours, and M826 reduced brain injury as noted by a reduction of DNA cleavage by ~30% and a reduction in tissue loss by 30%–35%. However, it should be noted that cell death due to necrosis was observed, and this was thought to be due to calpain-related necrotic cell death. These results were similar to those seen for Z-D(OCH$_3$)-FMK in a similar H-I study.[47] Therefore, a selective reversible caspase inhibitor such as M826 can protect the brain from injury in a neonatal H-I model.

Méthot and coworkers[43] reported the use of M867 in a rat cecal ligation and puncture (CLP)–induced sepsis model. It was demonstrated that continuous i.v. infusion of M867 (2 mg/kg/h) for 24 hours in a rat CLP model reduced αII-spectrin cleavage and DNA fragmentation by 88% and 57%, respectively. They also developed an *in vitro* cellular assay using cultured primary rat thymocytes, which normally die at a defined rate when cultured in suspension via apoptosis, and found that this model predicted the ability of caspase inhibitors to block apoptosis resulting from sepsis.

Han and coworkers[43] showed that a pyrazone could be successfully incorporated into a caspase inhibitor scaffold to afford potent selective caspase-3 inhibitors. Selectivity was achieved by modifying the P-4 amine, while potency was increased through the modification of the warhead. An increase in potency and selectivity was achieved by the addition of a t-butyl group of the pyrazone in the 5-position when they used aminomethyl ketone warheads instead of benzylthiomethyl ketones.

Hotchkiss and coworkers[49] reported the use of caspase inhibitors to improve survival during sepsis. Using a murine CLP-induced model of sepsis, they investigated two different types of caspase inhibitors: Z-VAD(OCH$_3$)-FMK, M920, pan caspase inhibitors; and M791, a selective caspase-3-inhibitor in comparison to results from caspase-3$^{-/-}$ mice. Initially they compared Z-VAD(OCH$_3$)-FMK against M920 in the CLP model of sepsis and found that when administered 90 minutes post-insult (i.p. every 12 hours for 96 hours) the inhibitors had similar efficacy in reducing apoptosis in a dose-dependent manner (0.1, 1, and 10 mg/kg) after ~24 hours. M791 was shown to be equipotent at reducing apoptosis in this model. Intraperitoneal administration of 10 mg/kg M920 resulted in 80% survival after 48 hours compared to ~50% survival of vehicle-treated mice, whereas a higher dose of 30 mg/kg resulted in 100% survival at 48 hours. M791 exhibited survival rates similar to those of M920 over 6 days. M920 was also shown to reduce blood bacterial counts and protect lymphocytes from apoptosis during sepsis. Overall, it was demonstrated that although the caspase-3[−/−] mice showed a decrease, but not total block of apoptosis, the caspase inhibitors tested reduced lymphocyte apoptosis and improved survival. Given that lymphocytes act rapidly (within 24 hours) to control infection, it was proposed that caspase inhibitors enhance immunity by preventing lymphocyte apoptosis.

Isabel and coworkers[50] developed pyridyl aspartyl ketones as reversible caspase-3 inhibitors. Initially benzamide and pyridyl amides of aspartyl aldehydes (Figure 7.34) were investigated (Table 7.32), and they found that the nicotinamide (**200**) proved to most potent (IC_{50} = 10 μM), while the iso-nicatinamide (**201**) was inactive. Substitution with bromine in the 5-position of the nicatinamide (**202**) slightly increased activity (IC_{50} = 6 μM). However, these compounds were much less active than the preferred tetrapeptide Ac-DEVD-H (IC_{50} = 0.027 μM).

FIGURE 7.34

TABLE 7.32
Arylamide Aspartyl Aldehydes

Cmpd	Ar	Csp-3 IC_{50} (μM)
198	Ph	23
199	2-Pyr	55
200	3-Pyr	10
201	4-Py	200
202	5-Br-3-Pyr	6

Building on this, the warhead pocket was probed using **202**, the 5-bromonicatinamide-Asp-H (Figure 7.35), as a benchmark compound (Table 7.33). Results from those efforts showed that the benzylthiomethyl ketone (**205**) was more potent than the benzyl ketone (**203**). The use of 4-fluorobenzylthioether (**206**) as a warhead was equipotent to its unsubstituted parent.

Using the 4-fluorobenzylthiomethyl ketone warhead scaffold, they then probed the P-3 pocket by modifying the pyridine 5-position (Table 7.27). Dialkyl amides (**207, 208**) were shown to be potent caspase-3 inhibitors with good cellular activity. Similar dialkyl sulfonamines (**209**) proved to be more potent in enzyme and cellular assays. Monoalkyl sulfonamides were also prepared and screened in the cellular assay. Lastly, Isabel and coworkers compared various warheads on compound **212**, which revealed that the phenoxymethyl ketone and the acyloxymethyl ketone (**213**

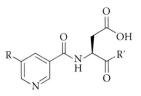

FIGURE 7.35

TABLE 7.33
5-Substituted Nicotinamide Aspartyl Aldehydes and Methyl Ketones

Cmpd	R	R'	IC$_{50}$ (μM) Csp-3	NT2 Cell
200	H	H	10	NT
202	Br	H	6	NT
203	Br	CH$_2$Ph	63.7	NT
204	Br	CH$_2$SPh	11.7	NT
205	Br	CH$_2$SBn	1.1	NT
206	Br	CH$_2$SCH$_2$-(4-F-Ph)	1.2	NT
207	Et$_2$NCO	CH$_2$SCH$_2$-(4-F-Ph)	0.62	>100
208	Et(iPr)NCO	CH$_2$SCH$_2$-(4-F-Ph)	0.25	16
209	Et$_2$NSO$_2$	CH$_2$SCH$_2$-(4-F-Ph)	0.17	12
210	c-PrNHSO$_2$	CH$_2$SCH$_2$-(4-F-Ph)	0.40	8
211	c-Pr(CH$_3$)NSO$_2$	CH$_2$SCH$_2$-(4-F-Ph)	0.27	22
212	c-PrNHSO$_2$	H	1.07	57
213	c-PrNHSO$_2$	CH$_2$OPh	0.39	4.3
214	c-PrNHSO$_2$	CH$_2$OCO-1-Np	0.13	1.3

and **214**, respectively) were more potent than the parent aldehyde (**212**) or the 4-fluorobenzylthio analogue (**211**) against caspase-3 and also in an NT2 cell assay. This assay examines camptotherin-induced DNA fragmentation in neuronal precursor cells (NT2) as a way of assessing apoptosis. In particular, the 1-naphthoylmethyl ketone was identified as a potent irreversible inhibitor.

Thus it was shown that a tetrapeptide could be truncated to an aspartate (Table 7.33) leading to the least peptidic caspase inhibitor that still contains an aspartyl group. These simple molecules were not as potent as peptide inhibitors, but the incorporation of a warhead enhanced activity and led to compounds with cellular activity. It should be noted that aspartyl sulfonamides,[51] utilizing a similar methodology, have been reported as caspase-1 inhibitors.

In order to identify selective caspase-1 and -3 inhibitors, Kim and coworkers[52] described the use of isoxazoles as a P-2/3 replacement (Table 7.34). Their initial focus was the optimization of the P-4 R group (Figure 7.36) using a variety of aryl and heteroaryl amides, with a 1-naphthyl amide (**216**, Xyz033mp) demonstrating the best caspase-1 activity.

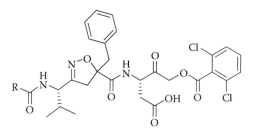

FIGURE 7.36

TABLE 7.34
Isoxazole as a P-2/3 Replacement: P-4 SAR

Cmpd	R	k_3/K_i (M^{-1} s^{-1}) Csp-1	k_3/K_i (M^{-1} s^{-1}) Csp-3
215	$PhCH_2CH_2$	13,000	114
216	1-Np	807,000	500
217	$Indol-1-CH_2$	294,000	132
218	2-Quin	408,000	160
219	2-Indol	707,000	1,230
220	3-Indol	357,000	1,560
221	$HO_2CCH_2CH_2$	7,560	28,000
222	2-NpCO	—	—

Truncating the R_2 group (Figure 7.37) from benzyl to methyl (LB84068MP) resulted in a minor loss of caspase-1 activity with a slight increase in caspase-3 activity (k_3/K_i = 616,000 and 2,390 M^{-1} s^{-1}, respectively). Further selectivity was achieved by the use of a succinate as an R group (Table 7.35), which mimics the Asp side chain of the optimal peptide sequence (DEVD) for caspase-3. Optimization of the R_2 group showed that alkyl groups were preferred for caspase-3 activity, with activity increasing from methyl (**223**) to n-propyl (**224**) to the n-butyl derivative (**225**), which was the most active. Further elongation of the alkyl chain resulted in a drop in activity, e.g., the n-pentyl (**228**), or the use of an ether (**227**), which also resulted in loss of activity.

Kim et al. also reported data using a cell survival assay that utilized WI38 (human embryonal lung fibroblast) cells. Three compounds (see Figure 7.38 for compound **229**) were compared against two known caspase inhibitors (Table 7.36), and it was reported that all three compounds were more effective than the two known caspase inhibitors in protecting cells from apoptosis, with Xyz033mp (**216**) providing complete protection at a concentration of 50 μM.

Kang and coworkers[53] reported the use of Xyz033mp (**216**) in a concavalin A-induced (Con A) hepatitis mouse model. Xyz033mp (20 and 100 mg/kg) was shown to reduce lethality in mice in this model of liver disease by 47 and 65%, respectively,

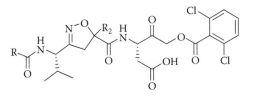

FIGURE 7.37

TABLE 7.35
Isoxazole as a P-2/3 Replacement: P-2 SAR

Cmpd	R	R₂	k_3/K_i (M⁻¹ s⁻¹) Csp-1	Csp-3
223	HO₂CCH₂CH₂	Me	14,500	40,900
224	HO₂CCH₂CH₂	n-Pr	14,000	197,000
225	HO₂CCH₂CH₂	n-Bu	14,900	1,140,000
226	HO₂CCH₂CH₂	Bn	7,560	28,000
227	HO₂CCH₂CH₂	CH₂OCH₃	3,400	27,400
228	HO₂CCH₂CH₂	n-Pent	NR	80,100

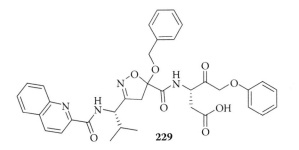

229

FIGURE 7.38

TABLE 7.36
Percent Cell Survival, WI38 (Human Embryonal Lung Fibroblast)

Conc. (μM)	Ac-DEVD-H	Z-DEVD-CH₂Cl	Cmpd 229	222	Xyz033mp
0.4	12.16	15.13	14.44	14.67	15.51
2	14.75	16.27	18.32	17.33	64.44
10	19.16	27.38	35.37	82.40	92.68
50	35.14	47.32	78.29	90.39	100.21

but no effect was seen when mice were treated with 4 mg/kg of Xyz 033mp (**216**). They also explored the optimal timing of treatment, and found that administration of Xyz033mp 1 hour pre- and 4 hours post-insult reduced ALT levels in a dose-dependent manner with the 100 mg/kg dose bringing levels down to near-baseline levels.

Continuing work in this series, Chang and coworkers[54] described the use of isoxazoles as a P-2 replacement by truncating the original isoxazole-peptidomimetic. The bulk of the data reported were focused on caspase-8 inhibition (Table 7.37), potentially due to the correlations noted among enzymatic inhibition, viability of Jurkat cells following Fas treatment, and extrapolations to the extrinsic pathway and activity in liver models of disease. They focused on modifying the 3- and 5-positions of the isoxazole ring and extensively used the fluoromethyl ketone warhead (Figure 7.39). Simple alkanes were investigated at the 5-position with either an ethyl or isopropyl group, whereas various groups were investigated at the 3-position with a focus on heteroaryls. The use of substituted phenyl groups (**231–236**) provided moderate caspase-8 inhibitors with good activity in a Jurkat cellular model. The 2-pyridyl analogue (**237**) exhibited a loss in potency in caspase-8 inhibition and a corresponding loss of activity in Jurkat cells. 4-Substitution on the pyridine ring (**238–246**) enhanced potency over the unsubstituted pyridyl parent, with the t-butyl and cyclohexyl analogues (**238** and **245** respectively) exhibiting the best activity. A few indoles were also investigated, which would presumably bind in a manner similar to that of IDN-1965.[15] The 1,5-dimethyl analogue (**249**) exhibited potent inhibitor activity; however, the (iso)quinolyl analogues (**250–259**) also proved to be potent caspase-8 inhibitors.

Chang and coworkers investigated various warheads, but focused on 1-isoquinoline analogues (Figure 7.40, Table 7.38). Common warheads (**261–263**) were incorporated with little success, with the exception being the DPP (**261**) analogue, which exhibited modest activity. The use of a TFP warhead[25] (**260**), introduced by Idun[28,31] also provided modest activity but no improvement over the FMK, and the use of Michael acceptors (**267–269**) proved to be ineffectual. They also reported on chiral fluoromethyl ketones, showing that the natural isomer (**274**) exhibited greatly enhanced caspase activity. Finally, they noted that the R isomer at the 5-position of the oxazole was preferred.

Using compound **272**, Park and coworkers reported on their experiments demonstrating its efficacy in modulating apoptosis. First, in a carrageenan-induced rat model of inflammation, compound **272** was shown to reduce edema in a dose-dependent manner up to 23% at the high dose of 10 mg/kg. Second, in a bile duct–ligated (BDL) rat model, compound **272** reduced liver apoptosis by 75% in an oral dosing regiment of 3 mg/kg twice daily, and reduced fibrosis to near-baseline levels. Third, in a collagen-induced arthritis (CIA) model of rheumatoid arthritis, compound **272** reduced swelling by 70% when dosed 10 mg/kg in a 14-day study. A comparison of compound **272** versus their previously reported inhibitor LB84068MP[53] was also reported (Table 7.39).

Researchers from LG Chemical reported on their clinical candidate, LB854318, in the form of a poster at the 2004 American Association for the Study of Liver Diseases.[55] Inactivation rates of LB84318 were described (Table 7.40), as well

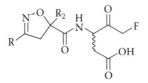

FIGURE 7.39

TABLE 7.37
2-Isoxazoline-5-Carboxamide Aspartylfluoromethyl Ketones

Cmpd	R	R_2	5-Postion Stereochemistry	Csp-8 k_3/K_i ($M^{-1} s^{-1}$)	Jurkat} IC_{50} (μM)
230	2-Bz-thiophenyl	Et	Racemic	120,000	1.1
231	2-iPr-Ph	Et	Racemic	180,000	0.33
232	3-t-Bu-Ph	Et	Racemic	320,000	0.56
233	4-t-Bu-Ph	Et	Racemic	140,000	2.65
234	2-iPr-Ph	i-Pr	R	500,000	0.67
235	3-t-Bu-Ph	i-Pr	R	300,000	0.3
236	2-Ph-Ph	i-Pr	Racemic	200,000	2.45
237	2-Pyridyl	Et	Racemic	44,000	2.27
238	4-tBu-2-Pyr	Et	Racemic	450,000	0.7
239	4-tBu-2-Pyr	i-Pr	R	230,000	1.24
240	4-iBu-2-Pyr	Et	Racemic	150,000	1.02
241	4-Ac-2-Pyr	Et	Racemic	64,000	31
242	4-cPropyl-2-Py	Et	Racemic	190,000	0.7
243	4-cPentyl-2-Py	Et	Racemic	310,000	0.35
244	4-cPentyl-2-Py	i-Pr	R	410,000	0.79
245	4-cHexyl-2-Pyr	Et	Racemic	430,000	0.72
246	4-Ph-2-Py	i-Pr	Racemic	180,000	1.14
247	1,3-$(CH_3)_2$-2-indoyl	Et	Racemic	160,000	2
248	5-Cl-1-CH_3-2-indoyl	i-Pr	Racemic	820,000	1.3
249	1,5-$(CH_3)_2$-2-indoyl	i-Pr	Racemic	1,800,000	0.6
250	1-Isoquin	Et	Racemic	1,500,000	0.17
251	4-Quinolinyl	Et	Racemic	190,000	4.7
252	1-Isoquin	Pr	S	430,000	0.16
253	1-Isoquin	Et	S	290,000	0.98
254	2-Quinolyl	Et	S	1,500,000	0.15
255	3-Isoquinolyl	Et	R	1,100,000	1.9
256	8-Quinolyl	Et	Racemic	40,000	26.5
257	3-Quinolyl	Et	Racemic	310,000	31.2
258	2-Quinolyl	i-Pr	R	420,000	1.25
259	5,6,7,8-H_4-1-Isoquin	Et	Racemic	620,000	0.73

as a comparison with IDN6556 in two murine models of liver disease (Table 7.41). In an LPS/D-Gln-induced model of liver failure, LB84318 protected mice from death when administered either pre- or 6 hours post-insult. According to Table 7.41, LB84318 was demonstrated to be more efficacious in the models described.

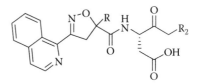

FIGURE 7.40

TABLE 7.38
P-2 and Warhead SAR of 2-(1-Isoquinyl)Isoxazoline-5-Carboxamide Aspartylmethyl Ketones

Cmpd	R	5-Postion Stereochemistry	R_2	Caspase k_3/K_i (M⁻¹ s⁻¹) 1	8	9	Jurkat IC₅₀ (μM)
260	Et	Racemic	OPhF₄		500,000		1.2
261	i-Pr	R	DPP		520,000		1.5
262	i-Pr	R	PTP		16,000		NA
263	i-Pr	R	BTA		37,000		NA
264	i-Pr	R	OiBu		NA		NA
265	i-Pr	R	O-2-Pyr		29,000		NA
266	i-Pr	R	(structure)		170,000		NA
267	i-Pr	R	=CHCO₂H		NA		
268	i-Pr	R	=CHCO₂Et				NA
269	i-Pr	R	=CHSO₂CH₃				NA
270	n-Pr	S	F		430,000		0.16
271	Et	S	F		290,000		0.98
272	i-Pr	R	F	1,100,000	13,000,000	200,000	0.1
273	i-Pr	S	F		190,000		
274	i-Pr	R	F (S)		23,000,000		
275	i-Pr	S	F (R)		470,000		

Note: NA = not active.

TABLE 7.39
In Vitro **Data for LG Chemical Leads**

Cmpd	LPS ED₅₀ (mg/kg)	Jurkat IC₅₀ (μM)	Caspase-8 k_3/K_i (M⁻¹ min⁻¹)
272	0.01	≤0.1	13,000,000
LB84068MP	0.99	1.6	1,020,000

TABLE 7.40
LB84318 Caspase Activity

Caspase	1	3	7	8	9
k_3/K_i (M^{-1} s^{-1})	1,100,000	61,000	320,000	14,000,000	200,000

TABLE 7.41
In Vivo* Comparison of LB84318 and *Emricasan

	ED_{50}, p.o.	
Cmpd	LPS/D-Gln	α-Fas
LB84318	0.022	0.018
Emricasan	0.148	0.357

The incorporation of oxazoles in tri- and dipeptide cores resulted in several potent caspase inhibitors. This heterocycle could be modified to incorporate specificity for caspase-1 or caspase-3. Caspase-3 specificity was also achieved by the use of a glutarate in the P-4 region. Both tri- and dipeptide cores of this type exhibited potent enzyme activity as well as *in vivo* potency. Unfortunately, there are few examples where data from these series have been published in peer-reviewed journals; therefore, most of the publicly available data are obtained from the patent literature and the poster cited above. Their lead caspase inhibitor from these series, LB84318, is currently under clinical investigation for the treatment of liver disease, and is believed to be the only pan-caspase inhibitor undergoing clinical evaluation other than emricasan.

7.5 AZAPEPTIDES

Aza-amino acids[56] (aAA) have been successfully incorporated into drug scaffolds to increase potency and metabolic stability. Graybill and coworkers[57] first applied this approach to caspase inhibitors as replacements for the P-1 aspartic acid moiety of Z-VAD-CH₂X to give a diacyl hydrazine (aD) (Figure 7.41). Their ease of synthesis

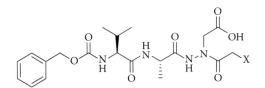

FIGURE 7.41 Diacyl hydrazine derivative of Z-VAD.

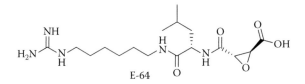

FIGURE 7.42 Aza-peptide E-64: Epoxysuccinate warhead.

and inability to epimerize at the P-1 position made them attractive from a drug development viewpoint. Unfortunately, these compounds proved to be considerably less active than their carbon analogues.

In 2002, Powers and coworkers[58] first described the fusion of azapeptides with epoxides to create new selective caspase inhibitors. The inspiration for this work was the discovery of E-64 (Figure 7.42), a natural product isolated from *Aspergillus japonicus*,[59] which employs an epoxysuccinate moiety. Derivatives of E-64 have also been developed as inhibitors of other cysteine proteases.[60] Incorporating this new warhead onto caspase-specific peptide sequences (Figure 7.43) yielded potent caspase inhibitors (Table 7.42). The use of the VAaD backbone with two different epoxides resulted in selective caspase-1 inhibitors (**276** and **277**). As the ester proved to be more active than the phenethyl analogue, Powers and coworkers investigated

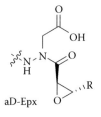

FIGURE 7.43

TABLE 7.42
Aza-peptides: Epoxysuccinate Warhead Analogues

Cmpd	P-2	R	k_3/K_i (M^{-1} s^{-1})			
			Csp-1	Csp-3	Csp-6	Csp-8
276	Ph(CH$_2$)$_2$CO-VA	trans-CO$_2$Bn	36,200	58	52	63
277	Ph(CH$_2$)$_2$CO-VA	trans-(CH$_2$)$_2$Ph	6,130	<10	<10	<10
278	Z-DEV	S,S-CO$_2$Et	—	205,300	7,100	6,600
279	Z-DEV	R,R-CO$_2$Et	—	30,560	<10	<10
280	Z-LET	S,S-CO$_2$Et	—	<10	7,720	8,620

TABLE 7.43
Aza-peptides: Caspase-Specific Epoxysuccinate Warhead Analogues

Cmpd	P-2	R	k_3/K_i (M^{-1} s^{-1})			
			Csp-1	Csp-3	Csp-6	Csp-8
281	Z-V	S,S-CO$_2$Et	—	3,650	350	1,260
282	Z-V	S,S-CO$_2$H	—	1,110	90	370
283	Z-DEV	S,S-CO$_2$Bn	54,700	1,910,000	12,700	188,000
284	Z-DEV	S,S-CO$_2$Et	11,800	1,070,000	5,440	95,500
285	Z-DEV	S,S-CONHBn	25,400	1,090,000	6,000	84,400
286	Z-DEV	S,S-CO-PheNH$_2$	9,250	722,000	6,140	29,600

the use of ethyl esters. The DEVaD backbone (**278** and **279**) provided selectivity for caspase-3, and the LETaD backbone (**280**) was selective for caspases-6 and -8. The S,S epoxides were more potent than their R,R counterparts. However, the use of epoxides in caspase inhibitors resulted in the introduction of two new stereocenters, which negates one of the key advantages of using an aza-peptide.

Powers and coworkers[61] further developed the SAR of these aza-peptide epoxides. The epoxide ester was investigated on different peptide backbones (Table 7.43), which were designed for caspase specificity, but they observed that the use of a Z-VaD dipeptide backbone proved to be less potent than the preferred tetrapeptides. A comparison of warhead esters suggested that hydrophic moieties provided important interactions with the P prime pocket (phenethyl ester > benzyl ester > ethyl ester). Amides were also investigated, and the benzyl amide (**285**) exhibited similar activity to that of the esters, whereas the use of peptide amides resulted in a loss in potency (**286**). Both phenyl epoxides and phenethylepoxides proved to be ineffective. Selectivity for caspase-3 was achieved using the Z-DEVaD backbone, as exemplified in Table 7.43. Selectivity for caspase-1 was maintained by utilizing the previously mentioned Ph(CH$_2$)$_2$COVAaD[61] backbone (not shown). The use of Z-IETaD backbone (data not shown) provided selectivity for caspases-6 and -8. The S,S diastereomers were again shown to be more potent than their R,R counterparts.

Moving away from epoxides, Powers and coworkers[62] next focused on azapeptide Michael acceptors. This new class of caspase inhibitors was derived from a fumarate cysteine protease inhibitor, E-64c (Figure 7.44).[60] Following the path

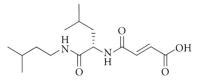

Fumarate derivative of E-64c

FIGURE 7.44 Fumarate cysteine protease inhibitor, E-64c.

TABLE 7.44
Aza-Peptides: Malonate Warheads

Cmpd	Backbone	R	k_3/K_i (M^{-1} s^{-1})		
			Csp-3	Csp-6	Csp-8
287	Z-DEV	CO$_2$Et	2,130,000	35,600	273,000
288	Z-DEV	CO$_2$Bn	1,700,000	8,470	121,000
289	Z-DEV	CONHBn	1,750,000	3,210	78,200
290	Z-DEV	CON(Me)Bn	2,640,000	9,500	90,300
291	Z-LET	CO$_2$Et	5,560	18,700	237,000
292	Z-LET	CO$_2$Bn	4,600	47,600	98,400
293	Z-LET	CON(Me)Bn	6,000	10,800	169,000

aD-Ma

FIGURE 7.45

TABLE 7.45
Aza-Peptides: Malonate Warheads

Cmpd	Backbone	R	k_3/K_i (M^{-1} s^{-1})		
			Csp-3	Csp-6	Csp-8
294	Z-DEV	CONHBn	1,750,000	3,210	78,200
295	Z-DEV	CONHCH$_2$-4-F-Ph	2,100,000	4,400	85,100
296	Z-DEV	CONH(CH$_2$)$_2$Ph	1,950,000	3,470	129,000
297	Z-DEV	CON(Me)Bn	2,640,000	9,500	90,300
298	Z-DEV	CON(CH$_3$)EtPh	1,180,000	4,000	31,900
299	Z-DEV	CONBn$_2$	3,000,000	5,100	8,600
300	Z-DEV	CON(CH$_2$-1-Np)$_2$	5,620,000	29,700	9,460
301	Z-LET	CONHCH$_2$-4-F-Ph	4,490	3,100	171,000
302	Z-LET	CONHPh	4,700	11,400	176,000
303	Z-LET	CONHBn	1,120	1,550	70,170
304	Z-LET	CONBn$_2$	8,630	14,100	129,000
305	Z-LET	CON(CH$_3$)CH$_2$-1-Np	11,200	21,700	179,000
306	Z-LET	CO-THQ	12,100	13,000	216,000

Note: THQ = tetrahydoquinoline.

that was mapped out with the aza-epoxides, they incorporated Michael acceptors such as vinyl ketoesters and amides (Figure 7.45) onto DEVaD and LETaD scaffolds to achieve selective inhibitors of caspase-3 and caspases-6 and -8, respectively (Table 7.44). The ethyl ester (**287**) proved to be more potent than the benzyl ester (**288**), which is contrary to the trend observed in the epoxide series. Powers and coworkers showed that the ester, as part of this Michael acceptor moiety, could be converted to an amide and maintain activity, which led to the methylbenzyl amide (**290**) that proved to be not only the most active inhibitor, but also the most selective toward caspase-3.

Inhibitors of this structural type were examined more closely[63] by focusing substitution efforts on the Michael acceptor region. Initial analysis suggested that the carbonyl moiety (R) was necessary for activity. Using the z-DEVaD (caspase-3-specific) backbone, it was shown that a phenyl ketone was a less effective inhibitor than the lead ethyl ester, and that the cis isomer was less active than the trans isomer.

SAR was also directed toward modification of the amide moiety (Figure 7.45, Table 7.45). Addition of fluorine to the 4-position of the benzylic group had little effect on inhibitor activity (**295**), nor did extension of the benzylamide to a phenethylamide (**296**). As Powers et al. previously found that the use of a disubstituted amide afforded an increase in activity,[62] they incorporated a methyl benzylamide (**297**) or a dibenzylamide (**299**), both of which increased activity as anticipated. That led to the most active inhibitor of caspase-3 in this study, which utilizes a di(methyl-1-naphthyl)amide (**300**).

Caspase-8-selective compounds could be readily obtained by changing the backbone from z-DEVaD to z-LETaD (**301–306**). They investigated a small number of amides in the P-1′ position and found that the phenyl amide (**302**) was more potent than the benzyl amide (**303**). The disubstitued amides proved to be more active than their monosubstituted parents, with the tetrahydroquinoline (**306**) being the most effective.

These two new warhead moieties described by Powers and coworkers are an interesting development in our understanding of the determinants that a potent and selective caspase inhibitor must possess. Unfortunately, data providing insight into the ability of these inhibitors to modulate apoptosis in cells were not reported. It should also be noted that truncation to dipeptides (**281, 282**) resulted in a substantial loss in potency. The optimized epoxide and Michael acceptor warheads may prove to be potent and efficacious when incorporated onto inhibitor scaffolds utilizing an aspartate as opposed to the aza-amino acid.

7.6 CONCLUSIONS

All art is derivative, and drug design is no different. Tetra- and tripeptide caspase inhibitors were derived from the natural peptide sequences identified by Thornberry and coworkers.[4] The initial warheads evaluated were gleaned from previous reports of serine and cysteine protease inhibitors. Even the latest warheads[58,62] were developed from natural products that were found to be cysteine protease inhibitors. The strategies that have been reported vary greatly in their approaches. However, one key to designing a successful caspase inhibitor is to work within the realm of peptidomimetics. This can be accomplished by making use of the selectivity provided by the P-1 pocket for aspartic acid, at the same time introducing sufficient modifications to generate novelty (and thus intellectual property), and utilizing key peptide sequence interactions with the protein. There are two strategies for constraining backbones (which are an effective replacement for peptide backbones): the use of cyclic backbones and the incorporation of a heterocycle into a core's backbone. Both strategies have been proven effective *in vivo.*

The truncation of tetrapeptides to dipeptides and peptidomimetics was the main focus of most research groups. This strategy gave rise to several potent caspase inhibitors with *in vivo* efficacy. The P-3 amino acid has been successfully replaced with oxamides, carbamates, indole amides, and phenylacetamides. The differing backbone strategies outlined also utilized different warhead strategies. FMK and DCB warheads were utilized successfully to develop irreversible caspase inhibitors. The phenoxymethyl ketone moiety was optimized to the irreversible tetrafluorophenoxymethyl ketone (TFP) warhead, which was shown to be potent *in vivo* on several different structural classes of caspase inhibitors. There are three main types of reversible warheads that have been most commonly used: aldehyde-based, benzylthiomethyl, and aminomethyl ketones, all of which are potent caspase inhibitors. Most recently, epoxides and Michael acceptors have been developed as potent irreversible warheads for caspase inhibitors. It is still unclear whether a reversible or an irreversible caspase inhibitor will provide a sufficient therapeutic index to be useful in a clinical setting. The biology underlying the caspase activation cascade suggests that an extremely potent reversible inhibitor would be required to generate sustained inhibition of caspase activity. Consistent with this notion, the majority of *in vivo* efficacy data has been generated with irreversible inhibitors. There is also the opinion that a reversible inhibitor may have less toxicity than an irreversible inhibitor. Nevertheless, research continues with reversible inhibitors, such as Merck's use of aminomethyl ketones (M826, **196**). Despite the interest in reversible inhibitors, there are currently two irreversible inhibitors in clinical trials, Emiricasan and LB84318, for the treatment of liver disease.

REFERENCES

1. Alnemri, E. S., et al. 1996. Human ICE/CED-3 protease nomenclature. *Cell* 87:171.
2. Chapman, K. T. 1992. Synthesis of a potent, reversible inhibitor of interleukin-1β converting enzyme. *Bioorg. Med. Chem. Lett.* 2:613. Fletcher, D.S., et al. 1995. A synthetic inhibitor of interleukin-1 beta converting enzyme prevents endotoxin-induced interleukin-1 beta production in vitro and in vivo. *J. Interferon Cytokine Res.* 15:243.
3. Chapman, K. T., et al. 1997. A combinatorial approach for determining protease specificities: Application to interleukin-1β converting enzyme (ICE). *Chem. Biol.* 4:149.
4. Thornberry, N. A., et al. 1997. A combinatorial approach defines specificities of members of the caspase family and granzyme B. *J. Biol. Chem.* 272:17907.

5. Brady, K. D. 1998. Bimodal inhibition of caspase-1 by aryloxymethyl and acyloxy-methyl ketones. *Biochemistry* 37:8508. Brady, K. D., et al. 1999. A catalytic mechanism for caspase-1 and for bimodal inhibition of caspase-1 by activated aspartyl ketones. *Bioorg. Med. Chem.* 7:621.

6. Mjalli, A. M. M., Chapman, K.T., Mac Coss, M., and Thomberry, N.A. 1993. *Bioorg. Med. Chem. Lett.* 3 (12):2689–2692. Phenylalkyl ketones as potent reversible inhibitors of interleukin-1B-convertins enzyme.

7. Mjalli, A. M. M., et al. 1994. Activated ketones as potent reversible inhibitors of interleukin-1β converting enzyme. *Bioorg. Med. Chem. Lett.* 4:1965.

8. Mjalli, A. M. M., et al. 1995. Inhibition of interleukin-1β converting enzyme by N-acylaspartic acid ketones. *Bioorg. Med. Chem. Lett.* 5:1405.

9. Dolle, R. E., et al. 1995. Aspartyl α-((diphenylphosphinyl)oxy)methyl ketones as novel inhibitors of interleukin-1β converting enzyme. Utility of the diphenylphosphinic acid leaving group for the inhibition of cysteine proteases. *J. Med. Chem.* 38:220 (and references therein).

10. Smith, R. A., et al. 1988. New inhibitors of cysteine proteinases. Peptidyl acyloxy-methyl ketones and the quiescent nucleofuge strategy. *J. Am. Chem. Soc.* 110:4429.

11. Grimm, E. L., et al. 2004. Solid phase synthesis of selective caspase-3 peptide inhibitors. *Bioorg. Med. Chem. Lett.* 12:845.

12. Garcia-Calvo, M., et al. 1998. Inhibition of human caspase by peptide-based and macromolecular inhibitors. *J. Biol. Chem.* 273:32608.

13. Holly, T. A., et al. 1999. Caspase inhibition reduces myocyte cell death induced by myocardial ischemia and reperfusion in vivo. *J. Mol. Cell. Cardiol.* 31:1709.

14. Karanewsky, D., and Bai, X. 1998. PCT Int. Appl. WO 9811129. Karanewsky, D., and Bai, X. 1999. U.S. Patent 5877197. Karanewsky, D., and Ternansky, R. 2000. PCT Int. Appl. WO 0001666. Revesz, L., et al. 1994. Synthesis of p1 as parte-based peptide acylovy methyl and fluoromethyl ketones as inhibitons of interleukin-1β-converting enzyme. *Terahedron Lett.* 35(52):9693–9696. Van Noorden, C. J. 2001. The history of Z-VAD-FMK, a tool for understanding the significance of caspase inhibition. *Acta Histochem.* 103:241.

15. Karanewsky, D., and Bai, X. 1999. U.S. Patent 5869519.

16. Gores, G. J., et al. 1999. Apoptosis of sinusoidal endothelial cells occurs during liver preservation injury by a caspase-dependent mechanism. *Transplantation* 68:89.

17. Hoglen, N. C., et al. 2001. Characterization of the caspase inhibitor IDN-6556 in a model of apoptosis-associated liver injury. *J. Pharmacol. Exp. Ther.* 297:811.

18. Cai, S. X., et al. 2004. Dipeptidyl aspartyl fluoromethylketones as potent caspase-3 inhibitors: SAR of the P2 amino acid. *Bioorg. Med. Chem. Lett.* 14:1269.

19. Cai, S. X., et al. 2003. MX1013, a dipeptide caspase inhibitor with potent *in vivo* antiapoptotic activity. *Br. J. Pharmacol.* 1401:402.

20. Hara, H., et al. 1997. Inhibition of interleukin 1β converting enzyme family proteases reduces ischemic and excitotoxic neuronal damage. *Proc. Natl. Acad. Sci. USA* 94:2007.

21. Yaoita, H., et al. 1998. Attenuation of ischemia/reperfusion injury in rats by a caspase inhibitor. *Circulation* 97:276.

22. Cai, S. X., et al. 2004. Dipeptidyl aspartyl fluoromethylketones as potent caspase inhibitors: SAR of the N-protecting group. *Bioorg. Med. Chem. Lett.* 14:5295.

23. Cai, S. X., et al. 2005. Dipeptidyl aspartyl fluoromethylketones as potent caspase inhibitors: Peptidomimetic replacement of the P2 α-amino acid by a α-hydroxy acid. *Bioorg. Med. Chem. Lett.* 15:1379.

24. Linton, S. D., et al. 2002. Acyl dipeptides as reversible caspase inhibitors. Part 1. Further optimization. *Bioorg. Med. Chem. Lett.* 12:2969.

25. Ullman, B. R., et al. 2003. Structure-activity relationships within a series of caspase inhibitors: Effect of leaving group modifications. *Bioorg. Med. Chem. Lett.* 13:3623.

26. Ullman, B. R., et al. 2005. Structure-activity relationships within a series of caspase inhibitors. Part 2. Heterocyclic warheads. *Bioorg. Med. Chem. Lett.* 15:3632.

27. Linton, S. D., et al. 2002. Acyl dipeptides as reversible caspase inhibitors. Part 2. Further optimization. *Bioorg. Med. Chem. Lett.* 12:2973.

28. Linton, S. D., et al. 2004. Oxamyl dipeptide caspase inhibitors developed for the treatment of stroke. *Bioorg. Med. Chem. Lett.* 14:2985.

29. Hoglen, N. C., et al. 2004. *J. Pharm. Exp. Ther.* 309:634. Linton, S. D., et al. 2005. First-in-class pan caspase inhibitor developed for the treatment of liver disease. *J. Med. Chem.* 48:6779.

30. Han, Y., et al. 2004. Discovery of novel aspartyl ketone dipeptides as potent and selective caspase inhibitors. *Bioorg. Med. Chem. Lett.* 14:805.

31. Dolle, R. E., et al. 1996. First examples of peptidomimetic inhibitors of interleukin-1β converting enzyme. *J. Med. Chem.* 39:2438.

32. Brown, F. J., Andisik, D. W., Bernstein, P. R., Bryant, C. B., Ceccarelli, C., Damewood, J. R. Jr., Edward, P. D., Earley, R. A., Feeney, S., et al. 1994. Design of orally active, non-peptide inhibitors of human leukocyle elastase. *J. Med. Chem.* 37(9):1259–1261.

33. Wu, J., et al. 2003. Structural and functional analysis of caspase active sites. *Biochemistry* 42:4151.

34. Semple, G., et al. 1997. Pyridone-based peptidomimetic inhibitors of interleukin-1β-converting enzyme (ICE). *Bioorg. Med. Chem. Lett.* 7:1337. Semple, G., et al. 1998. Peptidomimetic aminomethylene ketone inhibitors of interleukin-1β-converting enzyme (ICE). *Bioorg. Med. Chem. Lett.* 8:959.

35. Sanderson, P. E. J., et al. 1998. Efficacious, orally bioavailable thrombin inhibitors based on 3-aminopyridinone or 3-aminopyrazinone acetamide peptidomimetic templates. *J. Med Chem.* 41:4466.

36. Karanewsky, D. S., et al. 1998. Conformationally constrained inhibitors of caspase-1 (interleukin-1β converting enzyme) and of the human CED-3 homologue caspase-3 (CPP32, apopain). *Bioorg. Med. Chem. Lett.* 8:2757.

37. Dolle, R. E., et al. 1997. Pyridazinodiazepines as a high-affinity, P2-P3 peptidomimetic class of interleukin-1β-converting enzyme inhibitor. *J. Med Chem.* 40:1941.

38. Karanewsky, D. S., and Linton, S. D. 1999. U.S. Patent 5,968,927.

39. Roggo, S., et al. 2001. *Abstracts of Papers, 221st American Chemical Society National Meeting*, San Diego, April 1–5. Roggo, S., Hintermann, S., Rasetti, V., and Von Krosigk, U. 2002. Hexahydro-4-oxaazepino[3,2,1-HI]indoles. WO02/034751.

40. Becker, J. W., et al. 2004. Reducing peptidyl features of caspase-3 inhibitors: A structural analysis. *J. Med. Chem.* 47:2466.

41. Willoughby, C. A., et al. 2002. Discovery of potent, selective human granzyme B inhibitors that inhibit CTL mediated apoptosis. *Biorg. Med. Chem Lett.* 12:2197.

42. Deckworth, T. L., et al. 2001. Long-term protection of brain tissue from cerebral ischemia by peripherally administered peptidomimetic caspase inhibitors. *Drug Dev. Res.* 52:579.

43. Han, Y., et al. 2005. Novel pyrazinone mono-amides as potent and reversible caspase-3 inhibitors. *Bioorg. Med. Chem. Lett.* 15:1173.

44. Toulmond, S. et al. 2004. Neuroprotective effects of M826, a reversible caspase-3 inhibitor, in the rat malonate model of Huntington's disease. *Br. J. Pharmacol.* 141:689.

45. Holtzmann, D. M., et al. 2002. Selective, reversible caspase-3 inhibitor is neuroprotective and reveals distinct pathways of cell death after neonatal hypoxia-iscemic brain injury. *J. Biol. Chem.* 277:30128.

46. Vannucci, R. C. 1990. Experimental biology of cerebral hypoxia-ischemia: Relation to perinatal brain damage. *Pediatr. Res.* 27:317.

47. Holtzmann, D. M., et al. 1998. Caspase inhibitor affords neuroprotection with delayed administration in a rat model of neonatal hypoxic-ischemic brain injury. *J. Clin. Invest.* 101:1992.

48. Méthot, N., et al. 2004. Differential efficacy of caspase inhibitors on apoptosis markers during sepsis in rats and implication for fractional inhibition requirements for therapeutics. *J. Exp. Med.* 199:199.

49. Hotchkiss, R. S., et al. 2000. Caspase inhibitors improve survival in sepsis: A critical role of the lymphocyte. *Nat. Immunol.* 1:496.

50. Isabel, E., et al. 2003. Nicotinyl aspartyl ketones as inhibitors of caspase-3. *Bioorg. Med. Chem. Lett.* 13:2137.

51. Shahripour, A. B., et al. 2002. Structure-based design of nonpeptide inhibitors of interleukin-1beta converting enzyme (ICE, caspase-1). *Bioorg. Med. Chem.* 10:31.

52. Kim, E., et al. 2001. PCT WO01/021600 A1.

53. Kang, C.-Y., et al. 2000. A broad-spectrum caspase inhibitor blocks concanavalin A-induced hepatitis in mice. *Clin. Immunol.* 97:221.

54. Change, H-K, et al. 2004. PCT WO2005/021516 A1.

55. Park, M., et al. 2004. LB84318, a novel and small-molecule caspase inhibitor, potently protects liver damage in mouse models of liver injury. In *Abstracts of Papers, 55th American Association for the Study of Liver Diseases*, Boston, October 29–November 2, abstract 1215.

56. Gante, J. 1994. Peptidomimetics-tailored enzyme inhibitors. *Angew. Chem. Int. Ed. Engl.* 33:1699.

57. Graybill, T. L., et al. 1995. Synthesis of diacylhydrazines as inhibitors of the interleukin-1β converting enzyme (ICE). *Bioorg. Med. Chem. Lett.* 5:1197.

58. Powers, J.C., et al. 2002. Aza-peptide epoxides: A new class of inhibitors selective for clan CD cysteine proteases. *J. Med. Chem.* 45:4958.

59. Handa, K., et al. 2004. Studies on thiol protease inhibitors. Part I. Isolation and characterization of E64, a new thiol protease inhibitor. *Agric. Biol. Chem.* 14:805.

60. Barrett, A. J., et al. 1982. L-*trans*-Epoxysuccinylleucylamido(4-guanidino)butane (E-64) and its analogues as inhibitors of cysteine proteases including cathepsins B, H and L. *Biochem. J.* 201:189. Parkes, C., Kembhavi, A. A., and Barrett, A. J. 1985. Calpain inhibition by peptide epoxides. *Biochem. J.* 230:509. Roush, W. R., et al. 2000. Design, synthesis and evaluation of D-homophenylalanine epoxysuccinate inhibitors of the trypanosomal cysteine protease cruzain. *Tetrahedron* 56:9747. Schaschke, N., et al. 1997. E-64 analogues as inhibitors of cathepsin B. On the role of the absolute configuration of the epoxysuccinyl group. *Bioorg. Med. Chem.* 5:1789.

61. Powers, J.C., et al. 2004. Design, synthesis, and evaluation of aza-peptide epoxides as selective and potent inhibitors of caspases-1, -3, -6, and -8. *J. Med. Chem.* 47:1553.

62. Powers, J.C., et al. 2004. Aza-peptide Michael acceptors: A new class of inhibitors specific for caspases and other clan CD cysteine proteases. *J. Med. Chem.* 47:1889.

63. Powers, J.C., et al. 2006. Design, synthesis, and evaluation of aza-peptide Michael acceptors as selective and potent inhibitors of caspases-2, -3, -6, -7, -8, -9, and –10. *J. Med. Chem.* 49:5728.

8 Discovery of a First-in-Class Apoptotic Caspase Inhibitor Emricasan (PF-03491390/IDN-6556)

Niel C. Hoglen

CONTENTS

8.1 INTRODUCTION

Apoptosis is a well-controlled form of cell death used to maintain the homeostasis of cell populations. Dysregulation of this fundamental process can lead to an increase in cell numbers (e.g., cancer) or a decrease in cell populations, impairing tissue or organ function. Indeed, with respect to tissue loss, there are a number of diseases where excessive apoptosis occurs, due to various insults or defects in the apoptotic machinery. A critical family of enzymes, the caspases, plays a vital role in the apoptotic process, from the amplification of the apoptotic stimulus (e.g., the initiator caspases-8 and -9) to the destruction of cellular components (i.e., the executioner caspases-3, -6, and -7).

The potential to inhibit cell death and tissue damage via caspase inhibition therefore is a viable target for the pharmaceutical industry. The initial use of prototypical caspase inhibitors led to the notion that inhibition of caspases may be effective for treating various disease states. Early work with caspase inhibitors showed protection of the liver in models of liver injury where apoptosis and, more importantly, caspase activation occurs.[1–3] Prototypical molecules, such as ZVAD-FMK, contain up to five amino acids, have potentially toxic active groups (i.e., fluoromethyl ketone), and possess few drug-like properties. While these molecules remain in use as tools to study the role of caspases both *in vitro* and *in vivo*, they can be considered templates for the development of smaller molecules with the potential for human use.

One of Idun Pharmaceutical's most successful strategies was to develop small molecule, irreversible caspase inhibitors with broad-spectrum activity similar to that of the prototypical peptide inhibitors. The choice of an irreversible inhibitor was due largely to the nature of the caspase activation cascade. Because apical caspases activate the executioner capases in addition to auto-activation, the most effective molecules will need to inhibit all caspases to stop the amplification of the apoptotic machinery.[4] Here I summarize the preclinical development path followed by Idun to select and evaluate emricasan, a broad-spectrum caspase inhibitor, as a clinical candidate for the treatment of liver diseases.

8.2 *IN VITRO* ACTIVITY OF EMRICASAN

Potency with respect to *in vitro* activity against several caspases was one of the first screens used to identify lead candidates and ultimately emricasan. Caspase inhibitors were first screened against purified human recombinant caspases-3, -6, -8, and -9 as well as murine caspase-1. As with most inhibitors, the IC_{50} was determined to assess initial potency. In these studies, emricasan inhibited (IC_{50}) at low to subnanomolar concentrations all caspases tested (Table 8.1). However, because the IC_{50} for irreversible inhibitors can be misleading, a more complete evaluation of the inhibitors was warranted. Therefore, K_i (the dissociation constant that measures reversibility of the molecule) and the first-order rate constant, K_3 (covalent binding), were also determined. Further, the ratio of K_3/K_i was also of interest, as it describes the irreversibility (or off-rate) of the inhibitor.[4] As seen in Table 8.1, variable K_i values were determined for individual caspases, ranging from subnanomolar to ~100 nM concentrations. Irreversible binding of the molecule was observed with all caspases, but particularly with the apical caspases-8 and -9 (Table 8.1). Also of note was the high potency toward caspase-1, suggesting that emricasan may possess some anti-inflammatory activity, since caspase-1 is responsible for maturation of the pro-inflammatory cytokines interleukin-1β and interleukin-18. To determine its broader selectivity profile, emricasan was screened against a panel of cysteine and serine proteases. No notable inhibition was seen against several proteases, including thrombin, factor Xa, and calpain II, but modest activity was observed against cathepsin B ($IC_{50} \sim 37$ μM). Lastly, when screened against a panel of receptors, ion channels, and enzymes, no significant activities were noted, suggesting that emricasan was selective for caspases only.[5]

The potency of caspase inhibitors, including emricasan, was also tested in cellular models of apoptosis. The major cell-based screening assay included the Fas-stimulated Jurkat E6.1 cell lymphoma model system. Stimulation of these cells with an agonistic Fas antibody induces apoptosis via the extrinsic pathway. A plethora of Idun's caspase inhibitors from a number of structural series were screened, and a portion of this work has been previously published, including comparisons to

TABLE 8.1
Potency and Selectivity of Emricasan for Caspases

Enzyme	IC$_{50}$ (μM)	Ki (μM)	K$_3$/K$_i$ (M^{-1} s^{-1})
Caspase-1	0.0004	0.046	689,000
Caspase-2	0.020	0.072	6,130
Caspase-3	0.002	0.104	75,700
Caspase-6	0.004	0.121	58,700
Caspase-7	0.006		
Caspase-8	0.006	0.001	2,940,000
Caspase-9	0.0003	0.0006	2,191,000
Calpain II, plasmin, factor Xa, thrombin	>200	—	—
Cathepsin B	36.5	—	—
Cell Type/Stimulus	**IC$_{50}$ (μM)**		
Jurkat E6.1/Fas	0.025	—	—
SKW/Fas	0.043	—	—
THP-1/LPS	0.27	—	—

Note: Emricasan potently inhibited caspases-1 to -9 as demonstrated by low to subnanomolar IC$_{50}$ and K$_i$. Large k$_3$/K$_i$ ratios demonstrate irreversibility. In contrast, IC$_{50}$ values against other cysteine and serine proteases are in the μM range, indicating that emricasan is highly selective for only caspases. Lastly, IC$_{50}$ values for emricasan in various cell lines are given.

emricasan.[6] With respect to emricasan, the IC$_{50}$ in this Jurkat model was 0.025 μM, which compared favorably to other lead molecules (see Chapter 7). The ability of these compounds to inhibit caspase activity in other cell lines was also assessed. For example, the SKW human B cell line (stimulated with Fas antibody) was used to screen anti-apoptotic activity, whereas the THP-1 human monocyte cell line (stimulated with lipopolysaccharide and interferon-γ) was used to screen inhibition of interleukin-1β secretion. From these assays, IC$_{50}$ values of 0.043 and 0.27 μM, respectively, were observed for emricasan, confirming potent anti-apoptotic activity and the potential to inhibit interleukin-1β and IL-18 maturation.[6]

In sum, this *in vitro* screening process aided in the determination of both the enzymatic and cellular potency of Idun's caspase inhibitor portfolio. Superb potency was demonstrated with molecules from several series, including emricasan; however, a challenging hurdle was to identify clinical candidates among the most potent molecules. An *in vivo* assessment in animal models known to have activated caspase signaling cascades was then employed to aid in the selection of a leading clinical candidate.

8.3 *IN VIVO* EFFICACY MODELS FOR EMRICASAN DISCOVERY AND CHARACTERIZATION

The *in vivo* efficacy of Idun's lead caspase inhibitors (including emricasan) was first tested in the α-Fas model of liver injury. In this model, mice are injected intravenously (IV) with α-Fas, an antibody agonist of the Fas receptor.[7] Within 1.5 hours of injection, hepatocytes and sinusoidal endothelial cells begin to undergo morphological signs of apoptosis, including organized nuclear condensation, membrane blebbing, and ultimately, cellular fragmentation. Further, caspases become activated (as determined by Ac-DEVD-AMC cleavage) in a similar timeframe.[8] Antibodies that recognize activated caspases (e.g., CM-1) can also confirm activation of caspases and their cellular localization. Plasma alanine aminotransferases (ALTs) and other liver enzymes become elevated within 3 hours and by 5 to 6 hours peak ALT activities are detected. Depending on the dose of α-Fas, ALT activities (IU/L) can range from the hundreds to several thousands (normal ALT activities range from 20 to 40 IU/L). Lethality from severe liver injury also may occur at high doses of α-Fas. Therefore, this model provides a relatively quick and effective *in vivo* model for screening caspase inhibitors with the primary endpoints of serum liver enzymes, mortality, and their effect on caspase activation.

In these initial experiments, the strategy was to screen lead candidates (i.e., the most potent compounds identified in the *in vitro* assays) for potency *in vivo* in this Fas model. The i.p. route of administration was chosen for its experimental simplicity, allowing for determination of ED_{50} values; in this manner, direct comparisons between caspase inhibitors could be performed. As advanced leads were chosen based on potency, other routes of administration were then tested. ED_{50} values after i.p. and i.v. doses were first established when drug was administered simultaneously with α-Fas. Advanced leads were further ranked based on ease of synthesis and drug-like characteristics, including solubility and stability. Emricasan was ultimately chosen, as it ranked highest in these criteria.

For emricasan, ED_{50} values around 200 μg/kg were determined for multiple routes of administration.[9] For example, ED_{50} values for i.m. and i.p. administration were as low as 0.04 (0.02–0.07) to 0.08 (0.06–0.12) mg/kg, while p.o. and i.v. administration resulted in ED_{50} values of 0.31 (0.24–0.42) and 0.38 (0.11–1.27) mg/kg, respectively (Table 8.2). An impressive observation was that near-complete protection of liver injury and lethality could be achieved with a single administration of emricasan at doses between 1 and 3 mg/kg. Near-complete efficacy could also be achieved when drug was administered as late as 3 hours after administration of α-Fas.[9] These data were an important finding and suggested that the caspase inhibitor was not only working prophylactically, but was providing protection *after* the apoptotic machinery had been activated.

Efficacy with emricasan was then confirmed in an additional model of caspase (apoptotic)-mediated liver injury, the galactamine/lipopolysaccharide (D-gal/LPS) model. This is a well-established model of liver injury involving sensitization of the liver to tumor necrosis factor alpha (TNF-α) by depletion of uridine stores with d-galactosamine.[10] In this model, peak liver injury and associated lethality occur approximately 10 to 12 hours after D-gal/LPS administration, significantly later than in the α-Fas model. Due to the longer time to peak liver damage in the TNF-α model, caspase inhibitors, in general, were dosed 4 hours after administration of D-gal/LPS for greatest efficacy. This model also confirmed that post-insult efficacy with emricasan could be achieved. One other advantage of the model is that liver injury and associated lethality can be achieved in both rats and mice, so efficacy in two rodent species can be assessed.

TABLE 8.2
Potency of Emricasan in Two Murine Models of Liver Injury α-Fas

Model	Route of Administration	ED_{50}(mg/kg)	Confidence Intervals (mg/kg)
α-Fas	IP	0.08	0.06–0.12
α-Fas	IV	0.38	0.11–1.27
α-Fas	PO	0.31	0.24–0.42
α-Fas	IM	0.04	0.02–0.07
D-Gln/LPS	IP	0.17	0.09–0.32
D-Gln/LPS	IV	0.09	0.03–0.31
D-Gln/LPS	PO	<0.01	—

Source: Hoglen, N. C., et al., *J. Pharm. Exp. Ther.,* 309, 634, 2004. With permission.

Note: Mice were administered α-Fas (200 μg/kg i.v.) immediately followed by emricasan (0.03–3.0 mg/kg). Mice were euthanized 6 hours later and blood was taken via cardiac puncture. ALT activities were measured in plasma and are calculated as $(ALT_{\alpha\text{-Fas/drug}} - ALT_{vehicle})/$ $(ALT_{\alpha\text{-Fas/vehicle}} - ALT_{vehicle}) \times 100\%$. ED_{50} values are expressed with 95% confidence limits in parentheses. n = 6 to 10/group. D-Gln/LPS: BALB/c mice received D-Gln and LPS (700 mg/kg and 10 μg/kg, respectively) by i.p. injection dissolved in saline (5 ml/kg dosing volume). Four hours post administration of D-Gln/LPS, vehicle, or emricasan (0.01–3.0 mg/kg), was given by the specified route. Ten hours after administration of D-Gln/LPS, mice were euthanized with pentobarbital (50 mg/kg i.p.), plasma was harvested by cardiac puncture, and ALT activities were determined. Data are expressed as percentage of ALT of D-Gln/LPS-treated controls.

With respect to emricasan, ED_{50} values comparable to those seen in the α-Fas model were observed in the mouse D-gal/LPS model.[9] For example, ED_{50} values by multiple routes were generally in the 100 μg/kg range in the mouse (Table 9.2), and near-complete protection could be achieved when emricasan was given 4 hours post dose as determined by ALT activities and histopathology. Of note, oral administration was the most effective route of administration, with an ED_{50} of less than 0.01 mg/kg.

Emricasan was also efficacious in the rat, but with less potency than observed in the mouse, with an ED_{50} of approximately 3 mg/kg after oral administration. At 10 mg/kg, i.p. and p.o. dosing gave comparable responses with a decrease in ALT activities by approximately 80%.[9] Interestingly, complete protection of liver injury could not be achieved; the reasons for this are unknown, but it may be due to multiple mechanisms of liver injury in the rat. In any event, these studies demonstrated that emricasan is efficacious in both species and, as with the α-Fas model, post-insult efficacy was possible with emricasan.

By using the Fas and D-gal/LPS models of liver injury, a number of caspase inhibitors from various structural series were assessed relatively quickly in *in vivo* model systems. This helped identify emricasan over other leads due to its potency and marked efficacy by most routes of administration. Importantly, these models also allowed for characterization of the pharmacology of emricasan once chosen as a development candidate. While the α-Fas and D-gal/LPS models of liver injury are excellent for screening caspase inhibitors and a better understanding of their pharmacology, they are less clinically relevant than other models of liver disease. Given this, emricasan was then tested in a variety of other models more relevant to both acute and chronic liver diseases, to be discussed in detail later.

8.4 PHARMACOKINETICS AND ADME OF EMRICASAN

As potency was established in both rat and mouse models, an examination of the pharmacokinetic (PK) parameters was also initiated in the rat. Initial studies focused on exposure of emricasan after various routes of administration. PK analysis revealed that in the rat, emricasan was cleared rapidly from the plasma, with $t_{1/2}$ ranging from 46 to 51 minutes after i.p., s.c., and i.v. administration, or 32 minutes after oral administration.[9,11] Bioavailability after oral administration was only 4%. However, further studies revealed that liver and portal concentrations were greater than systemic concentrations after oral but not i.v. administration. This suggested that there was a significant first-pass effect by the liver, and that the fraction of drug absorbed was greater than expected given the apparent bioavailability. Therefore, the liver was exposed to relatively high concentrations of drug after oral administration. A secondary rise in liver and portal concentrations seen in the PK studies was suggestive of enterohepatic recirculation, which may be playing a role in increased exposure of emricasan to the liver.

An ADME study using radiolabeled emricasan was also performed after i.v. administration in the rat and was reported in abstract form.[11] In this study, most of the radioactivity (emricasan and equivalents) was eliminated in the feces (>50%) or in urine (~20%). The tissue with the highest concentrations of radiolabel was the liver, confirming the initial PK studies. Also of note was a secondary rise in radioactivity in blood around 8 to 12 hours, consistent with some enterohepatic recirculation seen in the PK studies. Collectively, the PK and ADME data explain, in part, why potency after oral administration was as effective as other routes of administration in the rodent models, and why the molecule is most potent in models of liver injury. Indeed, the molecule is effective when given by multiple routes of administration, having a much higher concentration in the liver than plasma. Importantly, due to extensive first-pass metabolism, systemic exposure is less in other tissues than in the liver; therefore, the likelihood of adverse effects may be less. It became clear from the efficacy models and pharmacokinetics that emricasan may be most effective for the treatment of patients with liver diseases characterized by excessive apoptosis.

8.5 ISCHEMIC LIVER INJURY AND LIVER TRANSPLANTATION

Liver transplantation was one of the first indications chosen for characterization of emricasan, a decision that was based on previously reported findings. Liver ischemia during cold storage preservation, followed by warm reperfusion, is characterized by sinusoidal endothelial cell (SEC) apoptosis, which occurs early after reperfusion.[12] Primary graft dysfunction associated with reperfusion is seen in a clinical setting and is also believed to be associated with this initial damage to the SEC. Therefore, it was hypothesized that exposure of livers to a caspase inhibitor during the ischemic or reperfusion period may lower the amount of apoptosis, particularly of endothelial cells, and ultimately lead to an increased incidence of successful graft. Initial data with first-generation caspase inhibitors had supported this hypothesis in animal models,[13,14] and therefore emricasan and other lead candidates were screened in a model of ischemia/reperfusion.

Gores and colleagues investigated whether five molecules from several series of caspase inhibitors would inhibit SEC apoptosis and caspase activation in livers undergoing an ischemic/reperfusion event.[15] In their model, livers were removed from donor rats and then flushed and stored in cold storage media (University of Wisconsin solution) for 24 hours. Afterwards, livers are reperfused in a Krebs-Ringers-based perfusion solution for 1 hour, and SEC apoptosis in the livers is estimated by TUNEL analysis. Their caspase inhibitors, which had similar potencies to emricasan in the α-Fas model, were administered to the donor rat (10 mg/kg, i.p., 10 minutes prior to initiation of the procedure), in the cold preservation media and in the reperfusion media (25 μM in both cases), i.e., providing maximum exposure over the course of the experiment. In this preliminary study, emricasan reduced TUNEL positive cells to the greatest extent compared to the other caspase inhibitors, with maximum inhibition being ~50%. Based on these data, emricasan was then further tested in three paradigms: (1) Emricasan was added to the cold storage media only (25 μM). (2) Emricasan was added to the cold storage and reperfusion solutions (25 μM). (3) Emricasan was present in both solutions and donor rats were pretreated as before. In these studies, the three different treatments with emricasan decreased the number of apoptotic SEC compared to the control treatment group. Interestingly, the presence of emricasan only in the preservation media was as beneficial as the other groups. This was explained in part by the fact that caspase-3 activities were reduced to near-maximum levels across all groups, suggesting that exposure during the cold ischemic phase was the most important time of exposure. Importantly, this study demonstrated that emricasan was more effective than other caspase inhibitors tested, and that it was effective in inhibiting its target (caspases) with minimal exposure during the procedure.

Further studies by the Dumont-UCLA transplant center evaluated the potential for emricasan in liver transplantation by employing an *ex vivo* model of ischemia/reperfusion as well as a syngeneic rat orthotopic model of liver transplantation.[16] In these studies, functional data would be assessed in addition to effects on caspase inhibition/cellular apoptosis. In the first set of experiments, a closed reperfusion system was used to determine the effects of emricasan on liver function after cold ischemia/warm reperfusion. In this model, livers from donor rats are stored in cold UW solution for 24 hours and then perfused in a closed system with a modified reperfusion buffer (Krebs-Ringer) containing 15% whole rat blood. Bile production, vascular resistance pressure, and oxygen consumption can be continuously monitored and are indicative of liver function. ALT activities as well as liver histology (graded based on the Suzuki scoring system) were also monitored to assess liver damage between groups. When present during cold storage and in the reperfusion media, emricasan (25–50 μM) improved bile flow, oxygen consumption and vascular resistance. Maximum efficacy appeared to be reached with the compound at concentrations of 25 μM in both the cold storage, and reperfusion media. ALT leakage in the media was also reduced compared to controls, and improvement was similar when the livers were only cold-stored with emricasan. Effects of emricasan on portal blood flow, bile flow, and ALT leakage are shown in Figure 8.1. Suzuki scores of

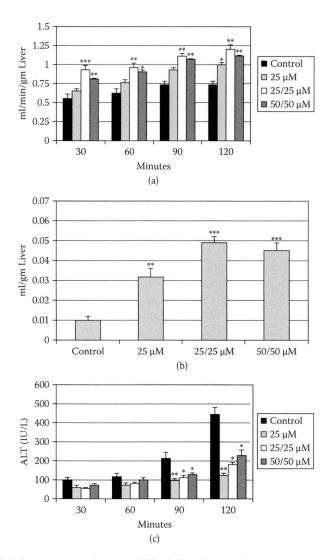

FIGURE 8.1 *Improvement of (a) portal blood flow, (b) total bile production, and (c) alanine aminotransferase (ALT) by emricasan.* The portal vein, common bile duct, and inferior vena cava from hepatectomized livers were cannulated, then perfused and stored in University of Wisconsin (UW) solution for 24 hours with or without addition of emricasan at a concentration of 25 or 50 μM. Livers were then perfused *ex vivo* for 2 hours on an isolated rat liver perfusion apparatus. Temperature was maintained at 37°C, pH was buffered at 7.4, and portal flow was adjusted to keep portal pressure constant at 13 cm H$_2$O. Treatment groups: control group, no addition of drug; 25 μM group, emricasan present in the UW solution for 24 hours; 25/25 μM, emricasan present during cold storage and reperfusion (25 μM); 50/50 μM, emricasan present during cold storage reperfusion (50 μM). Portal blood flow was recorded at 30-minute intervals (ml/min/g wet liver tissue). Total bile produced was collected and quantified over the 2-hour reperfusion period. ALT levels were recorded at 30-minute intervals (IU/L). N = 4–5 per group; data are expressed as mean ± SE. *, $p < .05$, **, $p < .0.1$, ***, $p < .001$. (From Hoglen, N. C. et al., *Liver Transplant.*, 13, 361, 2007. With permission.)

liver pathology were also improved, and decreases in apoptotic cells as determined by TUNEL assay were observed as well. These studies therefore demonstrated that the histopathology of livers improved, and more importantly, liver function was enhanced following treatment with emricasan.

The last set of experiments performed in this study tested emricasan in an orthotopic model of liver transplantation.[16] In these experiments, livers from donor Sprague-Dawley rats were cold-stored for 24 hours in UW with or without emricasan (50 μM). After 24 hours of cold storage, the livers were transplanted in donor Sprague-Dawley rats and monitored for survival over 14 days post transplant. In this model, approximately half of the control animals survive the transplantation. Most of the lethality occurs within the first 24 to 48 hours, and animals that die have marked liver injury characterized by sinusoidal congestion and liver damage with significant increases in ALT activities. Again, mortality is thought to be primarily due to graft nonfunction and loss of sinusoidal cells. Importantly, when livers of donor rats were exposed to emricasan during the cold ischemia phase and subsequently transplanted, all rats receiving those livers survived the experimentation period (14 days). These studies also provided confidence that emricasan may be beneficial to patients undergoing liver transplantation.

Data from preclinical studies with emricasan in the ischemia/transplant models were subsequently translated into the clinic, and recently this compound completed a phase II clinical trial in patients undergoing liver transplantation.[17] In this study, *post hoc* analysis revealed that livers of recipients exposed to emricasan (15 μg/ml) had decreased serum ALT and aspantate amino transferase (AST) activities when livers were stored with drug. Reductions in cells with activated caspase-3/7 as assessed by immunohistochemistry were also detected. A more thorough discussion of this trial is found in Chapter 10.

8.6 LUNG ISCHEMIA AND TRANSPLANTATION

Also worthy of mention is a study with emricasan in a preclinical model of lung ischemia/reperfusion/transplantation. A hypothesis similar to that in the liver exists in the lung ischemia/reperfusion setting: ischemia/reperfusion injury is associated with an increase in markers of apoptosis and caspase activation.[18] In the first set of experiments by Quadri et al., lungs from male Lewis rats were subjected to cold flush with a low-potassium dextran glucose preservation solution containing prostaglandin E1.[19] Caspase activities measured by DEVD-AMC, IETD-AMC, and LEHD-AMC cleavage activity (for caspases-3, -8 and -9, respectively) were elevated above controls after 6 and 18 hours of cold ischemia, with peak activities at 6 hours. Emricasan administration to the rat (i.p., 10 mg/kg) just prior to anesthesia and isolation of the lungs markedly decreased the elevation in caspase activation. This experiment suggested that emricasan can inhibit caspases expressed in lung.

In the next set of experiments, lungs from emricasan-treated rats were cold-stored for 6 or 18 hours following the same procedure, but emricasan at 25 μM was also added in the cold storage media.[19] After the cold storage period, one lung was used for TUNEL analysis and caspase assays, while the other lung was transplanted into a donor rat and reperfused for 2 hours. TUNEL staining demonstrated that emricasan markedly reduced the number of TUNEL positive cells compared to the control group at the 6-hour timepoint. After 18 hours of cold storage followed by transplantation, few TUNEL-positive cells in lungs of both control and treated rats were noted, and therefore there was no significant effect with emricasan. Lung function, assessed by partial oxygen pressure in the pulmonary vein, was significantly improved with emricasan when lungs were cold-stored for 18 hours. When assessed after 6 hours of cold storage, partial oxygen pressures were not affected, due to only a small decrease in function. Consistent with the studies' rat liver ischemia/reperfusion models, Quadri et al. concluded that reductions in apoptotic cell death in the lung from cold ischemia/reperfusion can improve function. These studies therefore demonstrated that emricasan may work in other organs, but variable results were noted in other ischemia/reperfusion models, including heart and brain (data not shown). While the reasons for this are not clear, they may be explained by differential uptake and tissue distribution of the molecule.

8.7   LIVER FIBROSIS

Recent data suggest that the progression of chronic liver diseases with fibrosis is initiated by continuous cell death consisting of both apoptotic and necrotic pathways of hepatocytes and other liver cells.[20,21] The result is a low-grade liver injury characterized by chronic inflammation resulting, at least in part, in activation of hepatic stellate cells, which in turn results in their production and deposition of collagen. In this manner, stellate cells can even contribute to the inflammation and apoptosis, further progressing the disease. Clinically, liver fibrosis is seen in patients with, for example, viral hepatitis (both HBV and HCV), alcoholic hepatitis, and cholestatic liver injury, and apoptosis of liver cells is clear in these diseases.[22] Experimentally, liver fibrosis can be induced by chronic administration of diverse chemicals, including carbon tetrachloride, thioacetamide, or concanavalin A, or by ligation of the common bile duct, the latter of which mimics cholestatic liver disease in humans. Therefore, testing whether caspase inhibition and apoptosis can reduce the progression of a fibrotic response is possible with these preclinical models.

For these reasons, Canbay et al. tested the ability of emricasan to attenuate liver injury and stellate cell activation in the bile duct ligation (BDL) model.[23] In their studies, C57/BL6 mice underwent bile duct ligation and emricasan was administered 3 hours after BDL and twice a day for a total of 3 days. After 3 days, liver injury was assessed histologically and by serum ALT activities, while fibrogenesis was determined by Sirius Red staining, which is indicative for collagen deposition. Compared to BDL animals receiving vehicle (saline) only, emricasan-treated mice had marked decreases in ALT activities and reduced the size and number of bile infarcts seen in the BDL control animals (Figure 8.2). Apoptosis and caspase activation in the livers of those animals were confirmed by an increase in TUNEL and caspase-3/7 positive cells, respectively (the latter determined immunohistochemically). The number of positive cells in livers of BDL mice treated with emricasan was substantially reduced. Importantly, minimal positive staining for collagen was noted in emricasan-treated BDL mice, unlike that seen in control BDL mice, where marked Sirius Red staining was noted in the periportal regions. Lastly, markers for stellate cell activation, including TGF-β and α-smooth muscle actin mRNA, were

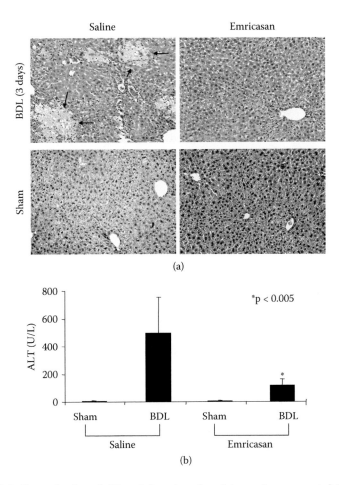

FIGURE 8.2 (See color insert) *Liver injury is reduced in emricasan-treated BDL mice.* Three days after the surgical procedure, the mice were anesthetized, and liver tissue and serum were obtained. (a) Fixed liver specimens from all mice were stained by conventional H&E. Bile duct proliferation, portal edema, and mild portal infiltrates, all features of extra-hepatic cholestasis, were present in all BDL mice. However, more bile infarcts (arrows), due to bile acid toxicity, occurred predominantly in saline-treated BDL mice. (b) Serum ALT values are significantly greater in saline-treated than in emricasan-treated BDL mice ($p < 0.005$, n = 4 for each experimental group). (From Canbay, A., et al., *J. Pharmacol. Exp. Ther.*, 308, 1191, 2004. With permission.)

elevated in BDL mice but were markedly reduced when BDL mice were treated with emricasan. This study supports the hypothesis that apoptotic cell death in the liver leads to stellate activation and collagen deposition. More importantly, agents such as emricasan that can block apoptotic cell death may diminish the initiation or progression of fibrosis in the liver.

As a side note, the same group examined the role of Kupffer cells in phagocytosis of apoptotic bodies in the BDL model and used emricasan as a tool to understand the role of apoptotic cells in Kupffer cell activation in the BDL model.[24] In their studies,

Kupffer cells increased the expression of TNF-α, Fas-L, TRAIL, and TGF-β when exposed to apoptotic bodies, suggesting that activated Kupffer cells may contribute to the initiation of apoptosis of neighboring cells. Further, Kupffer cells isolated from BDL mice expressed more TNF-α, Fas-L, TRAIL, and TGF-β than Kupffer cells isolated from sham-operated controls, suggesting that Kupffer cells play a role in the progression of cell death in the BDL model. If animals were pretreated with emricasan, thereby lowering the number of apoptotic cells to be engulfed, Kupffer cells expressed less of these death ligands, suggesting that apoptotic cells perpetuate the activation of Kupffer cells, hepatocellular injury, and fibrosis.

A phase II clinical trial was completed by Idun in patients with HCV and other diseases. The goal of this study was to determine whether treatment with emricasan for 14 days lowered ALT and AST activities in these patients. In summary, emricasan lowered serum liver enzymes in a dose-dependent manner without increases in adverse events. Details of this trial are summarized in Chapter 10 and have been recently reported by Pockros et al.[25]

8.8   TOXICOLOGY

Select preclinical safety information on emricasan has been reported in abstract form.[5] As part of the investigational new drug (IND) package submitted to the FDA, emricasan was evaluated in standard safety pharmacology assays as well as genotoxicity assays. No adverse events were reported in International Conference on Harmonization (ICH)–compliant safety pharmacology studies (GI, pulmonary, cardiovascular, respiratory, and CNS), and emricasan was negative in genotoxicity assays (Ames, mouse lymphoma, and mouse micronucleus assays).

The initial toxicology studies were performed following i.v. administration of emricasan. The i.v. route was chosen since the initial indication was to treat patients suffering from acute decompensated alcoholic hepatitis with infusions of emricasan for up to 14 days. Good laboratory practices (GLP) toxicology studies were performed in the rat and cynomolgus monkey for up to 28 days,[5] and infusions were given four times a day (q.i.d.). The major toxicological findings included vein irritation at the site of infusion in both species, with secondary reactions associated with the vein irritation. Interestingly, vein irritation in the toxicology studies was predictive of phlebitis noted in the phase 1 trials in humans.[26] In those studies, phlebitis was seen after either a single dose (10 mg/kg) or q.i.d. infusions, but was generally mild and did not result in premature withdrawal for the studies.

As previously discussed, efficacy obtained by oral administration in the preclinical models, in part, altered the strategy for emricasan, as well as changed the potential indication. As part of this change in strategy, oral indications and longer-term treatments were targeted. To maintain consistency with this altered clinical strategy, GLP toxicology studies were conducted in the monkey and rat after oral administration for up to 3 months. Data from the 28-day studies, with emricasan dosed three times daily, established a NOEL (no observed effect level) of 60 mg/kg/day in the rat and a NOAEL (no observed adverse effect level) of 45 mg/kg/day in the cynomolgus monkey. It appeared that p.o. administration of emricasan had a better safety profile in the nonclinical studies, and therefore a phase II clinical trial was initiated in patients with orally administered emricasan. Recently, results of this clinical trial were published and an excellent safety profile was reported.[25]

8.9 SUMMARY

The goal of this chapter was to highlight the preclinical development of emricasan. This molecule was selected, in part, from several thousand other inhibitors based on potency observed from *in vitro* screening against caspases, cell models of apoptosis, and efficacy in a number of *in vivo* models of tissue injury. Further, the toxicology of this molecule was also compared and tested by the proposed clinical routes of administration in order to understand its safety profile. Emricasan was efficacious in many models of liver injury where caspase-mediated apoptosis was well characterized and has demonstrated a necessary margin of safety to date. Recently, emricasan has been shown to be efficacious in patients with liver diseases and appears to be well tolerated to date. Further long-term clinical and preclinical studies will be needed to complete the registration of this exciting compound.

REFERENCES

1. Enari, M., Hug, H., and Nagata, S. 1995. Involvement of an ICE-like protease in Fas-mediated apoptosis. *Nature* 375:78.
2. Rouquet, N., et al. 1996. ICE inhibitor YVADcmk is a potent therapeutic agent against in vivo liver apoptosis. *Curr. Biol.* 6:1192.
3. Rodriguez, I., et al. 1996. Systemic injection of a tripeptide inhibits the intracellular activation of CPP32-like proteases in vivo and fully protects mice against Fas-mediated fulminant liver destruction and death. *J. Exp. Med.* 184:2067.
4. Wu, J. C., and Fritz, L. C. 1999. Irreversible caspase inhibitors: Tools for studying apoptosis. *Methods Enzymol.* 17:320.
5. Hoglen, N. C., et al. 2003. IDN-6556, the first anti-apoptotic caspase inhibitor in clinical trials: Preclinical efficacy and safety. *Hepatology* 38(Suppl. 1):579A (abstract).
6. Linton, S. D., et al. 2005. First-in-class pan caspase inhibitor developed for liver disease. *J. Med. Chem.* 48:6779.
7. Ogasawara, J., et al. 1993. Lethal effect of the anti-Fas antibody in mice. *Nature* 364:806.
8. Hoglen, N. C., et al. 2001. Characterization of the caspase inhibitor IDN-1965 in a model of apoptosis-associated liver injury. *J. Pharm. Exp. Ther.* 297:811.
9. Hoglen, N. C., et al. 2004. Characterization of IDN-6556 (3-{2-[(2-tert-butyl-phenyl-aminooxalyl)-amino]-propionylamino}-4-oxo-5-(2,3,5,6-tetrafluoro-phenoxy)-pentanoic acid): A liver-targeted caspase inhibitor. *J. Pharm. Exp. Ther.* 309:634.
10. Mignon, A., et al. 1999. LPS challenge in D-galactosamine-sensitized mice accounts for caspase-dependent fulminant hepatitis, not for septic shock. *Am. J. Respir. Crit. Care Med.* 159:1308.
11. Chen, L.-S., et al. 2003. IDN-6556, the first anti-apoptotic caspase inhibitor in clinical trials: Tissue distribution and pharmacokinetics. *Hepatology* 38(Suppl. 1):580A (abstract).
12. Gao, W., et al. 1998. Apoptosis of sinusoidal endothelial cells is a critical mechanism of preservation injury in rat liver transplantation. *Transplantation* 27:1652.
13. Natori, S., et al. 1999. Apoptosis of sinusoidal endothelial cells occurs during liver preservation injury by a caspase-dependent mechanism. *Transplantation* 68:89.

14. Cursio, R., et al. 2003. A caspase inhibitor fully protects rats against lethal normothermic liver ischemia by inhibition of liver apoptosis. *FASEB J.* 13:253.

15. Natori, S., et al. 2003. The caspase inhibitor IDN-6556 prevents caspase activation and apoptosis in sinusoidal endothelial cells during liver preservation injury. *Liver Transplant.* 9:278.

16. Hoglen, N. C., et al. 2007. A caspase inhibitor, IDN-6556, ameliorates early hepatic injury in an ex vivo rat model of warm and cold ischemia. *Liver Transplant.* 13:361.

17. Baskin-Bey, E. S., et al. 2007. Clinical trial of the pan-caspase inhibitor, IDN-6556, in human liver preservation injury. *Am. J. Transplant.* 7:218.

18. Fischer, S., et al. 2000. Dynamic changes in apoptotic and necrotic cell death correlate with severity of ischemia-reperfusion injury in lung transplantation. *Am. J. Respir. Crit. Care Med.* 162:1932.

19. Quadri, S. M., et al. 2005. Caspase inhibition improves ischemia-reperfusion injury after lung transplantation. *Am. J. Transplant.* 5:292.

20. Higuchi, H., and Gores, G. J. 2003. Mechanisms of liver injury: An overview. *Curr. Mol. Med.* 3:483.

21. Canbay, A., Friedman, S., and Gores, G. J. 2004. Apoptosis: The nexus of liver injury and fibrosis. *Hepatology* 39:273.

22. Guicciardi, M. E., and Gores, G. J. 2005. Apoptosis: A mechanism of acute and chronic liver injury. *Gut* 54:1024.

23. Canbay, A., et al. 2004. The caspase inhibitor IDN-6556 attenuates hepatic injury and fibrosis in the bile duct ligated mouse. *J Pharmacol. Exp. Ther.* 308:1191.

24. Canbay, A., et al. 2003. Kupffer cell engulfment of apoptotic bodies stimulates death ligand and cytokine expression. *Hepatology* 38:1188.

25. Pockros, P. J., et al. 2007. Oral IDN-6556, an antiapoptotic caspase inhibitor, may lower aminotransferase activity in patients with chronic hepatitis C. *Hepatology* 46:324.

26. Valentino, K. L., et al. 2003. First clinical trial of a novel caspase inhibitor: Anti-apoptotic caspase inhibitor, IDN-6556, improves liver enzymes. *Int. J. Clin. Pharm. Ther.* 41:441.

9 Novel Approaches to Caspase Inhibitor Discovery

Justin M. Scheer and Michael J. Romanowski

CONTENTS

9.1 INTRODUCTION

Most potent small-molecule inhibitors of caspases, regardless of whether they target the apoptotic or inflammatory enzymes, are peptide mimics of their natural substrates.[1] In spite of the highly conserved preference for an aspartic acid at the P1 position, inhibitor specificity has often been engineered. This is supported by reports that most caspases display substrate specificity *in vivo* and in combinatorial substrate identification studies.[2–6] The main determinants of inhibitor specificity, however, are not derived from the conserved inhibitor P1 position, where peptide bond hydrolysis occurs. This has guided design toward molecules with an aspartic acid or aspartic acid–like moiety that interacts with the S1 subsite of the substrate-binding groove and whose specificity determinants arise from contacts outside the S1 pocket. Other chapters of this volume and previous reviews evaluate in detail the advantages and disadvantages of inhibitors of this type.[1,7]

Despite their potency, some fundamental obstacles have been encountered with peptidomimetic caspase inhibitors, such as toxicity, poor pharmacokinetics, reduced oral bioavailability, difficulties with time-dependent inhibition, and at times, lack of specificity within the enzyme family. These issues are partly a consequence of the necessity of an aspartic acid–like P1 element. In addition, irreversible covalent warheads have often been necessary to provide increased potency to these molecules. Discovery tools that are distinct from the peptide mimetic approach will be discussed in this review, including recent technologies, such as fragment-based screening, that have yielded both covalent and noncovalent reversible inhibitors. Ending with possibilities for future research, a review of new approaches to inhibitor design that target a novel exosite, and the possibility of circumventing the aspartic acid requirement of active-site inhibitors will be presented.

9.2   NONPEPTIDIC INHIBITORS IDENTIFIED BY HIGH-THROUGHPUT SCREENING

Several series of nonpeptidic caspase inhibitors have been discovered and are reviewed in more detail in Chapter 5. In some cases, these compounds have been identified by a naive high-throughput screen (HTS) using *in vitro* enzymatic activity. Essentially three series have been identified through this approach: an isatin series, a series of barberine derivatives from which novel isoquinoline scaffolds were developed, and a quinoline series of compounds with a double electrophilic warhead based on either a 1,3-dioxo-2,3-dihydro-1*H*-pyrrolo[3,4-*c*]quinoline molecular scaffold or an anilquinazoline.[8–11] There was an additional steroid-like diacid identified in an HTS using caspase-8.[12] Although this compound is not competitive with a tetrapeptide substrate, it is possible that its diacid moiety is interacting with the S1 and S3 subsites of the enzyme, which prefer aspartic acid and glutamic acid, respectively. Interestingly, no other commercially available steroids tested in the assay showed inhibition.

The high-throughput screens reported to date appear to be limited to caspase-3, -7, and -8, although it is likely that such screens have been performed on other caspases. Although HTS has led to the successful identification of non-peptide-like scaffolds different from the bulk of the peptide-like inhibitors, it appears not to be the method of choice for developing caspase inhibitors. This is likely because direct application of substrate knowledge to the design and synthesis of potent inhibitors can rapidly facilitate the design of active-site peptide mimetics. The HTS approach may gain importance as recent structural data on caspases indicate that these enzymes undergo dramatic conformational changes when occupied by an active-site ligand. It is possible that the initial lack of this structural information has limited the diversity of compounds designed, and that high-throughput screens will reveal a broader range of chemotypes. Rational predictions of active-site structures may not always provide a complete complement of the relevant information on special structural details, suggesting that empirical approaches may have value in uncovering new opportunities.

9.3 FRAGMENT-BASED APPROACHES

Several technologies, methods, and techniques have been pioneered to identify small-molecular-mass fragments as starting points for the design of early hits. These fragment-based approaches typically seek to identify a small molecule with high ligand efficiency with the objective of rapidly surveying a large diversity space.[13–16] Some of the methods used include functional screens at very high concentrations of fragments, assaying for direct binding at very high concentrations, structure activity relationship (SAR) by nuclear magnetic resonance (NMR), screening by x-ray crystallography, and methods using mass spectrometry.[13,16] Direct binding uses surface plasmon resonance to detect proteins binding to fragments that have been immobilized on a chip. NMR-based screening (SAR by NMR) for fragments is performed by monitoring amide chemical shifts of N^{15}-labeled target proteins that occur when a fragment displays some binding affinity. This type of screen provides direct-binding-site information, as does screening by x-ray crystallography, the highest-resolution method available to date. The drawbacks to the x-ray crystallography-based method are the need for high protein concentration, the ability of a protein to generate diffraction quality crystals, and time. Another site-directed method is Tethering® (Sunesis Pharmaceuticals, Inc., South San Francisco, California), which probes free sulfhydrals on the protein surface using mass spectrometry-based identification of covalently linked compounds that have affinity for the site near the free sulfhydral. The interest in fragment-based methods rests partly on the premise that a 1-to-2-million-compound library of larger molecules can actually explore only a very limited portion of the total available small-molecule chemical space. By the recombination of small fragments, one can theoretically increase the chemical space surveyed and at the same time decrease the size of the library screened. Using an approach for identifying fragments followed by a linking strategy has proved successful in many cases. However, an investigation into the deconstruction of an inhibitor to its fragment cores suggests that the process of fragment linking and the process of molecular recognition may often evade the rationality of a fragment-based linking strategy.[17]

The fragment-based approach Tethering (and Tethering with Extenders) has been used to identify active-site inhibitors for caspases-3, -7, and -1. This technique was used at Sunesis Pharmaceuticals, Inc., where scientists developed the approach for discovering fragment ligands for multiple targets that include phosphatases, endopeptidases, protein-protein interactions, and more recently, kinases.[18–24] The hallmark of Tethering is the site-directed property in which a free cysteine, either naturally occurring or engineered, is screened against a compound library in which all the compounds have readily reducible disulfide bonds (Figure 9.1). Under mildly reducing conditions, the fragments undergo disulfide exchange with the protein, and those with affinity, albeit weak, can be covalently captured and identified by mass spectrometry. One major advantage of the technique is less purified protein is needed when compared with NMR- and x-ray-based approaches, and compound concentrations do not need to be very high, typically in the low micromolar range. Tethering with Extenders is referred to as a dynamic combinatorial technology that allows for the assembly of small-molecule fragments *in situ*, possibly optimizing the efficiency

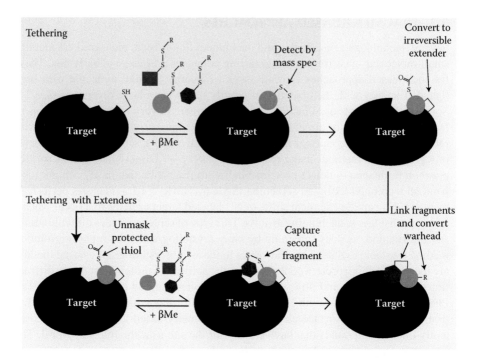

FIGURE 9.1 Tethering and Tethering with Extenders. The top panel (tan background) shows the Tethering method for fragment identification. A protein surface that displays a native cysteine at the site of interest or an engineered free thiol is targeted by the screen. Pools of disulfide-containing compounds are reacted with the protein in a mildly reducing environment to promote disulfide exchange. Fragments with affinity for a binding site near the thiol are captured by formation of a disulfide bond and can be rapidly detected by electrospray mass spectrometry as a change in the total mass of the protein. This mass change represents a compound mass that is unique in the pool facilitating positive identification. For Tethering with Extenders, the captured fragment is converted to a covalent extender that irreversibly modifies the protein free thiol. The process for Tethering with Extenders is shown in the lower half of the panel. After unmasking the new thiol, a similar mass spectrometry screen is done with the same disulfide compounds. After selection of a second fragment, the fragment and the extender are linked and the warhead is modified to make a reversible small-molecule inhibitor.

of fragment linking strategies.[25] Typically a core fragment is identified that can be converted to a compound with an irreversible covalent warhead and an additional masked free thiol. This moiety is termed the extender and is covalently and irreversibly attached to the protein surface. After covalent attachment to the site of interest on the protein, the blocked free thiol is unmasked to provide a new screening point (Figure 9.1, lower panel). From this point, a second screen can be performed to capture an adjacent binding element that can be converted to a potent inhibitor through a linking and merging strategy.

9.3.1 TETHERING WITH EXTENDERS FOR CASPASE-3 INHIBITORS

A series of potent inhibitors of caspase-3 with a K_i in the range of 20–50 nM was identified by optimizing an initial fragment hit with a salicylic acid core that binds to the S4 subsite of caspase-3. The approach used to identify the salicylic acid core by Tethering with Extenders was modular in nature.[26]

Extended-Tethering molecules were made by combining three components: an active-site warhead made of an arylacyloxymethyl ketone for attachment to the active-site cysteine, an aspartyl group for binding to the P1–aspartic acid-binding pocket, and a thioester that could be deprotected by reaction of the enzyme-inhibitor complex with hydroxylamine (Figure 9.2). This extender, or anchor, was bound irreversibly to the S1 subsite via covalent attachment to the active-site cysteine by an electrophilic warhead, and following deprotection of the extender, the enzyme-extender complex could be screened against a small library of disulfide-containing fragments. Two different extenders were designed based on this aspartic acid warhead to sample a different binding space due to the positioning of the thioester relative to the aspartyl component. In the case of extender A, the thioester was placed essentially in the P2 position of the peptide-inhibitor Ac-DEVD-CHO, while extender B was spaced with a benzyl group following a sulfonamide replacement of the amide bond between the P1 and P2 elements. These anchors, or extenders, can then be used to capture new S4-binding fragments by screening the disulfide-compound library with the extender probe anchored to the enzyme active site.

Protein modified with these two extenders was used in the mass spectrometry–based screening of 7,000 disulfide-containing fragments in pools of eight to twelve compounds to identify third-module fragments. From this screen, a salicylic acid sulfonamide containing a thiol linker (captured fragment A) was selected by the shorter extender A. The longer extender identified an unrelated thiophene-sulfone fragment (captured fragment B), supporting the notion that different extenders, and in particular those with different lengths, are sampling distinctly different regions on the protein surface. It is also likely that a combination of unique contributions provided by the linker and the fragment contacts influenced capture selectivity. One notable point is that the selected fragments were not trapped by the other extender in an isolated reaction containing only the selected fragment and the enzyme-extender complex. The authors also noted it was likely that these fragments would have been false negatives in functional screens since they did not inhibit the enzyme as isolated fragments, even at high concentrations.[25]

Analysis of the salicylic acid (captured A) + extender A and thiophene-sulfone (captured B) + extender B enzyme complexes by x-ray crystallography shows that the extenders bind in a canonical fashion to the P1-binding pocket and that the P4 fragments contact the enzyme through multiple hydrogen bonding interactions at the S4 subsite. However, the binding mode differs somewhat from that displayed by the tetrapeptide inhibitor. Typically, the S4 pocket is smaller, but in the salicylic acid + extender A complex there is apparent widening of this pocket, made possible by a conformational change resulting from the collapse of the S2 pocket. The S2 pocket collapses because the extender does not possess a bulky side chain, like a valine residue, that would typically occupy this site in a peptide-inhibitor complex. In the x-ray

Enzyme Subsite

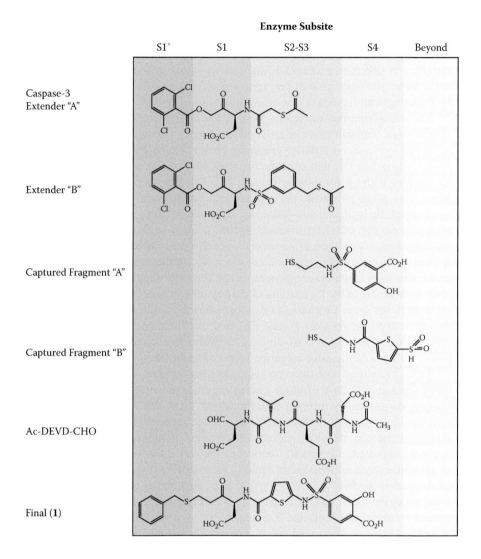

FIGURE 9.2 The process of discovering caspase-3 inhibitors. The caspase-3 extenders A and B were initially designed to screen for S3-S4-binding elements by extending from the P1 position. Two reported fragments selected from the disulfide-compound library using these extenders are shown: captured A that binds to extender A, and captured B that binds to extender B. The positions of the various regions of the compounds in the active site are shown in columns shaded with respect to the enzyme subsite that they occupy. The relationships among the P1′–P5 sites of the well-characterized peptidic inhibitor Ac-DEVD-CHO, the Tethering compounds, and the final linked and optimized molecule, compound 1, are also indicated by the columns.

structure of the protein-extender B complex, the larger benzene sulfonamide portion of the extender is seen in relative proximity to the S2 site, and may in turn influence the size of the S4 pocket, suggesting a reason for the selection of a completely different fragment. In this case, the thiophene sulfone makes interactions similar to those made by the salicylic acid, but is smaller in size and can be accommodated by the smaller S4 site. It has been found in combinatorial substrate studies that caspase-3 shows a very strong preference for aspartic acid in the P4 position.[2,4,5] Since the data reported in the substrate identification studies do not discriminate among effects on the P4 position in the presence of a particular change in the P2 position, conclusions about the dynamic interactions across the substrate-binding groove cannot be drawn. However, the results clearly show that there is flexibility in side chain preference in the P2 and P3 positions.[5]

The subsequent challenge of fragment-based design is to develop a potent and reversible small-molecule inhibitor from the fragment or extended fragment. According to a review on the subject, the transformation of a fragment hit, or even an advanced fragment compound that has been optimized by simple SAR experiments and limited medicinal chemistry, begins a longer, more challenging process that will turn the compound into a potent inhibitor.[13] The three basic considerations for advancing a fragment include fragment optimization, fragment merging and linking, and *in situ* fragment assembly.

Optimization and design of caspase-3 inhibitors from the extender-fragment complexes involved modification of the linker portion of the molecule by removal of the disulfide bond between the fragment and the extender, followed by conversion of the irreversible warhead to a reversible functionality. Replacement of the sulfurs with two methylenes and replacement of the arylacyloxymethyl ketone with an aldehyde resulted in a reversible inhibitor with a K_i of 2.8 µM. Both captured fragments A and B were converted to more potent and reversible inhibitors by these straightforward chemical changes, but displayed partial lack of selectivity when tested against a caspase panel. The most selective molecule came from linking extender B with its captured fragment (captured fragment B), which displayed greater than twenty-fold selectivity over most of the other caspases except caspase-7.

When tested in cell-based assays as aspartyl aldehyde inhibitors the compounds were found to be inactive. However, when converted back to an irreversible inhibitor by changing the warhead, the best molecule then displayed an EC_{50} of 18 µM for inhibiting poly (ADR ribose) polymerase (PARP) cleavage and 21 µM for preventing cell death in Jurkat cells induced with the death ligand Fas. This newly modified compound showed greater potency but less selectivity for caspase-3 than the precursor compound without an irreversible warhead. This may not be a serious liability since broader selectivity profiles may be useful in treating indications related to unwanted activation of cell death pathways in which more than one caspase is involved.

The next step in the development of these compounds involved chemical modifications of the inhibitor to increase potency and selectivity. Again, the approach took advantage of the modular nature of the compounds that could be broken down to the salicylic acid group (P4), the linker region (P2-P3), and the active-site warhead (P1'-P1) (Figure 9.2).[26] First, several changes were introduced to the salicylic acid group of the inhibitors that interacted with the residues in the S4 pocket of caspase-3. Removal of the carboxylate had a severe impact on inhibitory activity that was likely due to the loss of interaction with the indole of Trp214, whereas no other changes to the salicylic acid showed a real improvement in activity. Attempts to eliminate the charge were made in spite of the loss in biochemical potency in order to improve cell-based activity by increasing membrane permeability. Ultimately, no change that neutralized the charge without significant loss of activity could be identified, thus indicating the potential importance of charge interactions in the P4-binding pocket. This should not be surprising since interactions with the S4 site are important in establishing substrate specificity for acid side chains.[4,5] It is possible a prodrug approach to mask the acid group until cleavage by intracellular esterases could provide both the necessary cell permeability and subsequent enzyme affinity.

At the other end of the molecule, only a single β'-atom replacement of the covalent warhead with a lipophilic 2-chlorobenzyl thiomethyl ketone provided equal or better K_i. It appears that approaches used to enhance the potency by changing the warheads of other caspase inhibitors were not transferable to the caspase-3 inhibitors, suggesting a different structure at the P1′ position among the caspases, and therefore a different impact on inhibitor activity.[27] Modifications of the linker portion of the inhibitor were made with a selection of heterocycles in the context of both an aldehyde and aryl thiomethyl ketone warhead. As these derivations were made, the presence of a lipophilic warhead preserved the potency of the inhibitors, whereas an aldehyde version resulted in significant losses in potency. It was also observed that the size of the heterocycle appeared to affect the positioning of the S1 and S4 ligands; smaller furan rings at this position proved detrimental to potency. From the work presented, a compound selection was made for further modifications.[26]

Modification of the P1′ warhead leaving group entailed substitution of the group with a weakly activated phenylpropyl moiety. The new compound displayed activity nearly as potent as the aldehyde warhead, suggesting that a less reactive molecule at the P1′ is possible. Modification of the salicylic acid P4 group was not successful in finding a more potent replacement, while the sulfonamide linker connecting the warhead to the P4 element could be replaced with a tertiary amine. This preserved activity and allowed for reductive amination as a pathway to quickly sample chemical replacements at S4, finally resulting in compound **1** (Figure 9.2).[26]

The crystal structure of caspase-3 with compound **1** shows that the compound has a similar binding mode to peptide-based inhibitors such as Ac-DEVD-CHO. The selectivity of this and other related compounds could be expected to mimic that of the peptidic inhibitors. All the fragment-derived and optimized inhibitors that arose from this effort showed little cross-reactivity with the inflammatory caspases (-1, -4, and -5) and no reactivity with caspase-2. However, and perhaps of significance to the treatment of apoptosis-related disorders, these inhibitors appeared to be more potent pan-apoptotic inhibitors by nonselectively inhibiting caspases-3, -6, -7, and -8. This could be beneficial in stopping both the upstream activating caspases and the active downstream enzymes. It is interesting to note that two compounds displayed higher selectivity for caspase-3 than caspase-7, despite the identical substrate specificity of the two enzymes,[26] which is likely due to the difference in the linker region between the two compounds and the other inhibitors presented in the study. In these compounds, the linker region appears to bypass the P2-binding pocket. A second report on advancement of fragment-derived inhibitors for caspase-3 from the salicylic acid–sulfonamide compound shows that more potency can be derived by focusing attention on the region spanning this S2 subsite.[28] Alkyl analogs at the P2 position in the compounds were tested, as well as various phenyl derivatives and thiophene analogs. Larger alkyl chains in the P2 position were generally more active, which could be reasonably expected because of the increase in hydrophobic contact. With relation to other studies, however, one could expect that these molecules might concomitantly affect the impact that chemical moieties at P4 have on interactions with the protein. The phenyl group, identified as the most potent insertion at the P2, was selected as part of the scaffold for subsequent P4 replacements, but no chemical change at P4 improved the potency of the compound, and only one change in this position retained the same potency. Again, it is clear that an acid moiety is important in maintaining stability at this position.

The aldehyde warhead was also explored to identify functional groups that could improve potency. Within the series of less reactive ketones a thiomethyl ketone was identified that was sufficiently potent, whereas most other changes at this position had a drastically negative effect on inhibitor affinity for the enzyme. Finally, linker rigidification did not alter or enhance potency. Ultimately, using the modular-fragment approach as a progression from Tethering with Extenders, scientists rapidly sampled in a combinatorial-like fashion the potential of modifications along the length of the initial hit molecules. It seems the initial discovery of a salicylic acid group in the P4 position was quite an effective element in inhibitor design. Sampling of the S4 site by Tethering with Extenders made this possible since it allowed for a constricted S2 element and optimal sizing of the S4 for the group to bind. Quite interesting as well was that the molecule produced in this work formed interactions with the enzyme that mimic those of the peptidic inhibitors even though it was not based on a substrate peptide (except for the P1 aspartic acid).

9.3.2 TETHERING WITH EXTENDERS FOR CASPASE-1 INHIBITORS

A similar approach was taken to rapidly identify new inhibitors of caspase-1.[20] In summary, new fragments that bind to the S4 subsite of caspase-1 were identified and converted to aspartic acid-containing aldehyde inhibitors by simply linking the fragment and warhead moieties with an alkyl chain.[18,20] The strict conservation of the P1 element preference in caspases was an advantage since the irreversible warhead utilized for screening caspase-3 was directly transferable to the active site of caspase-1. This proved to be an equally effective fragment-capturing scaffold for caspase-1 from which ten new fragments were identified, two of which were reported (a tricyclic quinoxaline [fragment 1] and a hydroxyquinoline [fragment 2]; Figure 9.3). These P4-binding fragments shared little in common with the acid salicylate identified in caspase-3 (Figure 9.2), which is likely due to the differences in substrate preference. The P4 region of caspase-1 substrates is more likely to contain a hydrophobic side chain such as tyrosine or tryptophan, which is different from the acidic preference in the same site in caspase-3. Replacing the irreversible binding warhead with a reversible binding aldehyde derivative and linking the P1 and P4 fragments with a flexible linker generated a compound that showed a K_i of 0.15 µM (compound **2** in Figure 9.3) with selectivity over caspase-5. Based on the substrate specificity of caspase-1 in the P2 position for a hydrophobic group and the observations from the work with caspase-3, the addition of a hydrophobic side chain at the P2 position dramatically enhanced the activity of the initial compound, reducing the K_i to 0.007 µM in the most potent molecule (compound **3** in Figure 9.3). This represents

Enzyme Subsite

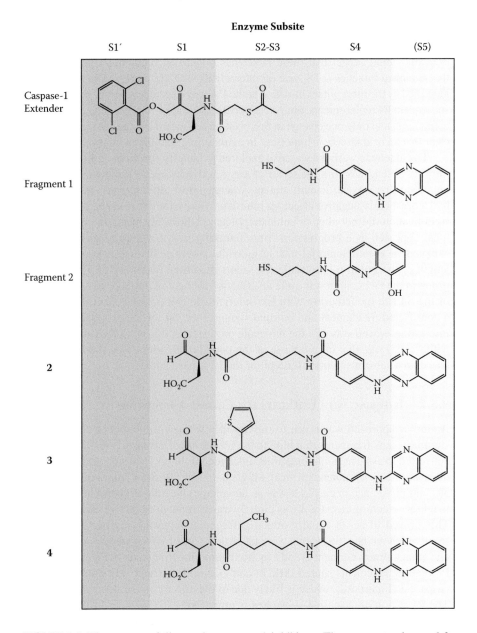

FIGURE 9.3 The process of discovering caspase-1 inhibitors. The same extender used for modification of the caspase-3 active site (extender A, Figure 9.2) was used for caspase-1. This extender bound equally well to caspase-1 because of the highly conserved preference for an aspartic acid in the P1 position and was used to screen the library of disulfide-containing fragments. Fragments 1 and 2 were identified in the screen and used to make compound **2** by linking fragment 1 to an aspartic acid aldehyde in the P1 position. Compound **2** was substituted at the P2 position with a hydrophobic constituent, resulting in compounds **3** and **4**. The hydrophobic side chains increase the potency of the compounds and help define a rigid S2 subsite in this caspase.

a potency increase of more than twenty-fold, indicating that structure-based rational observations can be effectively included into a fragment-based approach to inhibitor design. The hydroxyquinoline compound was similarly modified, yielding a more potent compound; however, the overall potency of this series was much less than that of the tricyclic quinoxaline series. It is interesting to note that these compounds showed extraordinary selectivity over caspases-1 and -5, both of which are involved in inflammation and display nearly identical substrate specificities. It is not clear, however, if selective inhibition of these caspases is favorable to pan-inhibition since both are often simultaneously activated during interleukin 1β processing. Structures of several of these compounds were reported in complex with caspase-1.[18]

In a co-crystal structure of the tricyclic quinoxaline compound irreversibly bound to the caspase-1, the quinoxaline ring is bent around Arg383, whose side chain constricts the region between the S4 and a large pocket beyond the S4 that could be considered an S5 subsite. In compound **4**, a related quinoxaline with a less rigid hydrophobic linker in the P2 position, the extended moiety was found to bend away from Arg383 in a different direction, but still within the same large pocket (Figure 9.3). The overall potency of compounds that include a functionality that extends beyond the S4 pocket in caspase-1 appears to be less influenced by the identity of the S2 side chain. To summarize these subsite relationships (Figure 9.4), caspase-3 appears to have a much more flexible S2

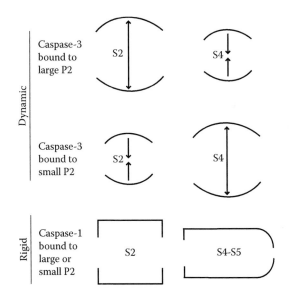

FIGURE 9.4 The influence of P2 elements in caspase-3 and caspase-1 inhibitors. In the presence of larger P2 elements, the S2 of caspase-3 opens wider to accommodate the larger molecule. This in turn reduces the size of the S4 subsite. The altered S4 resulted in the capture of different fragment chemotypes in an extender screen. By taking advantage of the subsite dynamics in caspase-3, modification of the extender in the P2 position could increase the diversity of molecules identified in fragment screening efforts. The S2 subsite in caspase-1 is more rigid and appears not to have the same type of influence on the S4-S5 regions as does the corresponding subsite in caspase-3.

that opens and closes depending upon the size of the occupying ligand. If the P2 position of the compound is bulky, the S2 site is found in a larger, open conformation that in turn affects the S4 site and the region beyond. In caspase-1, S2 appears to be open in all inhibitor-bound forms, regardless of the presence of a P2 element. Thus, the coordinating event between these active-site elements is subtly different across the enzyme family.

The most potent compound in a cell-based assay examining the secretion of IL-1β in response to lipopolysaccharide stimulation in human peripheral blood mononuclear cells (PBMCs) was the linked tricyclic fragment containing the alkyl chain with no P2 element (compound **2**, *in vitro* K_i of 0.15 µM and cell-based EC_{50} of 4.1 µM). This is, however, not the most potent *in vitro* compound. The comparison of compound potency in an assay that requires membrane permeability makes structure-function conclusions difficult to define, but clearly supports the notion that continued optimization of the compounds to improve cell potency is possible since many modifications of the compounds can produce submicromolar K_i compounds *in vitro*. One of the compelling conclusions from this work is the ease in which compounds can be identified by the Tethering and Tethering with Extenders approaches. Potent starting points can be transferred easily across an enzyme family to identify fragments with novel and important chemotypes that impart unpredicted interactions unlikely to be identified by any other method.

9.4 INHIBITION THROUGH NON-ACTIVE-SITE MECHANISMS

Dynamic conformational states of proteins have been recognized and exploited in drug discovery efforts on several enzyme classes, including kinases, phosphatases, and endopeptidases.[21,29–32] A notable example is the Abelson tyrosine kinase inhibitor, imatinib, which locks the kinase in an inactive conformation by an allosteric mechanism. Similarly, one interesting discovery in caspases during the last several years is that these enzymes can adopt several different conformations. These conformations can be trapped by various ligands that recapitulate structural features of the inactive zymogens.[33] It seems that two predominant conformations exist: an active conformation observed in active-site ligand-bound x-ray structures, and an inactive conformation observed in zymogen and active-site ligand-free structures.

Zymogen processing is commonly necessary to yield fully active caspases.[34–38] This may not be as important in caspase-9, in which the zymogen and the fully processed enzyme have relatively similar catalytic activities. However, relative to other family members, caspase-9 is the least efficient with a k_{cat}/K_m of 0.00013 × 10^6 M^{-1} s^{-1} (about 25,000-fold less than caspase-1)[39] and whose activity is enhanced by a large multiprotein complex in cells. For most other family members, processing of the zymogen is a critically important step.

In comparing active and inactive conformations of several caspases, a common series of alternate loop arrangements composing the active site has been observed (Figure 9.5). For example, the x-ray structure of the inactive procaspase-7 zymogen shows that the loops surrounding the active site are not arranged in a conformation that favors substrate binding.[35,40] The peptide-binding groove is not formed and a critical region of the N-terminal portion of the small subunit that helps to order the active site of the neighboring monomer is held away from this position by the

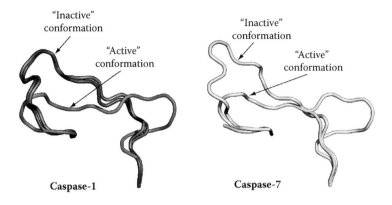

FIGURE 9.5 (see color insert) Comparison of active and inactive structures of caspase-1 and caspase-7. Three structures of caspase-1 are aligned and the backbone residues 330–348 are displayed as ribbons. The active conformation shown in red represents the enzyme bound to an active-site inhibitor, and in this conformation the loop along the active-site groove is flattened to allow for substrate binding. The two inactive conformations shown (ligand-free in blue and allosterically inhibited in purple) are nearly identical in the positioning of the loop in the active site. The arrangement of this loop in the inactive conformations disrupts the substrate-binding groove and repositions critical residues that compose the aspartic acid–binding pocket. Caspase-7 shows a similar conformational dynamic. The inactive conformations shown are zymogen (light blue) and the allosterically inhibited form (green). These loops also disrupt the substrate groove that is distinct from the flattened active site seen in the active structure (yellow). The loops in the active-site region in caspase-1 and caspase-7 undergo similar transitions between active and inactive conformations.

unprocessed amide bond. The conformation of this inactive ligand-free active site is most notably missing a formed S1, which accepts the substrate aspartic acid side chain. This pocket is formed by the positioning of Glu283, Arg189, and Arg341 whose side chains create a positively charged environment that strongly coordinates the carboxylic acid functionality of the P1 aspartate, and this coordination is apparent in all x-ray structures of caspase family members that are bound to an active-site ligand (see Romanowski[41]). However, in the zymogen structure of caspase-7, this highly conserved structure is completely displaced as the side chain of Arg341 is found nearly 20 Å away from the S1 subsite.[40] This displacement renders the active site incompetent for binding substrate. A similar conformational dynamic is seen in caspase-9. When the fully processed caspase-9 enzyme is bound to an active-site inhibitor, it is observed that one active site is fully formed, with inhibitor bound, while the other active site is in an inactive conformation that resembles the zymogen structure of caspase-7. There is one example, that of caspase-3, in which a ligand-free structure reveals a conformation similar to that of the active-site-bound structures. However, since these crystals were grown in the presence of a potent active-site inhibitor, the observed conformation of the active site may not predict what a true ligand-free structure of the enzyme might be in the absence of any active-site ligand in the crystallization buffer, given that even a very weak binding active-site ligand can have a dramatic effect on active-site conformation.[41]

One observation was that concomitant with the closing of the active site was the opening of a pocket at the dimer interface, possibly providing an opportunity to develop a new class of inhibitors that could act allosterically. In addition, it was postulated that the conformation of the active-site ligand-bound caspase-1 might differ from that of a ligand-free enzyme, as observed in caspase-7. This would clearly impact the effectiveness of rational drug design and attempts to replace the obligatory P1 element. To address the flexibility of the caspase-1 active site, the ligand-free form of the enzyme was crystallized and its structure was compared to that of a ligand-bound form. The ligand-free structure revealed a closed active site, which is very similar to that observed in the ligand-free form of caspase-7, the zymogen form of caspase-7, and the inactive site in caspase-9.

It was previously predicted by scientists at Vertex Pharmaceuticals that caspase-1 could be inhibited allosterically.[42] The authors described the results of mutations in the enzyme that affected residues distal to the active site. These substitutions caused considerable reduction in the activity of the enzyme as judged by a transient expression system that measured the level of proIL-1β processing induced by the mutants. Substitutions that reduced activity included His322 located near the dimer interface, Cys244 positioned 8–9 Å from the active site, and Cys136 located 30 Å from the active site. The triplet of Ser332, Ser333, and Ser339, which are close to the dimer interface, also had a mild impact on activity. A recent review of allosteric regulation highlighted that sites at subunit interfaces are often involved in allosteric regulation mechanisms.[43]

One interesting result from the Vertex study is that two noncompetitive inhibitors were identified, gold-thiomalate and auranofin (Table 9.1).[42] These compounds are anti-inflammatory agents that reduce the circulating levels of IL-1β and appear to react with cysteines 364 and 397 located near the surface-exposed dimer interface. The authors suggest that this site could be a physiological target for these anti-inflammatory agents functioning through an allosteric mechanism, albeit most likely non-selectively targeting caspase-1. The noncompetitive nature of the inhibitors suggests that they function through allosteric regulation, although it is difficult to predict from the available information the mechanism through which this may occur. One possible mechanism is that these compounds disrupt the dimeric structure of the protein, although no data were provided to support this. Nevertheless, these early observations suggest that there may be multiple ways of inhibiting caspases, and they have been addressed in several publications dealing with the inhibition of caspases by naturally occurring proteinaceous inhibitors.[44] A particularly relevant example is the inhibition of caspase-9 by the inhibitor XIAP, which binds at the dimer interface and prevents the association of dimers and the formation of a competent active site.[45] Given that an allosteric inhibitor would not have to bind in the active site, it is reasonable to speculate that small-molecule inhibitors could be identified that lack the aspartic acid moiety. The following sections will describe the use of Tethering to identify such inhibitors of caspases and other methods that have yielded non-aspartic acid-containing inhibitors.

9.4.1 Allosteric Inhibition of Caspase-3 and Caspase-7

Using fragment-based approaches to caspase drug discovery, published data demonstrated that caspase-3 and caspase-7 can be allosterically regulated.[30,33] In the

TABLE 9.1
Inhibitors That Bind Distal from the Active Site

Caspase	Compound	Structure	Site
Caspase-1	Compound 7		Cys331
	Gold-thiomalate		Cys364 and Cys397
	Auranofin		Cys364
	Thimerosal		Cys244
Caspase-3/7	DICA (compound 5)		Cys290
	FICA (compound 6)		Cys290
Caspase-9	Bir-2 domain of XIAP	Not shown	Dimer interface

previously described Tethering studies, a disulfide containing warhead was initially bound to the active site before the protein was screened to detect fragments that would then attach to the labeled warhead and extend into the S3 and S4 regions of the active site. As an alternative strategy, native protein was screened directly against the disulfide library in the hope of detecting fragments that would bind to the P1 and P2 regions and be trapped by disulfide linkage to the active-site cysteine residue. No fragments were detected by mass spectrometry that bound to the active-site cysteine residue located within the large subunit; however, fragments were identified that bound to a single cysteine residue on the small subunit. Peptide mapping and mutational analysis verified that these compounds were conjugated to Cys264, near the dimer interface. The position of this residue, however, is different from the locations of cysteines modified by gold-thiomalate and auranofin in caspase-1.[42] Rather than an exposed residue on the back of the enzyme, Cys264 is a component of the β-sheet that forms the core of the enzyme and makes direct interdimer contacts (Figure 9.6). The compounds inhibited enzyme activity in a competitive manner, as assessed by a fluorescence-based assay, and there was a linear correlation between the level of inhibition and the amount of compound covalently bound to Cys264.

The co-crystal structure of caspase-7 and both of these small, disulfide-linked fragments revealed that the conformation of the compound-bound enzyme was very similar to the zymogen form of the enzyme. In caspase-7, there are three different enzyme structural states known: the active structure, the zymogen structure, and the allosteric structure (Cys264 bound). Although these allosteric compounds bind to a region that is 14 Å from the active site, they exert a strong inhibitory function and seem to stabilize a zymogen-like structure.

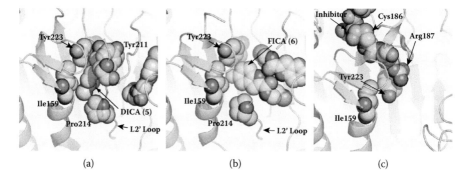

(a)   (b)   (c)

FIGURE 9.6 (See color insert.) Mechanism of allosteric inhibition of caspase-7 by DICA (**5**) and FICA (**6**). Caspases are composed of two identical monomers that bind together at an interface composed partly of an antiparallel β-sheet. The two monomers are differentially colored green and tan, with only side chains of the green monomer being displayed. In panel a, fragment **5** is shown with gray carbons, and in panel b, **6** is shown with yellow carbons (PDB ID codes: 1SHJ and 1SHL, respectively). Tyr223 is pushed out of position by the presence of allosteric compounds in both the **5** and **6** structures when compared to the active-site inhibitor-bound structure shown in panel c (PDB ID code: 1F1J). In the latter case, the active-site inhibitor (blue carbons, panel c) influences the structure of the active site such that Arg187 is positioned near Tyr223. Tyr223 moves closer to the dimer interface in this structure to make room for the side chain of Arg187. This arrangement is not possible in the allosteric structure because of the position of Tyr223. In addition to the Tyr223 restricted position, the L2′ loop closes in around **5** and **6** and is noticeably absent from the active structure in panel c. These structural changes appear to be the key elements in the inhibitory effect of compounds **5** and **6** and explain how binding ~15 Å from the active site can potently and competitively inhibit the enzyme.

The two compounds are small, ring-based structures (Table 9.1). One compound contains a dichlorophenoxy ring (DICA, compound **5**), and the other a fluoroindole (FICA, compound **6**). The conjugation of the molecules to the protein depends on the concentration of the reducing agent in the reaction; thus, a relative measure of affinity is the conjugation efficiency with increasing concentrations of reducing agent, an approach that has been used to rank compounds from disulfide-trapping Tethering screens.[22] It appears that even in the presence of high reducing agent concentrations these compounds stably bind to the dimeric interface. Moreover, it was found that binding of these compounds to the allosteric site was mutually exclusive with binding of an active-site inhibitor to the active site. Modification of the allosteric site by either compound **5** or **6** prevented binding of an active-site inhibitor, clearly suggesting an event that results in conformational changes.

A structural analysis elucidated a specific mechanism for allosteric inhibition at the dimer interface. This mechanism is presumably conserved between caspase-3 and caspase-7 since there is 100% conservation of the residues that line the allosteric binding pocket in the enzyme. The structure of both compounds **5** and **6** bound to the enzyme illustrate that despite a completely different binding orientation of the two molecules at the interface, the resulting structure of the enzyme is nearly identical. Compound **5** binds exclusively to the same monomer as the Cys264 to which it is tethered, making contacts between Tyr211 and Tyr223 and Pro214 and Ile159 (Figure 9.6). The second compound (**6**) is not quite as trapped, but it is buried in a hydrophobic pocket in the same region of the opposing monomer and again sits close to Tyr223, which may be a key residue in the mechanism of action of the two compounds. The distance between Tyr223 and Tyr211 and the compound is about 3.5 Å, and the para-chlorine of the compound is wedged between the δ-carbon of Pro214 and the γ-2 carbon of Ile159. Another factor contributing to inhibition is the position of loop L2′. This free strand of the enzyme comes from the N-terminus of the small subunit and is positioned toward the cavity along the dimer interface in the compound **5**- and **6**-bound structures. This loop is normally associated with the active site, but in these and the zymogen structures it is flipped onto the dimer interface. It may be that this loop is dynamic in solution and changes position between the allosteric cavity and the active site, with the allosteric compounds promoting the aggregation of the L2′ loop into the central cavity. The presence of a ligand in the active site or the central cavity may reduce the dynamic flexibility of the loop and stabilize it in one or the other conformation. Studies of loop dynamics are not reported, but these phenomena could be measured by NMR or evaluated by the addition of reporter species to the loop through engineering and chemical modification.

The binding of the compounds at the central cavity has two basic effects. One is to help stabilize the L2′ loop in an inactive position, and the second is to alter the position of the Tyr223 side chain. This side chain is positioned toward the cavity in the active structure of the enzyme. Occupying the space previously taken up by Tyr223 is the side chain of Arg187, which is adjacent to the catalytic Cys186. The role of repositioning Tyr223 after binding of compound **5** or **6** is likely to prevent Arg187 from burying its side chain closer to the central cavity, since it would be sterically disallowed (Figure 9.6). Thus, it is reasonable that this connection between the central cavity and the active site is a key aspect of the mechanism of action of the allosteric inhibitors.

This highlights the importance of Arg187 positioning in catalysis, presumably because of the influence on the alignment of Cys186 in the substrate recognition groove. However, other stabilizing factors may also be influenced by Arg187 positioning.

It is possible that binding to this pocket could form the basis of a potent therapeutic compound. As the current small molecules covalently bind to the cysteine at the dimer interface, further chemical optimization will be required to develop these as reversible binding small-molecule inhibitors. Although it might be possible to use Tethering with Extenders to generate more potent compounds, the site appears quite compact, which could make this approach difficult. Other fragment discovery approaches, such as SAR by NMR, could conceivably be used to obtain a noncovalent starting fragment. Nevertheless, a new mechanism of action, such as allosteric inhibition, may be one path to the discovery of potent nonpeptidic inhibitors.

9.4.2 ALLOSTERIC INHIBITION OF CASPASE-1

The conservation of Arg187 in the executioner and equivalent Arg286 in inflammatory caspases brings up the intriguing possibility that a conserved feature of the caspase family is a general mechanism for allosteric inhibition.[31] Caspase-1 was further investigated to explore whether such a mechanism existed in a much less conserved enzyme. Caspase sequences were aligned to determine if the cysteine at the dimer interface of caspase-7 was conserved in caspase-1. A straightforward sequence alignment suggested that a cysteine in this region of the protein was not conserved in the inflammatory caspases; however, an alignment of the three-dimensional structures of the enzyme showed that on the adjacent antiparallel β-strand there is a nearby cysteine. This cysteine, Cys331, is conserved in all human inflammatory caspases and is positioned such that modifications to the free thiol by small molecules could potentially exert a similar impact on enzyme activity. Indeed, from a screen of 8,000 disulfide-containing fragments, 14 compounds were identified that specifically interacted with Cys331. When covalently bound to caspase-1, the compounds inhibited activity in a competitive manner, as was observed with compounds **5** and **6** in caspase-3/7, and the inhibitory effect was dependent upon the concentration of reducing agents since increased concentrations of β-mercaptoethanol fully reversed the inhibitory effect. Structural and mutational studies revealed a mechanism of action of the caspase-1 allosteric inhibitors that is strongly dependent upon the very same arginine (Arg286) that is adjacent to the catalytic cysteine.

The x-ray structure of one of these compounds (compound **7**, Table 9.1) shows a pocket formed around two of the molecules by contributions from side chains on both sides of the dimer (see reference Scheer et al.[46]). Both molecules traverse the dimer interface in an orientation that is similar to the structure of compound **6** bound to caspase-7. Compound **7** interacts through a hydrogen bond between the amide of the linker portion of the molecule and Glu390 located at the dimer interface. This residue interacts with Arg286 through a salt bridge in the active structure of the enzyme. Thus, the repositioning of this critical arginine also appears important in the inflammatory caspases.

A structure-function analysis revealed a critical salt bridge formed between Arg286 and Glu390 during formation of the active conformation of the enzyme.

It is interesting to note that this salt bridge linking the active site to the central cavity is completely conserved among all inflammatory caspase homologs. Furthermore, the structures reveal that Glu390 in one monomer is coordinating the position of a water molecule directly at the dimer interface in cooperation with the symmetrical glutamic acid. Thus, there appears to be some connectivity between the two active sites that span the dimer interface via these two glutamates. Kinetic analysis of enzyme activity revealed a high degree of positive cooperativity in caspase-1 that is linked through this interaction, which supports the original hypothesis of allostery predicted by the first structural study of the enzyme.[42] It is possible that other naturally occurring molecules, whether proteins or smaller metabolites, might interact with this site.

The structure of the caspase-1/compound **7** complex exhibits striking similarity to the inactive, ligand-free conformation. When compared to the inactive conformation of caspase-1 adopted when the enzyme is not in complex with any ligand, there is a distinct conservation of the loop orientations and disruption of the interaction between Arg286 and Glu390. The authors suggest that these compounds are in essence trapping the enzyme in an inactive state. In a broader analysis, this inactive state is similar to the loop positions of the zymogen structure of caspase-7 and the structure of caspase-7 in complex with allosteric compounds **5** and **6**. The one difference between the caspase-7 and caspase-1 allosteric structures is that the L2' loop does not leave the active site in caspase-1. This provides the caspase-1 enzyme with a more exposed binding pocket at the dimer interface and the potential for accommodating larger, more potent small-molecule inhibitors. It seems as though there is a general mechanism of allosteric regulation that is inherent in the conformational changes that the enzyme undergoes as it becomes active. It is possible, although with no direct physiological evidence known, that this type of feature could allow for additional cellular regulation.

9.4.3 OTHER POSSIBLE SMALL-MOLECULE ALLOSTERIC INHIBITORS

Scientists from ChemDiv, Inc., have reported the identification of nonpeptidic inhibitors of caspase-3.[11] The authors screened a set of 15,000 compounds chosen from a larger library of 650,000 compounds. The sublibrary was enriched for molecules that showed properties expected for a thiol endopeptidase inhibitor. The effort combined elements of naive screening and rational drug design. The approach used to select this subset, called focused diversity, was based on two compound criteria: warhead electrophiles that would form reversible covalent bonds with cysteines, and diverse heterocyclic cores. A brief review of the subject is presented in Bamford et al.[47] The importance of a reversible warhead is highlighted in the context of achieving specificity in this enzyme class through the on/off functionality. Eleven scaffolds were identified in this screen, one of which was described as a sulfonamide-containing pyrrolo-quinoline (SPQ) scaffold, with two representative chemotypes presented in the report. This scaffold contained two electrophilic carbonyls (compound **8**, Figure 9.7) and was unique in this regard when compared to other caspase inhibitors. The potency of this compound and its derivatives ranged generally from 130 to 350 nM. The class was reversible and, interestingly, displayed noncompetitive binding. This led the authors to speculate that the compounds were likely acting through

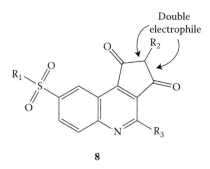

FIGURE 9.7 Compound 8.

other non-active-site cysteines, including Cys244, located very near the active site. The selectivity of the compound was for enzymes of the group-I and group-II caspases and, to a small degree, for caspase-6, which all present a conserved free thiol near the dimer interface. If indeed acting allosterically, it could be reacting with this well-characterized thiol.

Since it is difficult to predict a compound-binding site without structural data for the complex, mass spectrometry analysis can be particularly useful in deconvoluting a binding site when the interaction is covalent. Compounds reported to inhibit caspase-1 have been verified by mass spectrometry to bind at the active-site cysteine, the cysteine in the central cavity at the dimer interface, as with compound **6**, and on solvent-exposed surface cysteines observed with auranofin and gold-thiomalate (Table 9.1). Additionally, thimerosal was reported to be an uncompetitive inhibitor able to modify caspase-1 at Cys244 located very near the active-site groove.[42] Another molecule, disulfiram, was identified as an inhibitor of caspase-1 and caspase-3 and described as an active-site inhibitor, but with the possibility of additional sites of interaction.[48] Differences in sensitivity to disulfiram between caspase-1 and caspase-3, despite the highly conserved active-site sequence, suggested the latter possibility. Since both caspase-1 and caspase-3 contain several inactivating free thiols, mass spectrometry of these complexes coupled with either a peptide digest or systematic mutation of cysteines could definitively reveal the sites of interaction.

9.5 MALONATE ISOSTERES

In the structural study reporting the ligand-free form of caspase-1, the structure of the enzyme bound to the small molecule malonate was also reported.[41] This small, dicarboxylic acid–containing compound was observed bound to the active site of the enzyme in precisely the same location as the aspartic acid of peptide-mimetic substrates (Figure 9.8). Two significant conclusions drawn from this work are that noncovalent compounds are capable of stably binding to the active site and that binding to the S1 subsite may be the critical seed for converting the enzyme to the active conformation. It was postulated by the authors that the presence of a small ligand like malonate at the active site at high concentrations might act as a weak inhibitor and prevent a commonly observed phenomenon of autolysis that occurs at

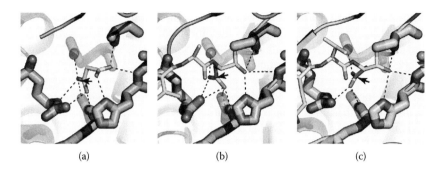

(a)	(b)	(c)

FIGURE 9.8 (See color insert.) Comparison of malonate, Ac-YVAD-CHO, and Ac-DEVD-CHO. Arrows indicate the carboxylate carbon. The structure of malonate is shown in yellow sticks bound noncovalently to the active site of caspase-1 in panel a. The hydrogen bonding interactions of one carboxylate of malonate mimic those of the side chain carboxylate of aspartic acid seen in panel b (Ac-YVAD-CHO, green sticks, bound to caspase-1) and panel c (Ac-DEVD-CHO, gray sticks, bound to caspase-7). The carbon of this carboxylate in the three inhibitors is indicated with an arrow. The second carboxylate of malonate provides additional hydrogen bond acceptors ~3 atoms away from the carbon of the first carboxylic acid. This second carboxylate is positioned in the oxyanion hole-like region, similar to the carbonyl of the aldehyde from the peptidic inhibitors. These structural features were used to select and screen for noncovalent weak inhibitors as structural mimics of malonate. PDB IDs for panels a, b, and c are 1SC3, 1ICE, and 1F1J, respectively.

room temperature and higher concentrations of enzyme. Indeed, in the presence of malonate and similar ligands such as aspartic acid, the enzyme was found to resist autolysis. This finding allowed for the development of a very rapid and simple assay for screening a set of malonate mimics for inhibitory activity that could serve essentially as new S1 constituents. A simple screen was devised that surveyed these types of compounds. To briefly describe the assay, the authors began with guanidine hydro-chloride–solubilized small and large subunits of caspase-1 that were expressed separately in *E. coli* as insoluble inclusion bodies. The guanidine-solubilized subunits were renatured together by dilution in a renaturation buffer containing 100 mM of the malonate mimetic. After room temperature incubation, samples were clarified by centrifugation and concentrated to about one-tenth the starting volume. The concentration step promoted the autolytic event and increased the time efficiency of the assay. Qualitative analysis of the protection assay on SDS-PAGE revealed either fully intact large and small subunits in the presence of a weak inhibitor like malonate, or the presence of an autolytic fragment mixture. In particular, the presence or absence of a common proteolytic product, a 7 kDa protein produced by cleavage at Asp381, was monitored.

A set of commercially available compounds was obtained with a similar structure to malonate and was analyzed in the assay. Of the compounds screened, five were identified that fully protected the enzyme from autolysis in the assay. The critical feature of these compounds was a carboxylic acid group that could mimic the carboxylate of the side chain of aspartic acid and an additional hydrogen bond acceptor located three or four atoms away from the carbon of the carboxylate. This additional hydrogen bond acceptor is hypothesized to form a hydrogen bond with elements of the protein that are normally associated with stabilizing the aldehyde oxygen in the transition-state complex.

In many ways this is a very canonical fragment identification screen. It is unlikely that any of these compounds would have been identified in a standard activity screen because of the difficult challenge of inhibiting high-affinity substrates such as the tetrapeptide substrates. It could be that a much lower K_m substrate would serve better in these types of screens. In fact, none of these compounds, including malonate, even at high (up to 100 mM) compound concentrations, inhibited the enzyme in a fluorescent-based assay using Ac-YVAD-AFC or Ac-WEHD-AFC.[41,49] Interestingly, succinate was not an inhibitor in the renaturation assay, even though it is structurally similar to aspartic acid, which is a good inhibitor in the assay. The absence of the amino group in succinate suggests that the amide in aspartic acid may help position the H bond acceptor of the carboxylate group that is not bound deep in the S1 subsite. Such a positioning element may be necessary in malonate isosteres that have a second H bond acceptor that is more than three carbons away from the side chain carboxylate.

This renaturation screening approach could help in identifying new noncovalent inhibitors of caspase-1. However, it is not clear if malonate works by inhibiting the enzyme in true competitive fashion or by promoting a structure that becomes more resistant to proteolysis, i.e., a refolding of the exposed loops into the active site might decrease the K_M of these loops as substrate. The common cleavage site observed in autolysis that yields the 7 kDa fragment is Asp381, which is on a loop adjacent to the active site. This loop region shows a much higher B-factor in the ligand-free structure, presumably because of its proximity to the destabilized active site, and this flexibility could increase the exposure of the loop to active enzyme, making it a more favorable substrate. In either case, this is a rapid and simple identifier for fragments that bind to the active site.

9.6 SUBITO

This approach, which is an abbreviation for structure-based iterative optimizations, includes a computational component coupled to an NMR-based functional screen.[50] The authors describe the use of this technique to quickly identify a small set of micromolar inhibitors from a very large set of commercially available compounds. The authors assembled a library of about 300 compounds that contain a unique drug scaffold. These compounds in turn represent a core scaffold for which extensive sets of closely related analogs are readily available. Thus, by screening only 300 compounds, a quick survey of diversity can be made and followed up through analysis of the readily available related compounds. Since the compounds are likely to be weak inhibitors, the authors included an NMR-based functional screen to allow high compound concentrations since conventional spectrophotometric assays at high concentrations are typically associated with problems such as compound fluorescence and compound aggregation effects. The screen was a ^{19}F NMR spectroscopy assay that monitors the cleavage of commonly used tetrapeptide-AFC (7-amino-4-trifluoromethyl-coumarin) substrates by observing the chemical shift of the trifluoromethyl group on coumarin after hydrolysis.[13,51,52] This screening approach has been described for several other enzymes, including trypsin and the serine/threonine kinase AKT1, and avoids the limitations typically associated with NMR: very high substrate, enzyme, and compound concentrations. Detection of the trifluoro group is sensitive enough to permit screening at concentrations of enzyme and substrate that are comparable to those used in standard high-throughput assays.[2,39,53] Authors of the initial report of this method describe using as low as single-digit nanomolar enzyme concentrations. One important detail is that the total volume of the assay must be about 500 μl for standard NMR, which can be reduced slightly with a flow-injection probe. The acquisition time is about 3 minutes, so the assay can be completed rapidly, but the throughput power is reduced compared to a plate-reader-based assay since only single samples can be simultaneously measured. The advantage is the identification of weak inhibitors, and possibly even fragments.

Caspase-8 (25–50 nM) was screened against their 300-compound library using 40 μM substrate (Ac-IETD-AFC). A benzodioxane-containing compound (BI-7E7) was identified by the screen and exhibited a micromolar IC_{50} (compound **9,** Figure 9.9). This compound bound reversibly and was used as a scaffold for the second step in the process of using computational methods to determine potential SAR from the ~500 compounds in the scaffold family. A docking analysis with these ~500 compounds was performed to identify analogs that would mimic the BI-7E7 benzodioxane moiety and alter the thiazole ring. This benzodioxane portion is postulated from docking experiments to bind to the S1 subsite similar to where the aspartic acid carboxylic side chain of the IETD-based inhibitor of caspase-8 is found. Ten compounds were identified as possible inhibitors and were tested in a standard fluorescence-based assay against caspase-8, and also the executioner caspases-3 and -7. One compound, BI-9B12, showed about ten-fold greater activity than the starting compound, with an IC_{50} around 10 μM, but with nonselectivity over caspases-3 and -7. Modeling suggested that the benzodioxane portion was binding in the S1 subsite, but it was also possible that these compounds were acting independent of the standard acidic P1 element seen in most caspase inhibitors. With no x-ray structure available, the authors left open the possibility that these compounds were not necessarily active-site inhibitors, and that they might act through an allosteric mechanism. Again, this is an example of a different screening approach that identified a new class of inhibitor that lacks an aspartic acid moiety.

Possible interaction of BI-7E7
with caspase-8 S1 subsite

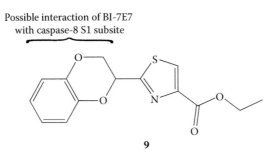

9

FIGURE 9.9 Compound 9.

9.7 CONCLUSION

Many new technologies have been pioneered to increase the repertoire of approaches to discover small-molecule inhibitors of protein targets. Fragment-based screens employing site-directed methods such as NMR, x-ray crystallography, and Tethering have been very powerful techniques that allowed the identification of new binding elements. For example, P4 replacements by modular substitution in Tethering with Extenders rapidly optimized the potency of inhibitors of caspase-1. In other cases, new binding sites have been identified that provide insight into the regulation and structure of the enzymes. These approaches complement the existing work on active-site inhibitors and also open entirely new avenues for discovery. The dynamic structure of the caspases provides an ideal model for the development of these and other exciting tools in small-molecule and protein research. Since no caspase-targeted therapeutic has yet been marketed, one of these methods may help contribute to the successful discovery of such compounds.

REFERENCES

1. O'Brien, T., and Lee, D. 2004. Prospects for caspase inhibitors. *Mini Rev Med Chem* 4:153–65.
2. Thornberry, N. A., Chapman, K. T., and Nicholson, D. W. 2000. Determination of caspase specificities using a peptide combinatorial library. *Methods Enzymol* 322:100–10.
3. Lee, D., et al. 1999. Substrate combinatorial array for caspases. *Bioorg Med Chem Lett* 9:1667–72.
4. Rano, T. A., et al. 1997. A combinatorial approach for determining protease specificities: Application to interleukin-1β converting enzyme. *Chem Biol* 4:149–55.
5. Thornberry, N. A., et al. 1997. A combinatorial approach defines specificities of members of the caspase family and granzyme B. *J Biol Chem* 272:17907–11.
6. Timmer, J. C., and Salvesen, G. S. 2007. Caspase substrates. *Cell Death Differ* 14:66–72.
7. Linton, S. D. 2005. Caspase inhibitors: A pharmaceutical industry perspective. *Curr Top Med Chem* 5:1697–717.
8. Lee, D., et al. 2001. Potent and selective nonpeptidic inhibitors of caspases 3 and 7. *J Med Chem* 44:2015–26.
9. Lee, D., et al. 2000. Potent and selective nonpeptide inhibitors of caspases-3 and -7 inhibit apoptosis and maintain cell functionality. *J Biol Chem* 275:16007–14.
10. Eun-sook Kim, S.-E. Y. et al. 2002. Design, synthesis and biological evaluations of nonpeptidic caspase 3 inhibitors. *Beull Kor Chem Soc* 23:7.
11. Okun, I., et al. 2006. Screening for caspase-3 inhibitors: A new class of potent small-molecule inhibitors of caspase-3. *J Biomol Screen* 11:277–85.
12. Smith, G. K., et al. 2002. Expression, preparation, and high-throughput screening of caspase-8: Discovery of redox-based and steroid diacid inhibition. *Arch Biochem Biophys* 399:195–205.
13. Erlanson, D. A., McDowell, R. S., and O'Brien, T. 2004. Fragment-based drug discovery. *J Med Chem* 47:3463–82.
14. Rees, D. C., et al. 2004. Fragment-based lead discovery. *Nat Rev Drug Discov* 3:660–72.
15. Hopkins, A. L., Groom, C. R., and Alex, A. 2004. Ligand efficiency: A useful metric for lead selection. *Drug Discov Today* 9:430–31.

16. Hajduk, P. J., and Greer, J. 2007. A decade of fragment-based drug design: Strategic advances and lessons learned. *Nat Rev Drug Discov* 6:211–19.

17. Babaoglu, K., and Shoichet, B. K. 2006. Deconstructing fragment-based inhibitor discovery. *Nat Chem Biol* 2:720–23.

18. O'Brien, T., et al. 2005. Structural analysis of caspase-1 inhibitors derived from Tethering. *Acta Crystallograph Sect F Struct Biol Cryst Commun* 61:451–58.

19. He, M. M., et al. 2005. Small-molecule inhibition of TNF-alpha. *Science* 310:1022–25.

20. Fahr, B. T., et al. 2006. Tethering identifies fragment that yields potent inhibitors of human caspase-1. *Bioorg Med Chem Lett* 16:559–62.

21. Hansen, S. K., et al. 2005. Allosteric inhibition of PTP1B activity by selective modification of a non-active site cysteine residue. *Biochemistry* 44:7704–12.

22. Erlanson, D. A., Wells, J. A., and Braisted, A. C. 2004. Tethering: Fragment-based drug discovery. *Annu Rev Biophys Biomol Struct* 33:199–223.

23. Erlanson, D. A., et al. 2003. Discovery of a new phosphotyrosine mimetic for PTP1B using breakaway tethering. *J Am Chem Soc* 125:5602–3.

24. Braisted, A. C., et al. 2003. Discovery of a potent small molecule IL-2 inhibitor through fragment assembly. *J Am Chem Soc* 125:3714–15.

25. Erlanson, D., et al. 2003. *In situ* assembly of enzyme inhibitors using extended tethering. *Nature Biotechnol* 21:308–14.

26. Choong, I. C., et al. 2002. Identification of potent and selective small-molecule inhibitors of caspase-3 through the use of extended tethering and structure-based drug design. *J Med Chem* 45:5005–22.

27. Vertex Pharmaceuticals, Inc. 2000. Inhibitors of interleukin-1b converting enzyme. U.S. Patent 6,103,711.

28. Allen, D. A., et al. 2003. Identification of potent and novel small-molecule inhibitors of caspase-3. *Bioorg Med Chem Lett* 13:3651–55.

29. Schindler, T., et al. 2000. Structural mechanism for STI-571 inhibition of abelson tyrosine kinase. *Science* 289:1938–42.

30. Hardy, J. A., and Wells, J. A. 2004. Searching for new allosteric sites in enzymes. *Curr Opin Struct Biol* 14:706–15.

31. Dennis, M. S., et al. 2000. Peptide exosite inhibitors of factor VIIa as anticoagulants. *Nature* 404:465–70.

32. Nagar, B., et al. 2002. Crystal structures of the kinase domain of c-Abl in complex with the small molecule inhibitors PD173955 and imatinib (STI-571). *Cancer Res* 62:4236–43.

33. Hardy, J. A., et al. 2004. Discovery of an allosteric site in the caspases. *Proc Natl Acad Sci USA* 101:12461–66.

34. Stennicke, H. R., et al. 1999. Caspase-9 can be activated without proteolytic processing. *J Biol Chem* 274:8359–62.

35. Riedl, S. J., et al. 2001. Structural basis for the activation of human procaspase-7. *Proc Natl Acad Sci USA* 98:14790–95.

36. Boatright, K. M., and Salvesen, G. S. 2003. Mechanisms of caspase activation. *Curr Opin Cell Biol* 15:725–31.

37. Shi, Y. 2004. Caspase activation, inhibition, and reactivation: A mechanistic view. *Protein Sci* 13:1979–87.

38. Shi, Y. 2004. Caspase activation: Revisiting the induced proximity model. *Cell* 117:855–8.

39. Garcia-Calvo, M., et al. 1999. Purification and catalytic properties of human caspase family members. *Cell Death Differ* 6:362–9.

40. Chai, J., et al. 2001. Crystal structure of a procaspase-7 zymogen: Mechanisms of activation and substrate binding. *Cell* 107:399–407.

41. Romanowski, M. J., et al. 2004. Crystal structures of a ligand-free and malonate-bound human caspase-1: Implications for the mechanism of substrate binding. *Structure (Camb)* 12:1361–71.

42. Wilson, K., et al. 1994. Structure and mechanism of interleukin-1β converting-enzyme. *Nature* 370:270–75.

43. Changeux, J. P., and Edelstein, S. J. 2005. Allosteric mechanisms of signal transduction. *Science* 308:1424–28.

44. Fuentes-Prior, P., and Salvesen, G. S. 2004. The protein structures that shape caspase activity, specificity, activation and inhibition. *Biochem J* 384:201–32.

45. Shiozaki, E. N., et al. 2003. Mechanism of XIAP-mediated inhibition of caspase-9. *Mol Cell* 11:519–27.

46. Scheer, J. M., Romanowski, M. J., and Wells, J. A. 2006. A common allosteric site and mechanism in caspases. *Proc Natl Acad Sci USA* 103:7595–600.

47. Bamford, M., Walkinshaw, G., and Brown, R. 2000. Therapeutic applications of apoptosis research. *Exp Cell Res* 256:1–11.

48. Nobel, C. S., et al. 1997. Disulfiram is a potent inhibitor of proteases of the caspase family. *Chem Res Toxicol* 10:1319–24.

49. Scheer, J. M., Wells, J. A., and Romanowski, M. J. 2005. Malonate-assisted purification of human caspases. *Protein Expr Purif* 41:148–53.

50. Fattorusso, R., et al. 2005. Discovery of a novel class of reversible non-peptide caspase inhibitors via a structure-based approach. *J Med Chem* 48:1649–56.

51. Maly, D. J., Choong, I. C., and Ellman, J. A. 2000. Combinatorial target-guided ligand assembly: Identification of potent subtype-selective c-Src inhibitors. *Proc Natl Acad Sci USA* 97:2419–24.

52. Dalvit, C., et al. 2003. A general NMR method for rapid, efficient, and reliable biochemical screening. *J Am Chem Soc* 125:14620–25.

53. Stennicke, H. R., and Salvesen, G. S. 2000. Caspase assays. *Methods Enzymol* 322:91–100.

10 Therapeutic Potential for Caspase Inhibitors
Present and Future

Hiroyuki Eda

CONTENTS

10.1 INTRODUCTION

Caspases, a family of cysteine proteases, play crucial roles in both maturation of inflammatory cytokines and apoptosis. After the caspase family was discovered, caspase activation was revealed to contribute to large numbers of pathological conditions, while caspase inhibition has been demonstrated to have enormous therapeutic potential in inflammatory diseases and apoptosis-driven disorders. One extremely active area of drug discovery and development targeting apoptosis is the identification of small molecule caspase inhibitors. A number of caspase inhibitors have demonstrated efficacy in a large variety of animal models, and several therapeutics have already progressed into clinical trials such as a caspase-1-specific inhibitor for the treatment of inflammatory diseases, and a pancaspase inhibitor for the treatment of liver diseases (liver preservation injury in liver transplantation and chronic liver disease, i.e., chronic hepatitis C). This chapter will review the present status and future potential of caspase inhibitors and explore the highlights of several caspase inhibitors currently in clinical trials. Focus will be attributed to apoptosis in liver diseases and discovery programs of pancaspase inhibitors against liver diseases. Lastly, this article will discuss the challenges associated with caspase inhibitors, especially mechanism-related toxicity associated with apoptotic caspase inhibition.

This chapter will describe the present status and future potential of the small molecule caspase inhibitors under clinical investigation. It will also address challenges facing the development of caspase inhibitors, especially long-term toxicity associated with caspase inhibition.

10.2 OVERVIEW OF CASPASES

Caspases (cysteinyl aspartate-specific proteases) are classified into two groups based on their biological function. The first group (caspase-1, -4, -5, -11, and -12) is responsible for the activation of the pro-inflammatory cytokines. The second (caspase-2, -3, -6, -7, -8, -9, -10, and -14) is responsible for apoptosis. Caspase-1, a member of group 1 caspases, which are primarily responsible for the cleavage of the inactive precursor of interleukin-1β (IL-1β) and interleukin-18 (IL-18), was initially designated IL-1β-converting enzyme (ICE). Since IL-1β and IL-18 contribute to inflammatory disorders in many organ systems,[1,2] caspase-1 inhibition therefore represents a promising target for a wide range of inflammatory diseases, such as rheumatoid arthritis, osteoarthritis, colitis, and psoriasis. Two caspase-1 inhibitors, VX-740 (pralnacasan) and VX-765 (VRT-043198), have progressed to clinical trials for the treatment of inflammatory disease.

Inhibition of apoptosis by inhibition of apoptosis-regulating caspases is an intriguing target for many disorders where inappropriate apoptosis occurs. These diseases include neurodegenerative disorders, such as Alzheimer's disease[3–5] and Parkinson's disease,[6,7] ischemic disorders,[8,9] sepsis,[10,11] and liver diseases.[12–24] A number of preclinical experiments have demonstrated the efficacy of caspase inhibitors against these types of diseases both *in vitro* and *in vivo*. For instance, pancaspase inhibitors are effective in animal models of traumatic brain injury,[25] neurodegenerative disorder,[26] cardiac dysfunction,[27,28] septic shock,[29] ischemia-reperfusion injury,[30–32] and liver diseases.[33–38]

Excessive apoptosis of hepatocytes is a distinguishing feature in all liver disease, such as viral hepatitis,[17,20–22] cholestatic liver disease,[12] alcoholic liver disease (ALD),[18,19,24] ischemia-reperfusion liver injury in transplantation surgery,[14–16] and nonalcoholic steatohepatitis (NASH).[23,24] In these liver diseases, apoptosis of hepatocytes is mediated via extrinsic or intrinsic pathways of apoptosis. The extrinsic pathway is triggered by death receptors such as CD95 and TNF-related apoptosis-inducing ligand (TRAIL) receptors. On the other hand, the intrinsic pathway is initiated from mitochondria by various triggers, such as reactive oxygen species (ROS). Both apoptotic pathways are intertwined in a complex fashion in promotion of liver diseases, although the intrinsic apoptotic pathway is more predominant in some liver diseases, whereas the extrinsic apoptosis pathway is more predominant in others. Therefore, liver disease could potentially benefit from a strategy that inhibits apoptotic caspases. Indeed, as described above, caspase inhibitors have shown to be effective against these liver diseases in preclinical animal models.[33–38] Currently, small molecule pancaspase inhibitors, emricasan (IDN-6556/PF-03491390) and LB84451, have been studied in clinical trials for the treatment of liver diseases (ischemia-reperfusion injury in liver transplantation and chronic liver diseases).[39,40]

Although the development of caspase inhibitors has opened up a new treatment modality in inflammatory and apoptosis-driven disorders, and some caspase inhibitors have shown promising results in early clinical trials, several issues critical for the development of caspase inhibitors remain to be resolved. Preclinical studies have revealed that although caspase inhibition is undoubtedly effective in a large variety of disorders, long-term inhibition of caspases can potentially cause autoimmune disorders or cancer. Furthermore, it is still unknown whether caspases have a physiological role besides the maturation of inflammatory cytokines and initiation and execution of apoptosis. Although clinical evidence has revealed that caspase inhibition is safe and tolerable in the short term,[40] long-term inhibition of caspases could lead to unexpected consequences.

10.3 CASPASE-1 INHIBITORS

The role of caspase-1 has been characterized in *in vivo* animal models for inflammation. For example, monocytes derived from mice harboring a disrupted caspase-1 gene reduced the production of pro-inflammatory cytokines IL-1β and IL-18 after stimulation with lipopolysaccharide (LPS), whereas thymocytes and macrophages underwent apoptosis normally upon *in vitro* stimulation by several different signals.[41] Furthermore, caspase-1-deficient mice were resistant to endotoxic shock.[41,42] This indicates that a small molecule inhibitor of caspase-1 would be a promising candidate for inflammatory disorders such as rheumatoid arthritis, osteoarthritis, colitis, and psoriasis.

10.3.1 VX-740 (PRALNACASAN)

Vertex Pharmaceuticals developed VX-740 (pralnacasan), a reversible caspase-1 inhibitor for the treatment of inflammatory diseases, especially osteoarthritis (OA) and rheumatoid arthritis (RA).[43] Pralnacasan was optimized by structure-based

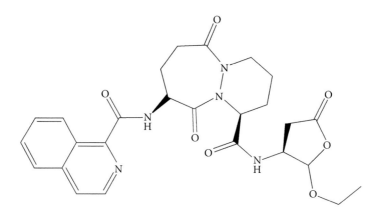

FIGURE 10.1 Chemical structure of VX-740/pralnacasan.

design starting with a chemical series in-licensed via a Sterling Winthrop patent estate. The structure of pralnacasan is shown in Figure 10.1. Preclinical studies showed that pralnacasan exhibited caspase-1-specific inhibitory activity with an IC_{50} of 1.3 nM, compared to 2.3 μM for caspase-3 and 0.12 μM for caspase-8.[44] In pre-clinical studies, pralnacasan inhibited type II collagen-induced arthritis in mice, and prophylactic treatment of pralnacasan (50 and 100 mg/kg, bid) delayed the onset of forepaw inflammation and reduced disease severity by up to 70%. Also, pralnacasan (100 mg/kg) inhibited the production of serum IL-1β by up to 80% in mice induced by LPS, and inhibited carageenan-induced increase of IL-1β in peritoneal exudates in mice by up to 75%.[44] Furthermore, pralnacasan reduced forepaw inflamma-tion when administered to mice with established arthritis (therapeutic treatment)[44] and showed anti-inflammatory activity in two OA models (collagenase-induced and STR/1N murine spontaneous OA model) and a colitis model (dextran sulfate sodium–induced colitis model).[45,46] In OA models, oral or food-drug mixture treat-ment of pralnacasan significantly reduced joint damage as indicated by histopatho-logical score of OA or by urinary levels of hydroxylysylpyridinoline (HP) cross-links and the ratio of hydroxylysylpyridinoline/lysylpyridinoline (HP/LP). HP and LP are trifunctional cross-links joining three adjacent collagen molecules, and the HP/LP ratio predominantly reflects cartilage degradation. In an experimental colitis model, intraperitoneal administration of pralnacasan reduced the clinical score of colitis and also reduced the expression of intracolonic IL-18. A long-term toxicity study revealed that pralnacasan showed no dose-limiting toxicity up to 6 months in rats or dogs.[43]

Phase I clinical studies demonstrated that pralnacasan was well tolerated and had an oral bioavailability of 50%.[43] A phase II clinical study was conducted in a 12-week, multicenter, double-blinded, randomized, placebo-controlled trial.[47] In this trial, 285 patients with RA received pralnacasan (100 or 400 mg tid, orally) or a placebo (tid) for at least 12 weeks. Although statistical significance was not achieved, there was a dose-dependent improvement in signs and symptoms of the disease, which were measured using the American College of Rheumatology 20

(ACR20) response rates in patients who received pralnacasan. A significant reduction of key biomarkers (C-reactive protein [CRP], erythrocyte sedimentation rate [ESR], serum amyloid A [SAA], matrix metalloproteinase-1 [MMP-1], and tissue inhibitor of metalloproteinases-1 [TIMP-1]) was observed in patients who received 400 mg tid of pralnacasan ($p < .05$ vs. placebo). An increase in ACR20 responses were observed when patients on methotrexate (MTX) for ≤6 months were excluded from the analysis, as well as in subsets of patients who received either concomitant stable (>6 months) MTX therapy or no MTX therapy. Adverse events were similarly distributed among the treatment group, with the most common adverse events being mild to moderate diarrhea and nausea, which was only seen in <5% of the study population. This trial demonstrated that pralnacasan was well tolerated and showed significant anti-inflammatory effects, indicating pralnacasan was a novel, orally active, anti-RA agent with no significant adverse effects. Phase II clinical trials of pralnacasan for OA were subsequently conducted,[43] but results of this clinical trial have not yet been made available.

Unfortunately, pralnacasan induced liver toxicity in animal studies with long-term treatment and was recently withdrawn from clinical trials.[48] According to the annual report from Vertex issued in 2006, Vertex suspended the clinical development of pralnacasan because of the toxicity in nonclinical animal studies. In their 9-month toxicology study, high doses of pralnacasan were associated with the development of fibrosis in circumscribed areas of the liver of one species of animal (Vertex Pharmaceuticals, Inc., annual report, 2006, http://investor.shareholder.com/vrtx/secfiling.cfm?filingID=1047469-06-3585). The adverse effect observed in a nonclinical 9-month toxicology study may not be attributable to the long-term inhibition of caspase-1, as caspase-1 knockout mice are overtly normal (see below). Also, no significant adverse events associated with liver toxicity have been reported in patients who were enrolled in clinical trials with pralnacasan.

10.3.2 VX-765

VX-765 is an orally active reversible caspase-1 inhibitor for inflammation developed by Vertex Pharmaceuticals (structure shown in Figure 10.2). *In vitro* pharmacological studies demonstrated that VX-765 is more potent than pralnacasan for inhibition of LPS-induced IL-1β release and *Staphylococcus aureus*–induced IL-1β and IL-18

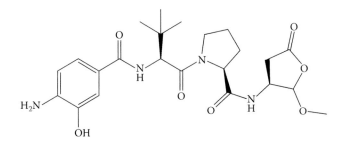

FIGURE 10.2 Chemical structure of VX-765.

TABLE 10.1
In Vitro **Profile of Pralnacasan and VX-765**

| | Inhibitory Concentration (IC$_{50}$) Value | |
	Pralnacasan	VX-765
LPS-induced IL-1β release	850 nM	470 nM
Staphylococcus aureus–induced IL-1β release	3.7 μM	0.84 μM
Staphylococcus aureus–induced IL-18 release	6.6 μM	2.5 μM

release (Table 10.1).[49–52] In animal models, VX-765 exhibited efficacy in a colla-genase-induced arthritis model, a STR/1N spontaneous OA model, and an oxazo-lone-induced allergic dermatitis model. In an oxazolone-induced allergic dermatitis model, VX-765 at an oral dose of 10–100 mg/kg bid dose-dependently reduced ear edema (up to 75% reduction). Also, significant reduction of thickening and cel-lular infiltration was observed in both the dermis and epidermis in mice treated with VX-765. Tissue levels of inflammatory and allergic cytokines and chemokines were reduced (74–99%), as well as levels of myeloperoxidase and nitric oxide.[49–52] Recently, VX-765 was reported to be effective in an *ex vivo* model for familial cold autoinflammatory syndrome (FCAS),[53] which is an autosomal dominant inherited inflammatory disorder with episodic symptoms, including rash, fever, and joint pain caused by exposure to cold. The pathological mechanism of FCAS is hypoth-esized to be driven by dysregulation of the pro-inflammatory cytokines, IL-1β and IL-18, because of a mutation in the CIAS1 gene encoding cryopyrin that is involved in the activation of caspase-1. The mutation in cryopyrin may result in hyperre-sponsiveness of IL-1β and IL-18 production to inflammatory stimuli, e.g., LPS stimulation. VX-765 blocked IL-1β secretion with equal potency in LPS-stimulated peripheral blood mononuclear cells (PBMCs) isolated from patients with FCAS and healthy subjects, indicating the clinical potential of VX-765 for autoinflammatory disorders.

Phase I clinical single- and multiple-dosing trials in healthy volunteers demon-strated a dose-dependent reduction of IL-18 level in plasma (Vertex Pharmaceuti-cals, Inc., annual report, 2006). Vertex has completed dosing in a 4-week, phase IIa safety and pharmacokinetic study of VX-765 in sixty-eight patients with psoriasis (Vertex press release, October 4, 2005, http://www.vpharm.com/Pressreleases2005/pr100405p.html). These results appear to support further clinical study; Vertex reported in 2006 that it may consider entering into a collaborative arrangement to advance clinical trials of VX-765.

10.3.3 Summary

Many studies both *in vitro* and *in vivo* have demonstrated that inhibition of cas-pase-1 reduces circulating levels of the pro-inflammatory cytokines, IL-1β and IL-18, and improves symptoms of inflammation in animal models. Clinical trials of pralnacasan and VX-765 have shown that orally available small molecule caspase-1 inhibitors are effective in patients with RA, OA, and psoriasis. Other preclinical investigations have also demonstrated that caspase-1 inhibitors are effective in a large variety of disease models, including inflammatory bowel disease (IBD)[46] and ischemic disorders of the brain and heart.[54] Caspase-1 knockout mice present with an almost complete absence of colitis induced by chronic treatment of dextran sulfate sodium,[55] indicating that caspase-1 inhibitors could ameliorate the symptoms of IBD in humans. Indeed, pralnacasan demonstrated efficacy in a rodent dextran sulfate sodium-induced colitis model.[46] Also, caspase-1-deficient mice were studied in a model of permanent cerebral ischemia by middle cerebral artery occlusion (MCAO). Brain edema at 4 hours after MCAO and a histologically defined brain lesion at 24 hours after MCAO were significantly reduced in caspase-1-deficient mice.[56] Furthermore, treatment with a caspase-1 inhibitor by continuous intracerebroventricular infusion to mice of Huntington's disease model slowed disease progression and resulted in a 25% increase in survival compared to controls.[57] Thus, although pralnacasan and VX-765 are the only compounds that have been evaluated in clinical studies for OA, RA, and psoriasis, small molecule caspase-1 inhibitors could be very beneficial as therapeutic agents for a large variety of diseases.

10.4 APOPTOSIS-REGULATING CASPASE INHIBITORS

Apoptosis is essential for many *in vivo* processes, such as development, regulation of the immune system, and maintenance of homeostasis. Dysregulation of apoptosis leads to a number of pathological conditions. Many of the key players have been identified, and some have been targeted by therapeutic strategies, including death receptors, Bcl-2 family members, and caspases. Although apoptosis is a very complex mechanism and varies among cell types and apoptotic stimuli, caspases are the key mediators for apoptotic cell death, and thus excessive caspase-dependent apoptosis leads to a number of diseases, such as Alzheimer's disease,[3-5] Parkinson's disease,[6,7] ischemic disorders,[8,9] sepsis,[10,11] and liver diseases.[12-24] An early report that systemic injection of caspase inhibitors to mice challenged with anti-Fas antibody protected them from fulminant liver failure was one of the first to demonstrate that *in vivo* inhibition of caspase activity had potential therapeutic implications.[58] Since then there have been many reports describing the positive effects of small molecule caspase inhibitors in animal models of human diseases, including Alzheimer's disease,[25] Parkinson's disease,[26] ischemic disorders,[27,28] sepsis,[29] and liver diseases.[33-38]

10.4.1 APOPTOSIS IN LIVER DISEASE

10.4.1.1 HCV Hepatitis

Hepatocyte injury and resulting cell death are major features in all liver diseases, and excessive hepatocyte apoptosis leads to exacerbation of disease symptoms. Viral hepatitis is the most common cause of liver disease. Among the hepatitis viruses, hepatitis B virus (HBV) and hepatitis C virus (HCV) are the major causes of chronic hepatitis. Infection of HCV is characterized by necroinflammatory liver injury and long viral persistence associated with an increased risk of developing liver fibrosis. Progression of liver fibrosis results in cirrhosis, portal hypertension, liver failure, and hepatocellular carcinoma. Hepatocyte death is mainly mediated by host immune systems, indicating hepatocyte apoptosis in hepatitis C is regarded as a protective mechanism. Recent evidence has shown that death receptors, especially CD95, play a crucial role in hepatocyte apoptosis. In fact, upregulation of CD95 in hepatocytes and CD95 ligand (CD95L) in T lymphocytes is reported to correlate with disease severity.[59–61] In hepatitis C, the prevalence of Fas antigen expression in HCV antigen-positive patients was higher than in HCV antigen-negative patients.[59] Additionally, serum-soluble Fas antigen levels in chronic hepatitis C patients were significantly higher than found in healthy subjects,[62] and Fas expression correlated not only with hepatocyte apoptosis but also with liver fibrosis.[63] These conclusions would suggest that hepatocyte apoptosis is mainly mediated by CD95/CD95L in hepatitis C. The role of CD95/CD95L in hepatocyte apoptosis was also demonstrated in hepatitis B transgenic mice.[64,65] A cytotoxic T lymphocyte (CTL) specific to hepatitis B surface antigen (HBsAg) killed hepatocytes expressing HBsAg in a Fas-dependent manner, indicating CD95 is essential for the development of hepatitis B.[65]

A positive correlation between hepatocyte apoptosis and apoptotic caspase activation was found in HCV-infected liver tissues.[66] Activation of caspase-3 and caspase-7, as well as cleavage of poly ADP-ribose polymerase (PARP), which is a specific substrate for caspases, was elevated in the liver of hepatitis C patients compared to those in healthy controls. The extent of caspase activation correlated with necroinflammatory activity in patients with chronic hepatitis C, and a corresponding correlation between hepatocyte apoptosis and necroinflammatory response in the liver was demonstrated in animal models. [67,68] In animal models, activation of caspase-3 is critical for the development of hepatocyte apoptosis, which in turn is an important signal for transmigration of primed neutrophils sequestered in sinusoids, thus leading to a necroinflammatory response in the liver. Recently, caspase activity was detected in sera from more than 50% of patients with chronic hepatitis C with normal serum alanine aminotransferase (ALT) levels but histological evidence of disease,[69] suggesting that chronic hepatitis C is an apoptotic disease and anti-apoptotic treatment via inhibiting the caspases would be beneficial for the amelioration of liver damage caused by HCV-related hepatocyte apoptosis.

10.4.1.2   Cholestatic Liver Disease

Besides viral hepatitis, other liver diseases, such as cholestatic liver disease, ALD, and NASH, lead to excessive apoptosis activation. In cholestatic liver disease, failure of bile salt excretion leads to high concentrations of toxic bile salts in hepatocytes, which then induces hepatocyte apoptosis.[12] Specifically, in an animal model, glycochenodeoxycholate, a toxic bile salt, induces hepatocyte apoptosis via ligand-independent oligomerization of Fas and activation of caspase-8.[70] In an experimental model, TNF-related apoptosis-inducing ligand (TRAIL) was also found to play a role in toxic bile acid–induced hepatocyte apoptosis,[71,72] including activation of caspase-8 and caspase-10.

10.4.1.3   Alcoholic Liver Disease (ALD)

Although the pathogenesis of alcoholic hepatitis (AH) remains poorly understood, hepatocyte apoptosis was significantly increased in alcoholic hepatitis.[18,19,24] Liver specimens from patients with AH revealed that caspase-3-positive hepatocytes were readily observed, and that hepatocyte apoptosis was significantly higher in patients with severe AH. Also, Fas receptor was strongly expressed in AH patient hepatocytes. One study in rats revealed that alcohol exposure increased hepatocyte apoptosis,[73,74] and consistent with this, increased numbers of apoptotic cells were observed in rats that developed ethanol-induced pathological liver injury. In another study, apoptotic cells were seen in the liver of ethanol-exposed mice, and this increase in apoptotic cells correlated with the duration of ethanol exposure; furthermore, the effect of ethanol on the level of apoptotic cell death was completely reversed by a period of abstinence. Indeed, exposure of human or rat hepatocytes to ethanol induced apoptosis in a dose-dependent manner.[75]

Several mechanisms seem to be involved in alcohol-induced hepatocyte apoptosis. Alcohol-induced hepatocyte apoptosis has been linked to an increase in reactive oxygen species (ROS),[76] and it has been shown that ROS promotes mitochondrial permeability and the release of cytochrome c.[104] Cytoplasmic cytochrome c interacts with the apoptotic-protein activation factor-1 (Apaf-1) and caspase-9 to form the apoptosome, which then activates the downstream effector caspases, caspase-3, -6, and -7. Potential sources of ROS are endogenous pro-inflammatory cells (Kupffer cells) and infiltrated inflammatory cells (neutrophils and lymphocytes). Ethanol exposure alters gut microflora, resulting in an increase of circulating endotoxins, which then activate Kupffer cells via Toll-like receptor 4 (TLR-4), resulting in an increase of ROS.[77,78] NADPH oxidase in Kupffer cells is proposed to be a major source of ROS oxidants when stimulated by ethanol. In NADPH oxidase–deficient mice treated with alcohol, there was no increase in free radical production or of TNF-α mRNA levels, which was consistent with the lack of evidence of liver pathology.[79] Additionally, alcohol-induced liver injury in TNF receptor-1-deficient mice was less severe than in wild mice. These results indicate not only that the oxidants generated by NADPH oxidase in Kupffer cells play a pivotal role in the development of alcohol-induced liver injury, but also that TNF-α production by activated Kupffer cells may also contribute to alcohol-induced liver injury.[80] Furthermore, liver cytochrome P450 2E1 (CYP2E1) induced by ethanol is a potential source of ROS that leads to liver injury. Indeed, CYP2E1 inhibitors ameliorated liver pathology in rats exposed to ethanol,[81,82] and liver injury induced by ethanol was more severe in transgenic mice expressing human CYP2E1 than in control mice.[83] CYP2E1 protein levels and enzyme activity were significantly higher in transgenic mice than in wild-type mice when fed an alcohol diet; conversely, CYP2E1 knockout mice developed liver injury similarly to control mice.[84] Although further studies are required to resolve these paradoxical results, one of the explanations is that the induction of antioxidant enzymes by ROS generated by CYP2E1 may attribute to the detoxification of CYP2E1-derived ROS.[85]

10.4.1.4 Nonalcoholic Steatohepatitis (NASH)

Nonalcoholic fatty liver disease (NAFLD) is the most common chronic liver disease in the United States, Asia, and Europe.[86–88] NAFLD encompasses a wide spectrum of conditions ranging from simple steatosis (fatty liver) to nonalcoholic steatohepatitis (NASH). The term NASH was coined in 1980, and its histopathology was described as lobular hepatitis, focal necrosis with mixed inflammatory infiltrates, and Mallory bodies.[89] The pathogenesis of NASH, however, remains to be defined, and it is still unknown why steatosis remains stable and nonprogressive in some patients with

NAFLD, whereas other patients progress to NASH and then NASH-fibrosis. Current concepts suggest that progression from NAFLD to NASH is a "two hit" process.[90]

The first hit is hepatic fat accumulation. Insulin resistance in adipocytes and myocytes leads to hepatic steatosis. Normally, insulin activates phosphatidyl inositol 3-kinase (PI-3K) and Akt/protein kinase B via the adipocyte and myocyte insulin receptors; as a consequence, GLUT-4, the insulin-responsive glucose transporter, is translocated to the cell membrane and maintains glucose homeostasis.[91] In contrast, adipocytes and myocytes accumulating fat possibly reduce PI-3K and lead to a decrease in the translocation of GLUT-4 to the cell membrane, which results in an increase of blood glucose and insulin levels.[91,92] Insulin resistance and hyperinsulinemia may contribute to steatosis by increasing the uptake of free fatty acids (FFAs) and increasing the synthesis of FFA mediated by enhanced levels of sterol regulatory element-binding protein-1c (SREBP-1c) and peroxisome proliferator-activated receptor γ (PPAR-γ) in hepatocytes.[93–95] Indeed, insulin resistance is much more frequent in patients with NASH than with other liver diseases.[92]

FFA accumulation in hepatocytes may result in an increased susceptibility to the second hit, which is possibly oxidative stress.[95,96] Hepatic lipid peroxidation by immunohistochemical staining for 3-nitrotyrosine, an indicator of lipid peroxidation, in the livers of patients with NASH was significantly higher than in those with fatty or normal livers.[97] Systemic levels of lipid peroxidation in patients with NASH were also higher than those in controls.[98] Oxidation of FFA accumulated in hepatocytes may ultimately lead to an increase in mitochondrial ROS production.[99] Furthermore, increased levels of CYP2E1 in the liver of patients with NASH could be a result of increased ROS levels.[100] In leptin-deficient (*ob/ob*) and hyperleptonemic (*db/db*) mice models, endotoxin levels in portal blood were higher than those of normal lean mice, indicating that Kupffer cells in the liver may be activated by endotoxins and produce ROS.[101] Taken together, the data suggest that ROS may play a central role in the pathogenesis of NASH. ROS induces hepatocyte cell death via both intrinsic and extrinsic apoptotic pathways and may increase levels of TNF-α, which may be involved in the pathogenesis of NASH. An increase of TNF-α and TNFR1 mRNA levels in hepatic tissue and peripheral fat, as well as an increase in the serum level of TNF-α, was observed in patients with NASH, compared to in obese subjects.[102,103] Indeed, hepatocyte apoptosis, indicated by caspase-3 activation, was markedly increased in NASH patients and directly correlates with disease severity, e.g., hepatic fibrosis and inflammatory activity.[23,24] Thus, it appears that apoptosis plays a critical role in both the pathogenesis and progression of NASH.

10.4.1.5 Ischemia-Reperfusion Injury

Hepatocyte death after ischemia and reperfusion in liver transplantation and resulting graft malfunction is the major contributor of clinical failure. Induction of apoptosis during ischemia-reperfusion injury has been identified as a critical cause of graft malfunction in both preclinical and clinical studies.[14–16] Sinusoidal endothelial cells also undergo apoptosis during ischemia-reperfusion injury, which may play a crucial role in progression of liver preservation injury, as endothelial cell apoptosis may lead to graft malfunction.[105–107] In addition to the induction of apoptotic cell death, Kupffer cells are activated during ischemia-reperfusion injury that release cytokines, including TNF-α, chemokines, and ROS, that contribute to the induction of apoptosis of other cells in the liver.[108,109] Preclinical studies revealed that caspase inhibition prevents apoptotic liver damage following ischemia-reperfusion, suggesting a therapeutic opportunity for apoptotic caspase inhibitors in this disease.[107,110]

Although a number of sources suggest that apoptosis in ischemia-reperfusion injury is critical for liver damage, the contribution of apoptosis in ischemia-reperfusion injury is still controversial.[111] In a rat ischemia-reperfusion injury model, most of the injured hepatocyte and sinusoidal endothelial cells undergo necrosis, but not apoptosis as measured by terminal deoxynucleotidyl transferase–mediated dUTP nick-end labeling (TUNEL) assay in combination with morphological observation. Thus, both necrosis and apoptsis may be important mechanisms of cell death in ischemia-reperfusion injury.[112] Nevertheless, apoptosis appears to play a role in ischemia-reperfusion injury, as inhibition of caspase activity can prevent liver damage induced by ischemia-reperfusion.[107,110]

10.4.1.6   Liver Fibrosis

Chronic liver diseases, such as chronic hepatitis B and C, ALD, NASH, autoimmune hepatitis, and cholestatic disorders, progress to liver fibrosis. Progression to liver fibrosis is a long-term process influenced by various factors. In this respect, liver fibrosis is not an independent disease *per se* but is caused by a variety of different liver diseases, all of which show significant pathological similarities, such as an inflammatory response and an increase in hepatocyte apoptosis. Once inflammation and a fibrogenic process are provoked, disease progression goes through a very similar path regardless of the initial insult.[113] Hepatic stellate cells (HSCs) are the key fibrogenic cell type and are nonparenchymal, mesenchymal cells localized in a perisinusoid in the liver. In liver disease, HSCs transform from a quiescent to an activated form (myofibroblastic transdifferentiation) and produce an excessive cellular matrix. More precisely, hepatocyte injury causes a release of inflammatory cytokines and other mediators, including growth factors from hepatocytes and other cells in the liver. These growth factors activate Kupffer cells, and also induce the infiltration or stimulation of T lymphocytes, which lead to the activation of HSCs.[114] In addition to mediators released from cells in the liver, phagocytosis of apoptotic bodies by HSCs leads to a fibrogenic response, as indicated by an increase in the expression of collagen α1(I) and transforming growth factor-β (TGF-β) mRNA, and is eventually associated with liver fibrosis.[115,116]

In conclusion, it would appear that hepatocyte apoptosis is an underlying cause of liver fibrosis; therefore, inhibition of hepatocyte apoptosis could be one potential therapeutic approach to prevent or reduce liver fibrosis.

10.4.2   PANCASPASE INHIBITORS IN CLINICAL TRIALS

10.4.2.1   Emricasan (IDN-6556/PF-03491390)

Idun Pharmaceuticals, Inc., has developed IDN-6556, a novel, irreversible, orally active, pancaspase inhibitor (structure shown in Figure 10.3). This compound was

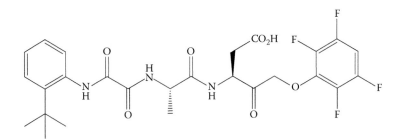

FIGURE 10.3 Chemical structure of emricasan (IDN-6556/PF-03491390).

evaluated as a therapeutic for the treatment of chronic hepatitis C and liver transplant rejection. After the acquisition of Idun by Pfizer in the first quarter of 2005, Pfizer proceeded to develop this compound as emricasan (PF-03491390) in the same indications. The current status of this compound is currently unclear, as Pfizer recently disclosed that it was discontinued from further development (Pfizer Pipeline, February 28, 2008). Preclinical studies are outlined below, as the preclinical development was described in detail in Chapter 8.

In a lung transplantation animal model both donor and recipient animals, along with the preservation solutions, were treated preoperatively with emricasan.[36,117] Emricasan reduced the number of apoptotic cells and improved lung function after lung transplantation and reperfusion, while in control mice an increase in endothelial cell and lymphocyte apoptosis in the lung during ischemia-reperfusion injury was observed.[117] Similarly, cell death in grafted liver after ischemia and reperfusion in liver transplantation leads to graft malfunction, which is one of the highest risk factors in a clinical setting. Emricasan was also shown to be effective against ischemia-reperfusion injury seen in liver transplantation,[36] as emricasan reduced sinusoidal endothelial cell apoptosis when added to liver preservation solution. This indicates that the compound appears to be effective against ischemia-reperfusion injury in liver transplantation.

Efficacy of emricasan against acute liver injury and liver fibrosis has also been evaluated.[37,38] Two different methods to induce liver injury were used in these studies, D-galactosamine/LPS and α-Fas. Emricasan dramatically reduced plasma ALT levels and caspase activity in the liver of mice challenged with either of these insults.[37] Interestingly, the concentration of emricasan in liver remained high for a prolonged period when administered orally compared to systemically, indicating that this compound has a high enterohepatic recirculation and first-pass effect. Consistent with this, emricasan attenuated hepatic injury and liver fibrosis in a bile duct–ligated mouse model,[38] where Fas-mediated apoptosis is one of the causes of hepatocyte death and resulting liver fibrosis.[118,119] Emricasan significantly reduced the number of apoptotic cells in the liver, and significantly reduced plasma ALT levels and mRNA levels of α-smooth muscle actin (a marker for activation of HSC), TGF-β (a potent fibrogenesis mediator), and collagen α1(I) (a major form of collagen in liver fibrosis). Additionally, liver collagen deposition was significantly reduced in emricasan-treated mice. These results indicate that emricasan can ameliorate liver fibrosis via inhibiting hepatocyte apoptosis.

Although emricasan has been profiled in several preclinical models by a variety of routes of administration, the only dose-limiting toxicity noted was vein irritation following intravenous administration. However, oral administration of this compound for 28 days to rats led to no adverse effects at a dose of 60 mg/kg, and the safety margin was at least sixty-fold greater for oral administration than intravenous administration.[120]

10.4.2.2 Emricasan: Clinical Trials for Liver Diseases

10.4.2.2.1 Phase I Clinical Trial

A phase I clinical trial was conducted to determine the pharmacokinetic profile and safety of emricasan in patients with hepatic dysfunction, including chronic hepatitis C, alcoholic liver disease, fatty liver, and unknown diagnosis.[121] The study was divided into 3 groups, and an outline of the treatment protocol is shown in Table 10.2. The first group (group 1) consisted of a single-dose escalation study (0.1, 0.5, 1, 5, and 10 mg/kg, 30-minute intravenous infusion) in normal volunteers (active drug, 25 subjects; placebo, 5 subjects). The second group (group 2) consisted of a multiple-dose escalation study (0.1, 0.5, 1, and 1.5 mg/kg qid, 30-minute intravenous infusions) for 7 days in normal volunteers (active drug, 25 subjects; placebo, 5 subjects). The third group (group 3) was a multiple-dose escalation study (0.1, 0.5, and 1 mg/kg qid, 30-minute intravenous infusions) for 7 days in patients with hepatic dysfunction (active drug, 11 subjects; placebo: 5 subjects). A significant reduction of ALT levels was observed during the 7-day treatment in group 3 (patients with hepatic dysfunction) at all doses tested compared to a placebo, while only one patient receiving 1.0 mg/kg of emricasan did not respond to treatment. At the end of the 7-day dosing period, ALT levels returned to baseline levels within 1–3 days. However, a transient increase above baseline in a liver function test was observed in some patients with hepatic dysfunction. ALT levels peaked 1–7 days after the treatment was stopped and returned to baseline values by 14 days. However, HCV assays in the four HCV-positive patients showed no clinically meaningful changes during the study.

Adverse events observed in this study were mild to moderate in severity, and the duration was relatively short, with the main adverse event being vein inflammation (phlebitis) after infusion of emricasan and leukocytosis (phlebitis was both drug and dose related). The incidence of phlebitis decreased when the infusion volume was

---

TABLE 10.2
Outline of Phase I Clinical Trial of Emricasan

Group	Regimen	Dose (mg/kg)	Dosing Schedule	Subjects Number (Active Drug, Placebo)
1	A single-dose escalation	0.1, 0.5, 1.0, 5.0, 10	30 min i.v. infusion	Normal (25, 5)
2	A multiple-dose escalation	0.1, 0.5, 1.0, 1.5	30 min qid i.v. infusion for 7 days	Normal (25, 5)
3	A multiple-dose escalation	0.1, 0.5, 1.0	30 min qid i.v. infusion for 7 days	Patients (11, 5)

increased or the dilution was increased. Interestingly, in the single-dosing group, phlebitis developed several hours after infusion but did not develop during the infusion, whereas in the multiple-dosing group, the onset of phlebitis occurred several days after the infusions started, suggesting that phlebitis is not caused by the compound. During the study, ALT levels in normal volunteers did not change, while a transient increase of white blood cell counts was observed in normal volunteers receiving a single dose of 5 or 10 mg/kg. Leukocytosis was observed only in subjects with phlebitis and was not observed in subjects absent of phlebitis; many subjects developed phlebitis without leukocytosis. The safety profile seen in this phase I trial was quite similar to the profile seen in preclinical toxicity studies,[120] as animals treated intravenously with emricasan incurred phlebitis and neutrophilic leukocytosis.

Maximum plasma concentration of emricasan (30-minute infusion) occurred at 0.58 hours in each of the three groups. In group 1, plasma concentration of emricasan declined with a mean apparent terminal half-life of 1.7–3.1 hours. $AUC_{(0-\infty)}$ and C_{max} increased dose-proportionately with a dose of 0.1 to 1 mg/kg of emricasan, and then non-dose-proportionately with a dose of 5 to 10 mg/kg. In group 2, $AUC_{(0-\infty)}$ and C_{max} increased dose-proportionately with all doses tested (0.1–1.5 mg/kg). In group 3, $AUC_{(0-\infty)}$ and C_{max} increased dose-proportionately with doses of 0.1 and 0.5 mg/kg, while slightly less than dose-proportionately from 0.5 to 1 mg/kg. In multiple dosing groups (group 2 and 3), no accumulation of emricasan was observed over a 7-day dosing period.

10.4.2.2.2 Phase IIa Clinical Trial

A multicenter, double-blinded, randomized, placebo-controlled phase II clinical trial was conducted.[122] Forty-nine patients were enrolled in this trial: 48 patients with chronic hepatitis C and one patient with NASH. Baseline ALT or AST levels in these patients ranged from 1.5- to 10-fold of upper limit of normal (ULN). Emricasan was administered orally once daily (25, 100, and 200 mg) or twice daily (5, 50, and 100 mg) for 14 days, with 21 days of follow-up. ALT levels at day 14 had significantly decreased compared to those at the baseline: –40% (25 mg qd), –33% (100 mg qd), –35% (200 mg qd), –49% (5 mg bid), –42% (50 mg bid), and –56% (100 mg bid). Similar changes of AST levels were observed (Table 10.3). The greatest reduction of ALT was observed at 100 mg bid. Both ALT and AST levels in all six patients that received emricasan at 100 mg bid decreased to a normal level (lower than ULN). Levels of ALT in placebo-controlled patients were not significantly changed (2%), and levels of AST were similar to those of ALT. Adverse events observed in this trial were all mild and transient. HCV mRNA titers did not change more than 1 log unit during the trial in 47 out of 48 chronic hepatitis C patients, but one patient receiving 100 mg bid of emricasan showed complete viral clearance at the end of follow-up.

Emricasan treatment of HBV patients was also part of this trial: 7 patients (4/7 patients were negative for HBV DNA, 3/4 were confirmed, and 1 patient was not confirmed) and 7 placebo patients. A dose of 100 mg bid of emricasan reduced ALT levels in HBV patients by an average of 50%.

10.4.2.2.3 Phase IIb Clinical Trial

A multicenter, double-blinded, randomized, placebo-controlled, parallel-group, dose-ranging phase II clinical trial was conducted in 204 patients with chronic

TABLE 10.3
ALT Reduction in Patients Receiving Emricasan

Treatment Schedule	Dose (mg)	% Reduction of ALT on Day 14[a]
Qd[b]	25	−40
	100	−33
	200	−35
Bid[c]	5	−49
	50	−42
	100	−56

[a] Percent changes from baseline.

[b] Once daily for 14 days.

[c] Twice daily for 14 days.

hepatitis C and liver fibrosis.[40] Patients enrolled in this trial were confirmed HCV positive who did not achieve a virological response, had relapsed, could not tolerate standard therapy, or failed to maintain a response. Patients who had decompensated cirrhosis, or severe liver disease, hepatocellular carcinoma, or concomitant infections, e.g., HBV or HIV, or a history of alcohol or drug abuse were excluded from the trial. Patients were randomized to receive either placebo or emricasan at doses of 5, 25, or 50 mg orally twice a day (bid) for 10 weeks. If ALT and AST levels in patients were still elevated at week 10, then a double-dose administration was continued to week 12. Patients were observed for an additional 4 weeks after discontinuation of treatment at week 12. Median absolute ALT and AST levels at week 10 significantly decreased in all treatment groups compared to baseline levels (Table 10.4). Reduction of ALT levels from the baseline were −34% (5 mg bid), −41% (25 mg bid), and −41% (50 mg bid); reduction of AST levels from the baseline were −17% (5 mg bid), −23% (25 mg bid), and −25% (50 mg bid). Levels of both ALT and AST in placebo-controlled patients did not change significantly (−2%), whereas ALT levels in some patients who received emricasan returned to normal at week 10: 15% (5 mg bid), 35% (25 mg bid), and 19% (50 mg bid), compared with 3% of patients who received a placebo. The degrees of reduction of ALT and AST levels were quite similar across all treatment doses, and reductions in all treatment doses were statistically significant ($p < .0001$ vs. placebo). Reductions of ALT and AST levels were observed during the first week and were maintained during the entire treatment period, and then returned to baseline levels after stopping treatment. Adverse events seen in this trial were mild to moderate, and incidences of adverse events were similar among treatment and placebo arms. The most frequently observed emricasan treatment-related adverse events were headaches in 24 patients and fatigue in 22 patients. During the treatment, no change in the mean log of HCV RNA titer was observed in any of the treatment groups.

TABLE 10.4
Emricasan Reduced Serum AST and ALT by Week 10 in the
Intent-to-Treat Population

	Placebo	5 mg Bid	25 mg Bid	50 mg Bid
Number of patients	51	55	50	48
Baseline AST (IU/L)	60	69	58	73
AST (IU/L) reduction from baseline at week 10	–2	–17*	–23*	–25*
Baseline ALT (IU/L)	101	103	98	115
ALT (IU/L) reduction from baseline at week 10	–2	–34*	–41*	–49*

Source: Reprinted from reference 40 by courtesy of John Wiley & Sons, Inc.

*$p < .0001$ versus placebo.

This clinical trial clearly indicates that oral administration of emricasan was well tolerated and reduced both ALT and AST levels in patients with chronic hepatitis C. Furthermore, ALT and AST reductions were maintained throughout the 12 weeks of treatment without any severe adverse events, and also without increasing HCV RNA levels. Further clinical trails, especially long-term trials, are necessary to determine whether emricasan is an effective treatment for liver fibrosis.

10.4.2.3 Emricasan: Clinical Trials for Liver Preservation Injury

Hepatocyte death after cold ischemia and warm reperfusion in liver transplantation remains a critical therapeutic problem. Induction of apoptosis during cold ischemia and warm reperfusion injury has been identified as one of the causes of graft malfunction in preclinical and clinical studies.[14–16,107,110] In one preclinical study, emricasan was shown to be effective in an ischemia-reperfusion liver injury model.[36] A multicenter, randomized, placebo-controlled, double-blinded, parallel-group, phase II trial to assess emricasan on cold ischemia and warm reperfusion injury during liver transplantation was conducted.[39,123] A total of ninety-nine subjects were assigned to four groups:

Group 1: Organ storage/flush, placebo; recipient, placebo (n = 23)
Group 2: Organ storage/flush, 15 μg/ml of emricasan; recipient, placebo (n = 23)
Group 3: Organ storage/flush, 5 μg/ml of emricasan; recipient, 0.5 mg/kg of emricasan (n = 27)
Group 4: Organ storage/flush, 15 μg/ml of emricasan; recipient, 0.5 mg/kg of emricasan (n = 26)

All 7-day biopsies of 35 patients were performed at one center (Mayo Clinic). Emricasan was supplemented into the cold storage and flush solution of the organ and was administered to the recipients intravenously every 6 hours for 24 hours after transplantation. Apoptosis was assessed by measuring the serum levels of CK18Asp396, which is a neopeptide cleaved from cytokeratin 18 by caspases,[124] TUNEL, and caspase-3/7 immunohistochemistry of liver specimens. Liver injury was assessed by measuring serum ALT and AST levels.

Liver apoptosis, as indicated by caspase-3/7 immunohistochemistry, was increased in intraoperative post-reperfusion liver biopsies compared to donor pre-perfusion liver specimens in group 1 (7% vs. 2%, $p < 0.03$), indicating a rapid increase in apoptosis in the liver that occurred during the reperfusion of the injury. This increase in apoptosis was significantly prevented in patients treated with emricasan (groups 2, 3, and 4). In these groups, there was no difference in the number of caspase-3/7-positive cells in the liver specimen between pre-perfusion and post-perfusion. Additionally, serum levels of CK18Asp396 were measured every 6 hours for 24 hours after transplantation and were found to be higher in the placebo group (group 1) than in the treatment groups when measured after venous perfusion and prior to intravenous infusion of placebo or emricasan. Also, levels of CK18Asp396 in serum of groups 1 and 2 were similar and higher than those of groups 3 and 4 throughout the 24 hours after transplantation. In contrast, levels of CK18Asp396 in serum of groups 3 and 4 did not increase and were maintained lower than that of the placebo group (group 3 vs. group 1, $p < 0.0001$; group 4 vs. group 1, $p < 0.001$). These observations indicate that emricasan reduced reperfusion-mediated cell apoptosis in the liver, and treatment of the recipient with emricasan led to a reduction of apoptosis during the 24 hours after transplantation.

ALT and AST levels in serum were measured from day 1 post liver transplantation to day 7. ALT and AST levels in group 1 were elevated immediately after liver transplantation. In contrast, ALT and AST levels in group 2 were significantly lower than those in group 1 for 7 days ($p < 0.01$). In groups 3 and 4, however, reduction of ALT and AST was not observed and their serum levels were similar to those of group 1. In all groups, ALT and AST levels decreased to normal levels by day 7. Emricasan did not lead to adverse renal or biliary functions, and there was no difference in serum levels of creatinine or bilirubin among all groups. By day 7, serum ALT and AST levels and the percent of TUNEL-positive and caspase-3/7-positive cells were lower in group 2 than in group 1, and these numbers in group 3 were also smaller than those of group 1 (Table 10.5). In contrast, the percent of TUNEL-positive and caspase-3/7 immunostaining-positive rates in group 4 did not decrease (Table 10.5).

They also determined if there was a correlation between inflammation (assessed by neutrophil accumulation) and levels of ALT/AST and apoptosis among the groups. Neutrophil accumulation in liver allografts 7 days after transplantation in group 2 was less than that found in other groups, whereas neutrophil accumulation and distribution in group 4 was similar to that in group 1. Additionally, in group 4, the number of myeloperoxidase immunostain-positive cells was significantly higher than that in group 2 ($p < 0.04$). During the first 30 days after the liver transplantation there was no difference in the incidence of adverse events among the groups, while no delayed graft function and primary nonfunction were observed in group 2. Also, there was no difference in the rate of acute cellular rejection in all groups.

TABLE 10.5

Levels of Apoptosis Indicated by TUNEL Assay
and Caspase-3/7 Immunostaining in Liver Tissue

Group	TUNEL (%)[a]	Caspase-3/7 (%)[b]
Group 1	10.3 ± 5	14 ± 6
Group 2	7.6 ± 2.5	8.0 ± 3.0
Group 3	7.7 ± 3.3	9.0 ± 8
Group 4	9.6 ± 5	26 ± 3

[a] Percent of TUNEL-positive cells/total hepatocytes.

[b] Percent of caspase-3/7-positive cells/total hepatocytes.

Taken together, the addition of emricasan into the storage and flush solution was well tolerated and reduced reperfusion-mediated cell apoptosis in the liver up to 24 hours after transplantation and also reduced neutrophil infiltration to the liver. This indicates that emricasan could be a promising agent to minimize injury induced by cold ischemia and warm reperfusion in liver transplantation when added to cold storage and flush solutions. Further clinical studies are required to determine whether this beneficial effect on the allograft dysfunction is sufficient enough for clinical use.

Surprisingly, intravenous treatment of emricasan to recipients negated beneficial anti-apoptotic effects when the liver was treated with emricasan in the storage and flush solution. The mechanism for this unanticipated result is unknown. Neutrophil infiltration was observed in the livers of recipients who received emricasan every 6 hours for 24 hours after the transplantation. It is possible that neutrophil accumulation may attribute to the adverse effect of emricasan since the impairment of neutrophil apoptosis can alter its longevity, and consequently may result in the development and promotion of an inflammatory condition.[125–127]

10.4.2.4 LB84318 and LB84451

10.4.2.4.1 LB84318

LG Life Sciences, Ltd., has developed LB84318, an orally active irreversible pancaspase inhibitor, for the treatment of various liver diseases, including hepatitis C virus (HCV) infection, NASH, alcoholic hepatitis, fulminant hepatitis, primary biliary cirrhosis, and organ transplantation. LB84318 inhibits caspase-1, -3, -7, -8, and -9 in an irreversible manner, while exhibiting no inhibitory activity against other proteases, such as cathepsin B, H, and thrombin.[128,129] In animal studies, LB84318 dramatically reduced serum ALT levels in a LPS/D-galactosamine (LPS/D-Gal)-induced mouse acute liver injury model in a dose-dependent manner ($ED_{50} = 0.008$ mg/kg [i.v.], 0.022 mg/kg [oral], and 0.013 mg/kg [i.p.], measured 8 hours after the insult), as well as in an α-Fas-induced mouse acute liver injury model ($ED_{50} = 0.018$ mg/kg, oral).[128,129]

In a survival study using a LPS/D-Gal model, mice treated with LB84318 (0.3 mg/kg, oral administration) survived 7 days after insult, whereas all mice treated with LPS/D-Gal died 8–24 hours after insult. Furthermore, LB84318 at a dose of 0.1 mg/kg (either intravenous or oral administration) protected mice from lethal toxicity of LPS/D-Gal even when administered 6 hours after insult. In this study, LB84318 reduced ALT levels and blocked apoptosis within 1 hour when administered intravenously 6 hours after the insult, indicating that LB84318 reduced hepatic apoptosis effectively and quickly. A preadministration study using an α-Fas-induced liver injury model showed that oral administration of LB84318 (1 mg/kg) significantly reduced ALT levels up to 6 hours before the insult,[128] suggesting LB84318 is retained in the liver for prolonged periods following oral administration. Preliminary toxicity studies revealed that LB84318 was safe (MTD > 600 mg/kg/day).[130] No further development of LB84318 has been reported, and thus its current status is not known.

10.4.2.4.2 LB84451

LG Life Sciences is developing what appears to be a second-generation caspase inhibitor, LB84451. The only publicly available information on this compound is provided by the LG Life Sciences Web site. LB84451 is an orally active caspase inhibitor with a unique scaffold that was shown to have a fast onset and long duration of action as a potential hepatoprotective agent targeted for indications such as HBV, HCV hepatitis, and NASH. A 28-day preclinical toxicology study appears to have demonstrated an acceptable safety profile. A phase I clinical trial was initiated in October 2005 in the Netherlands (LG Life Sciences IR presentation report, July 2006, http://www.lgls.co.kr/data/presentationData/060720_IR_PT_eng.pdf#search='IR%20presentati on%20LG%20life%20sciences, and R&D pipelines in LG Life Sciences presented on company Web site: http://www.lgls.co.kr/eng/rd/pipeline.jsp), and the company appears to have entered a phase II trial recently for liver disease (based upon information contained in its July 2007 investment relations presentation). In November 2007 LG Life Sciences entered into a license agreement with Gilead Sciences, Inc., to develop a series of caspase inhibitors for the treatment of fibrotic diseases, which gives Gilead rights to all their caspase inhibitors, including LB84451.

10.4.3 SUMMARY

Treatment of emricasan in patients with chronic hepatitis C who have failed standard therapy revealed that it was well tolerated and effectively reduced ALT and ASL levels without any severe adverse events. Importantly, this occurred without an increase in HCV RNA levels during the course of 12-week administration, indicating that treatment of emricasan has no influence on hepatitis C viral growth in patients with chronic hepatitis C. This result warrants further clinical investigation to determine whether emricasan improves liver histology/function in patients with liver fibrosis following long-term treatment. The current standard treatment for chronic hepatitis C is pegylated interferon-α (PEG-IFN-α) in combination with ribavirin. However, a sizable proportion of patients treated with PEG-IFN-α plus ribavirin fail to achieve a response, especially patients who are infected with HCV genotype 1b. More than 10% of patients who received this combination had to discontinue the therapy due to severe adverse events.[131–133] Although these problems may be solved by the development of novel antiviral agents, emricasan may have a potential benefit to these patients if it improves liver histology in the absence of a virological response. Prevention of progressive liver damage and regression of liver fibrosis are extremely significant secondary treatment goals.

ALT levels in patients treated with emricasan decreased by up to 30%–50% compared to placebo-treated patients, and ALT levels in 15%–30% of those patients were normalized. Normalization of ALT levels in all patients by treatment of emricasan would be difficult to achieve as it is not an antiviral agent. What degree of reduction of ALT levels will result in the improvement of liver histology in patients with liver fibrosis is still unclear, and there is little information regarding the correlation between ALT reduction and histological improvement even following treatment with agents other than the standard therapy, i.e., IFN-based therapies. In this respect, determining whether long-term treatment with emricasan will improve liver histology in patients with liver fibrosis will be an important question to answer in the near future.

Phase IIb results also suggest that emricasan could be effective in other chronic liver diseases, e.g., NASH. Hepatocyte apoptosis is significantly increased in NASH patients and correlates with disease severity and plasma caspase-3-generated cytokeratin-18 fragments compared to steatosis or healthy subjects, indicating that hepatocyte apoptosis mediated by caspase activation is one of the pathological characteristics of NASH.[23,134] Interestingly, emricasan reduced hepatocyte lipoapoptosis induced by free fatty acids (FFAs) *in vitro*.[135] Given that the accumulation of oxidized FFAs in hepatocytes could lead to an increase in mitochondrial ROS production, and therefore increased hepatocyte death via both intrinsic and extrinsic apoptosis pathways, emricasan could possibly ameliorate liver damage by reducing hepatocyte apoptosis mediated by caspase activation in NASH.

Emricasan reduced liver preservation injury after cold ischemia and warm reperfusion in liver transplantation. Supplementing the storage and flush solution with emricasan reduced reperfusion-mediated cell apoptosis in the liver. However, intravenous treatment of emricasan to recipients reversed this beneficial anti-apoptotic effect in the liver allograft. The mechanism for this is still unknown, but neutrophil infiltration in the liver following administration of emricasan to the recipient could be attributed to an unexpected effect. Neutrophil accumulation and activation is one of the critical causes of ischemia-reperfusion liver injury, and functional inactivation of neutrophils protects the liver against ischemia-reperfusion injury.[136–138] In animal models, pretreatment of caspase inhibitors or small interfering RNA to animals 2 minutes or 1 hour before ischemia is reported to lead to a decrease in the infiltration of neutrophils into the liver during ischemia-reperfusion, and consequently protects the liver from injury.[139,140] These studies suggest that caspase inhibition results in a decrease of neutrophil accumulation and, as a consequence, inhibits liver injury. Indeed, inhibition of caspase-3 and caspase-8 by treatment of siRNA in mice resulted in a decrease of polymorphonuclear leukocyte infiltration into the liver, along with a corresponding preservation of liver architecture. These results are consistent with clinical results. In the clinical trials described above, adding emricasan to the storage and flush solution reduced neutrophil infiltration to the liver and reperfusion-mediated cell apoptosis in the liver for 24 hours after transplantation compared to liver allograft in the placebo group. This indicates that emricasan has at least some therapeutic potential by protecting the liver from cold ischemia-reperfusion-related apoptosis and injury. However, the reason why prolonged treatment of emricasan to the recipient resulted in an increase of neutrophil infiltration into liver allografts is unknown. Inhibition of neutrophil apoptosis by treatment of emricasan may promote an inflammatory condition, and consequently induce caspase-independent cell death, i.e., necrosis.[111,125–127,141]

Although apoptosis is a pivotal trigger for ischemia-reperfusion liver injury, it may not be the key instigator in the progression of liver damage. Further studies are required to determine whether the effects of emricasan on allograft dysfunction shown in these clinical studies are beneficial for recipients. Furthermore, the appropriate treatment regimen to maximize the efficacy of emricasan on ischemia-reperfusion liver injury will need to be determined.

10.5 CHALLENGES TO THE DEVELOPMENT OF CASPASE INHIBITORS

10.5.1 LONG-TERM INHIBITION OF CASPASE-1

Despite encouraging clinical data, theoretical concerns resulting from long-term inhibition of caspases still remain to be resolved. Caspase-1 is involved in the inflammatory response by activating the pro-inflammatory cytokines, IL-1β and IL-18, but is not involved in the apoptotic caspase cascade. Thus, it is believed that chronic inhibition of caspase-1 may not cause adverse effects related to apoptosis inhibition. Indeed, caspase-1 knockout mice are overtly normal,[41,42] and histopathological evaluation of major organs in these mice revealed no hematological abnormality, e.g., the numbers of peripheral blood cell and proportion of various B and T cell subsets were unchanged. Also, there has been no evidence of an increase in the spontaneous tumor rate, at least at up to 25 weeks.[41] Consistent with the fact that caspase-1 does not play a role in the apoptotic cascade, it was found that apoptosis occurs normally in these mice (e.g., macrophages undergo ATP-induced apoptosis and thymocytes undergo apoptosis induced by dexamethasone and γ-irradiation, but are resistant to anti-Fas-induced apoptosis). On the other hand, caspase-1 knockout mice exhibit dramatically reduced production of IL-1β and IL-18 in response to inflammatory challenges and are resistant to endotoxin shock. These mice are highly resistant to a lethal dose of LPS that normally kills all wild mice. Furthermore, caspase-1 knockout mice are resistant to models of inflammation (pancreatitis model[142]), ischemia (acute renal failure model[143]), hypoxic-ischemia,[144] and pneumococcal meningitis.[145] Therefore, the evidence would suggest that chronic inhibition of caspase-1, i.e., long-term treatment with a caspase-1 inhibitor, should not evoke harmful adverse effects. Nevertheless, there may be unforeseen consequences of chronically inhibiting caspase-1; for example, it still remains to be determined if the toxicity observed with VX-740 is target or compound related.

10.5.2 LONG-TERM INHIBITION OF OTHER CASPASES

10.5.2.1 Caspases and Tumorigenesis

In contrast to inhibition of caspase-1, inhibition of apoptosis could raise the potential risk of tumorigenesis due to the inhibition of cell death. Apoptosis is thought to be an essential process in the development and homeostasis of multicellular organisms and acts as an elimination system to prevent aberrant cells from transforming into cancer cells. Inhibition of apoptotic caspases could interrupt this system and consequently result in neoplasia. However, there is as yet no clear evidence that apoptotic caspase inhibition leads to enhanced tumorigenesis. Instead, progress has been made in recent years in the investigation of complex mechanisms of apoptosis and also the redundancy of apoptosis regulation. In other words, although it is obvious that apoptotic caspases are essential for apoptosis, caspase-independent cell death may act as a fail-safe mechanism to protect an organism from aberrant cells and tumorigenesis.[146–151]

Many studies have shown that caspase inhibition using caspase inhibitors often does not prevent cell death induced by a variety of cell death stimuli *in vivo* as well as *in vitro*, suggesting that inhibition of caspase activation could promote alternative pathways of cell death.[152–159] One of the alternative mechanisms of cell death is autophagic cell death or type II cell death. Under normal physiological conditions, autophagy is the survival response cells use to remove damaged organelles for reuse of their amino acids or other building blocks during normal organelle turnover or in a nutrient-deprived condition of healthy cells.[160] In an experimental study using rat primary hepatocytes, induction of mitochondrial permeability transition stimulates autophagy, indicating autophagy may be a protective mechanism against apoptosis, especially mitochondria-mediated apoptosis.[161] However, excessive autophagy leads to cell death in a caspase-independent manner. Indeed, caspase inhibitors fail to inhibit autophagic cell death.[159,162,163] Interestingly, a number of studies have supported a suppressive role of autophagy in tumorigenesis, as dysfunction of the autophagic machinery leads to the development of tumors.[163–166] Introduction of Beclin 1 (BECN 1), an autophagy-inducible protein that interacts with Bcl-2 in mammalian cells, into MCF-7 human breast cancer cells results in the inhibition of clonigenicity *in vitro* and tumorigenesis *in vivo*. Moreover, a decrease in the expression level of Beclin 1 is observed in breast carcinoma tissues compared to normal breast tissue in humans.[165] In contrast, there are reports that describe how autophagy promotes tumor cell growth by providing nutrients from recycling of proteins and intracellular components to aid survival under low nutrient conditions, i.e., tumor cells located in the center of tumor tissue, and how autophagy protects cells from death by eliminating damaged intracellular organelles induced by anticancer treatments.[167,168] Thus, although recent studies have suggested that autophagy may have a suppressive role in tumorigenesis, further studies are required to resolve this.[169–172]

Another caspase-independent cell death mechanism is subcellular organelle-mediated cell death. Endoplasmic reticulum (ER) stress, due to calcium depletion or the accumulation of excess proteins, induces caspase-12-mediated apoptosis.[173] However, calcium-activated neutral protease calpain induces cell death without the activation of caspases, and not surprisingly, caspase inhibitors fail to inhibit calpain-mediated cell death. This indicates that ER-mediated stress leads to cell death in both a caspase-dependent and -independent manner.[174,175]

Mitochondria also play a crucial role in regulating cell death. Mitochondrial membrane permealization is increased by stress resulting in the release of pro-apoptotic proteins, such as cytochrome c, Smac/DIABLO, apoptosis-inducing factor

(AIF), OMI/HtrA2, and endonuclease G.[146,176,177] These pro-apoptotic proteins induce apoptosis in both a caspase-dependent and -independent manner. AIF is normally localized in the intermembrane mitochondrial space and translocates to the nucleus in response to such stress, and therefore induces apoptotic cell death. Although the function of AIF is still largely unexplored, AIF is the mediator of caspase-independent apoptosis and the apoptotic effect mediated by AIF cannot be prevented by caspase inhibitors, indicating that AIF plays a role in caspase-independent apoptotic cell death by inducing chromatin condensation and large-scale fragmentation of DNA.[153,156,178,179] Omi/HtrA2 is also confined in mitochondria under normal physiological conditions but is released upon an apoptotic stimuli and induces apoptosis in a caspase-independent manner through its serine protease activity.[180,181] Endonuclease G is a sequence-unspecific DNase that translocates from the mitochondria to the nucleus and induces cell death by degradation of nuclear DNA.[182] Thus, these pro-apoptotic proteins in the mitochondria can induce apoptosis not only in a caspase-independent manner, but also in a caspase-dependent manner in certain physiological conditions.[146,176,177]

Death receptor-mediated caspase-independent cell death has also been reported. Fas- or TNF receptor-mediated apoptosis can be switched to non-apoptotic necrotic cell death when caspases are inhibited in experimental models.[154,157,183–186] Also, receptor-mediated necrotic cell death in the presence of caspase inhibitors can be prevented by oxygen radical scavengers, indicating that necrotic cell death may be attributed to ROS.[185,186] Consistent with this, cellular ROS levels are elevated following TNF-induced necrosis in the presence of caspase inhibitors but not in cells lacking receptor-interacting protein (RIP), tumor necrosis factor receptor-associated factor 2 (TRAF2), or Fas-associated death domain (FADD), which are critical components of the TNF-induced signaling cascade for necrosis.[186] Therefore, caspase-dependent apoptosis is not an exclusive mechanism that eliminates aberrant cells that could potentially develop into tumors, but rather is one of many backup mechanisms that are caspase independent, and which serve as an alternative system to eliminate aberrant or preneoplastic cells.

It has been suggested that the anti-inflammatory activity of caspase-1 inhibitors might contribute to the inhibition of tumorigenesis. A connection between inflammation and cancer was noted by Rudolf Virchow in the nineteenth century, and recent studies support the notion that chronic inflammation increases the risk of cancers.[187–189] IL-1β, one of the pro-inflammatory cytokines that is activated by caspase-1, is reported to be involved in tumorigenesis, proliferation of tumor cells, and angiogenesis.[190–192] *In vivo* experimental studies have shown that tumor cells transfected with the IL-1β gene secrete angiogenic factors, such as vascular endothelial growth factor (VEGF) and macrophage-inflammatory protein-2 (CXCL2), and grow rapidly with high vasculature *in vivo*.[190] In contrast, neither local tumor development nor metastases were observed in IL-1β knockout mice when tumor cells were injected into mice.[191] These results indicate that IL-1β may play a role in tumor development and invasion via regulating tumor angiogenesis. Furthermore, IL-1β induces VEGF via activation of NF-κB through the phosphatidylinositol 3-kinase/mammalian target of rapamycin (PI-3K/mTOR) pathway and induction of hypoxia-inducible factor-1α (HIF-1α) through the NF-κB/cyclooxygenase-2 (COX-2) pathway.[192] COX-2 and

HIF-1α are key players linking inflammation and tumorigenesis.[189] These results suggest that long-term treatment of caspase-1 inhibitors and pancaspase inhibitors that also inhibit caspase-1 could inhibit tumor development and tumor-related angiogenesis via reduction of IL-1β production.

Although a long list of investigations support the notion that long-term inhibition of caspases may not raise the risk of tumorigenesis, there are a number of reports describing the correlation between dysfunction of caspases and tumor growth in both preclinical and clinical studies, but the results are still controversial. For example, a deletion or loss of expression by methylation of the promoter region of caspase-8 has been observed in both pediatric tumors[193-197] (e.g., neuroblastoma, Ewing tumor, rhabdomyoblastoma, medulloblastoma, retinoblastoma), and nonpediatric tumors (e.g., lung cancer).[198-201] For example, caspase-8 is deleted or silenced by methylation in approximately 50% of neuroblastoma specimens, and a caspase-8 mutation is strongly correlated with amplification of the MycN oncogene, indicating that caspase-8 may act as a tumor suppressor in neuroblastoma.[193,194,196] In contrast, a large cohort study revealed no correlation between loss of caspase-8 expression and MycN oncogene amplification or prognosis.[202] Caspase-8 is absent in neural progenitor cells,[202,203] indicating that an absence of caspase-8 may only reflect the neuroblast developmental status, and the absence of caspase-8 by itself may not solely cause tumorigenesis.

The expression level of caspases in hematopoietic malignancies has been investigated, but the results are also controversial.[204-208] Interestingly, higher levels of caspase-2 and -3 are observed in patients with acute myelogenous leukemia than in normal subjects.[204] In acute lymphoblastic leukemia (ALL), procaspase-3 was detected at either the initial diagnosis or relapse stages, whereas the activated form of caspase-3 was detected only at the initial diagnosis; thus, processing of caspase-3 was absent in relapsed ALL patients.[206] A correlation between the dysfunction of caspases and tumors in clinical samples is still controversial. Dysfunction of other apoptosis pathways is also intricately linked with tumorigenesis.[209-211]

In summary, recent progress has shown that caspase-independent cell death acts as a fail-safe mechanism to protect organisms from aberrant cell growth and tumorigenesis, and that the anti-inflammatory activity of pancaspase inhibitors may contribute to the inhibition of tumor development. The theoretical concern that inhibition of apoptotic caspases could increase the potential risk of tumorigenesis due to inhibition of cell death may therefore be too simplistic.

10.5.2.2 Non-Apoptotic Functions of Caspases

Recent progress has shown that apoptotic caspases also function in non-apoptotic processes, for example, cell development, activation, proliferation, and differentiation.[212-215] *In vitro* experiments have shown that active caspase-3 plays a role in erythroid maturation, and maturation is inhibited by treatment with a caspase inhibitor or using siRNA directed to caspase-3.[216,217] Also, caspase-3 and caspase-9 induce differentiation of monocytes into macrophages in the presence of macrophage colony-stimulating factor (M-CSF), while a caspase inhibitor abolishes this function.[218] Similar functions of caspase-3 and caspase-9 are suggested by the formation of proplatelets from mature megakaryocytes.[219] However, failure of maturation of proplatelets is not observed in caspase-3 knockout mice.[220–223] Instead, caspase-3 knockout mice show reduced activation-induced cell death of peripheral T cells and hyperproliferation of peripheral B cells.[220–223] Caspase-8 is reported to have an essential role in lymphocyte activation and proliferation, and inactivation of caspase-8 by conditional knockout or expression of a dominant negative caspase-8 revealed that caspase-8 activity is required for the proliferation of hematopoietic progenitor cells.[224,225] Spontaneous thymocyte apoptosis *in vitro*, which partly mimics death by neglect, one of the selection processes of thymocyte, is partially inhibited by caspase inhibitors, indicating that caspases may be involved in the death-by-neglect process.[226–228] On the other hand, experimental studies using caspase-9 knockout mice or a caspase-9 inhibitor revealed that caspases are not essential for negative selection in thymocyte development.[226,229] Activity of caspase-8 is required for CD-3-induced proliferation and interleukin 2 production of human T cells, and these reactions are blocked by caspase inhibitors.[230,231]

Inherited human caspase-8 or caspase-10 mutations have been identified, although the number of patients is quite small.[232–235] The activity of these mutated caspase-8 or caspase-10 proteins is reduced or unable to function compared to that of wild-type proteins. Individuals with a homozygous mutation in caspase-8 (point mutation of Arg248Trp in the p18 large subunit) show lymphadenopathy and splenomegaly. These conditions are typically associated with autoimmunity; however, the main symptoms of these patients are related to immunodeficiency (see below).[233] In these patients, interleukin-2 production stimulated by T cell receptors (TCRs) and CD25 induction on both CD4+ and CD8+ T lymphocytes is dramatically decreased. Also, functions of natural killer (NK) cells and B cells are impaired, as indicated by recurrent sinopulmonary and herpes simplex virus infection and a decrease in serum immunoglobulin concentrations—features very similar to those observed in the *in vitro* experimental studies previously mentioned. Conversely, individuals who carry heterozygous mutation of caspase-8 are asymptomatic. Development of a conditional caspase-8 deletion mutant restricted to T cell lineage revealed that caspase-8 mutant mice manifest a range of symptoms, including lymphadenopathy, splenomegaly, and immunodeficiency, which are quite similar to those observed in patients carrying homozygous caspase-8 mutations.[236,237] Note that these mice carry a deletion mutant of caspase-8, not a full-length "protease dead" caspase-8. The expression level of mutant caspase-8 protein and its recruitment to the death-inducing signaling complex (DISC) were both markedly decreased in patients with a homozygous point mutation, suggesting that mutant caspase-8, albeit the full-length protease-dead caspase-8, may not have allosteric regulatory effects, such as caspase-8 independent death receptor–mediated NF-κB activation (see below).[233]

Individuals carrying inherited caspase-10 mutations manifest autoimmune lymphoproliferative syndrome (ALPS) as defined by chronic nonmalignant lymphadenopathy or splenomegaly, increased number of double-negative T lymphocytes (DNT cells), and impaired lymphocyte apoptosis.[232,234,235] Heterozygous mutations in caspase-10 are sufficient to promote the development of ALPS. Surprisingly, this mutated caspase-10 has a dominant negative effect when co-transfected with a wild-type caspase-10 gene into lymphocyte cell lines.[235] These studies indicate that the effects of caspase inhibition by a small molecule would be different from those observed in individuals carrying caspase-8 or -10 mutations.

Furthermore, death receptor–mediated NF-κB activation by an enzymatically inactive mutant of caspase-8 was observed, indicating that the death receptor domains (DEDs), but not proteolytic activity, of caspase-8 are required for the activation of NF-κB.[238–240] In contrast to these results, proteolytic activity of caspase-8 is required for NF-κB activation after TCR activation in T cells.[241] These results indicate that activation of NF-κB via either a caspase-8 enzymatic activity-dependent or -independent manner depends on upstream signals. Therefore, the conformation of caspase-8 or its association with other proteins in the DISC complex, along with its enzymatic activity, may contribute to its pleiotropic functions.

Recent advances have unveiled the pleiotropic function of apoptotic caspases, as they are involved not only in apoptosis but also in non-apoptotic cellular processes, such as immune cell development, activation, proliferation, and differentiation. Nevertheless, many questions still remain to be resolved concerning the non-apoptotic cellular functions of caspases. Detailed knowledge of their functions, regulatory mechanisms, specific targets, and cell specificity is required to understand the pleiotropic function of apoptotic caspases.

10.6 CONCLUSIONS

A large variety of experimental investigations of caspases have opened up the possibility of caspase inhibitors as a new therapeutic intervention. Caspase activation has been demonstrated to contribute to large numbers of pathological conditions, e.g., inflammatory diseases and apoptosis-driven disorders, and experimental studies have revealed that caspase inhibitors are effective in a large variety of disease models. Recently, several caspase inhibitors have entered clinical trials, most notably, VX-765, a caspase-1-specific inhibitor for the treatment of inflammatory diseases, and emricasan, a pancaspase inhibitor for the treatment of liver diseases (liver preservation injury in liver transplantation and chronic liver disease, i.e., chronic hepatitis C). The complexities involved in moving a caspase inhibitor successfully through clinical trails were recently highlighted by the withdrawal of emricasan from clinical development for liver fibrosis (Pfizer has not disclosed any information regarding this decision). Therefore, numerous issues critical for the development of caspase inhibitors still remain to be resolved; for instance, long-term inhibition of caspase activity could theoretically induce autoimmune disorders or cancer. Recent advances indicate that caspase-dependent apoptosis is not an exclusive mechanism for elimination of aberrant cells, but that there are many other backup mechanisms to promote cell death, which may serve as alternative systems. Additionally, there is a pleiotropic function of apoptotic caspases in which apoptotic caspases are involved in a non-apoptotic cellular process, such as immune cell development, activation, proliferation, and differentiation. Also, further investigations of the pleiotropic function of caspases are required to understand their regulatory mechanisms, specific targets, and cell specificity. Finally, the effect, if any, of long-term caspase inhibition on the immune system remains to be determined.

Nevertheless, caspases are clearly a viable target for a large variety of diseases, and there are likely to be potential beneficial effects of inhibiting caspase-1 in a range of diseases. Novel compounds arising from these efforts will provide the basis for the future direction of development for caspase inhibitors.

ACKNOWLEDGMENTS

I apologize to the authors who made important contributions to this field but have not been cited due to limitations of space. I thank Dr. Shinichi Koizumi and Dr. Katsuhiro Shinjo for critically reading the manuscript.

REFERENCES

1. Dinarello, C. A. 1996. Biological basis for interleukin-1 in disease. *Blood* 87:2095.
2. Dinarello, C. A. 1999. IL-18: A T_{H1}-inducing, proinflammatory cytokine and new member of the IL-1 family. *J. Allergy Clin. Immunol.* 103:11.
3. Kim, T. W., et al. 1997. Alternative cleavage of Alzheimer-associated presenilins during apoptosis by a caspase-3 family protease. *Science* 277:373.
4. Barnes, N. Y., et al. 1998. Increased production of amyloid precursor protein provides a substrate for caspase-3 in dying motoneurons. *J. Neurosci.* 18:5869.
5. Walter, J., et al. 1999. Phosphorylation of presenilin-2 regulates its cleavage by caspases and regards progression of apoptosis. *Proc. Natl. Acad. Sci. USA* 96:1391.
6. Hartmann, A., et al. 2000. Caspase-3: A vulnerability factor and final effector in apoptotic death of dopaminergic neurons in Parkinson's disease. *Proc. Natl. Acad. Sci. USA* 97:2875.
7. Hartmann, A., et al. 2001. Caspase-8 is an effector in apoptotic death of dopaminergic neurons in Parkinson's disease, but pathway inhibition results in neuronal necrosis. *J. Neurosci.* 21:2247.
8. Yue, T. L., et al. 1998. Staurosporine-induced apoptosis in cardiomyocytes: A potential role of caspase-3. *J. Mol. Cell Cardiol.* 30:495.
9. Narula, J., et al. 1999. Apoptosis in heart failure: Release of cytochrome c from mitochondria and activation of caspase-3 in human cardiomyopathy. *Proc. Natl. Acad. Sci. USA* 96:8144.
10. Jaeschke, H., et al. 1998. Activation of caspase 3 (CPP32)-like proteases is essential for TNF-alpha-induced hepatic parenchymal cell apoptosis and neutrophil-mediated necrosis in a murine endotoxin shock model. *J. Immunol.* 160:3480.
11. Tinsley, K. W., et al. 2000. Caspases -2, -3, -6, and -9, but not caspase-1, are activated in sepsis-induced thymocyte apoptosis. *Shock* 13:1.
12. Patel, T., Bronk, S. F., and Gores, G. J. 1994. Increases of intracellular magnesium promote glycodeoxycholate-induced apoptosis in rat hepatocytes. *J. Clin. Invest.* 94:2183.
13. Que, F. Q., and Gores, G. J. 1996. Cell death by apoptosis: Basic concepts and disease relevance for the gastroenterologist. *Gastroenterology* 110:1238.
14. Sasaki, H., et al. 1996. Activation of apoptosis during the reperfusion phase after rat liver ischemia. *Transplant. Proc.* 28:1908.
15. Borghi-Scoazec, G., et al. 1997. Apoptosis after ischemia-reperfusion in human liver allografts. *Liver Transplant. Surg.* 3:407.
16. Kuo, P. C., et al. 1998. Apoptosis and hepatic allograft reperfusion injury. *Clin. Transplant.* 12:219.
17. Schuchmann, M., and Galle, P. R. 2001. Apoptosis in liver disease. *Eur. J. Gastroenterol. Hepatol.* 13:785.
18. Natori, S., et al. 2001. Hepatocyte apoptosis is a pathologic feature of human alcoholic hepatitis. *J. Hepatol.* 34:248.
19. Ziol, M., et al. 2001. Clinical and biological relevance of hepatocyte apoptosis in alcoholic hepatitis. *J. Hepatol.* 34:254.

20. Bantel, H., et al. 2001. Caspase activation correlates with the degree of inflammatory liver injury in chronic hepatitis C virus infection. *Hepatology* 34:758.

21. Bantel, H., and Schulze-Osthoff, K. 2003. Apoptosis in hepatitis C virus infection. *Cell Death Differ.* 10:S48.

22. Kountouras, J., Zavos, C., and Chatzopoulos, D. 2003. Apoptosis in hepatitis C. *J. Viral. Hepatol.* 10:335.

23. Feldstein, A. E., et al. 2003. Hepatocyte apoptosis and Fas expression are prominent features of human nonalcoholic steatohepatitis. *Gastroenterology* 125:437.

24. Ribeiro, P. S., et al. 2004. Hepatocyte apoptosis, expression of death receptors, and activation of NF-κB in the liver of nonalcoholic and alcoholic steatohepatitis patients. *Am. J. Gastroenterol.* 99:1708.

25. Abrahamson, E. E., et al. 2006. Caspase inhibition therapy abolishes brain trauma-induced increases in Abeta peptide: Implications for clinical outcome. *Exp. Neurol.* 197:437.

26. Yang, L., et al. 2004. A novel systemically active caspase inhibitor attenuates the toxicities of MPTP, malonate, and 3NP *in vivo*. *Neurobiol. Dis.* 17:250.

27. Neviere, R., et al. 2001. Caspase inhibition prevents cardiac dysfunction and heart apoptosis in a rat model of sepsis. *Am. J. Respir. Crit. Care Med.* 163:218.

28. Hayakawa, Y., et al. 2003. Inhibition of cardiac myocyte apoptosis improves cardiac function and abolishes mortality in the peripartum cardiomyopathy of Gαq transgenic mice. *Circulation* 108:3036.

29. Hotchkiss, R. S., et al. 1999. Prevention of lymphocyte cell death in sepsis improves survival in mice. *Proc. Natl. Acad. Sci. USA* 96:14541.

30. Schulz, J. B., Weller, M., and Moskowitz, M. A. 1999. Caspases as treatment targets in stroke and neurodegenerative diseases. *Ann. Neurol.* 45:421.

31. Braun, J. S., Tuomanen, E. I., and Cleveland, J. L. 1999. Neuroprotection by caspase inhibitors. *Expert Opin. Invest. Drugs* 8:1599.

32. Sastry, P. S., and Rao, K. S. 2000. Apoptosis and the nervous system. *J. Neurochem.* 74:1.

33. Jaeschke, H., et al. 2000. Protection against TNF-induced liver parenchymal cell apoptosis during endotoxemia by a novel caspase inhibitor in mice. *Toxicol. Appl. Pharmacol.* 169:77.

34. Kim, K. M., et al. 2000. A broad-spectrum caspase inhibitor blocks concanavalin A-induced hepatitis in mice. *Clin. Immunol.* 97:221.

35. Yang, W., et al. 2003. MX1013, a dipeptide caspase inhibitor with potent *in vivo* antiapoptotic activity. *Br. J. Pharmacol.* 140:402.

36. Natori, S., et al. 2003. The caspase inhibitor IDN-6556 prevents caspase activation and apoptosis in sunisoidal endothelial cells during liver preservation injury. *Liver Transplant.* 9:278.

37. Niel, C., et al. 2003. Characterization of IDN-6556 (3-{2-(2-*tert*-butyl-phenylaminooxalyl)-amino]-propionylamino}-4-oxo-5-(2,3,5,6-tetrafluoro-phenyl)-pentanoic acid): A liver-targeted caspase inhibitor. *J. Pharmacol. Exp. Ther.* 309:634.

38. Canbay, A., et al. 2004. The caspase inhibitor IDN-6556 attenuates hepatic injury and fibrosis in the bile duct ligated mouse. *J. Pharmacol. Exp. Ther.* 308:1191.

39. Baskin-Bey, E. S., et al. 2006. The pan-caspase inhibitor, IDN-6556, attenuates human liver preservation injury. *Hepatology* 44(Suppl. 1):190A.

40. Shiffman, M. L., et al. 2006. PF-03491390 (formerly IDN-6556), a pancaspase inhibitor, is well-tolerated and effectively reduces raised aminotransferases (ALT and AST) in chronic active hepatitis C (HCV) patients (PTS). *Hepatology* 44(Suppl. 1):224A.

41. Li, P., et al. 1995. Mice deficient in IL-1 beta-converting enzyme are defective in production of mature IL-1 beta and resistant to endotoxic shock. *Cell* 80:401.

42. Kuida, K., et al. 1995. Altered cytokine export and apoptosis in mice deficient in interleukin-1 beta converting enzyme. *Science* 267:2000.

43. Siegmund, B., and Zeits, M. 2003. Pralnacasan Vertex Pharmaceuticals. *Idrugs* 6:154.
44. Ku, G., et al. 2001. Selective interleukin-1β converting enzyme (ICE/Caspase-1) inhibition with pralnacasan (HMR 3480/VX-740) reduces inflammation and joint destruction in murine type II collagen-induced arthritis (CIA). *Arthritis Rheum.* 44: S241.
45. Rudolphi, K., et al. 2003. Pralnacasan, an inhibitor of interleukin-1 beta converting enzyme, reduces joint damage in two murine models of osteoarthritis. *Osteoarthritis Cartilage* 11:738.
46. Loher, F., et al. 2004. The interleukin-1 beta converting enzyme inhibitor pralnacasan reduces dextran sulfate sodium-induced murine colitis and T helper 1 T-cell activation. *J. Pharmacol. Exp. Ther.* 308:583.
47. Pavelka, K., et al. 2002. Clinical effects of pralnacasan (PRAL), an orally-active interleukin-1 beta converting enzyme (ICE) inhibitor, in a 285 patient PhII trial in rheumatoid arthritis (RA). In *66th Annual Science Meeting of the American College of Rheumatology*, abstract LB02.
48. Fischer, U., and Schulze-Osthoff, K. 2005. Apoptosis-based therapies and drug targets. *Cell Death Differ.* 12:942.
49. Randle, J., Ku, G., and Qadri, A. 2001. Interleukin-1β converting enzyme (caspase-1) inhibition with VX-765 reduces inflammation and cytokine levels in murine oxazolone-induced dermatitis. *J. Invest. Dermatol.* 117:532.
50. Braddock, M. 2003. Inflammatory process in drug discovery—SRI conference. *IDrugs* 6:1049.
51. Braddock, M., and Quinn, A. 2004. Targeting IL-1 in inflammatory disease: New opportunities for therapeutic intervention. *Nat. Rev. Drug Discovery* 3:1.
52. Le, G. T., and Abbenante, G. 2005. Inhibitors of TACE and caspase-1 as anti-inflammatory drugs. *Curr. Med. Chem.* 12: 2963.
53. Stack, J. H., et al. 2005. IL-converting enzyme/caspase-1 inhibitor VX-765 blocks the hypersensitive response to an inflammatory stimulus in monocytes from familial cold autoinflammatory syndrome patients. *J. Immunol.* 175:2630.
54. Faubel, S., and Edelstein, C. L. 2005. Caspases as drug targets in ischemic organ injury. *Curr. Drug Targets Immune Endocr. Metabol. Disord.* 5:269.
55. Siegmund, B., et al. 2001. IL-1β-converting enzyme (caspase-1) in intestinal inflammation. *Proc. Natl. Acad. Sci. USA* 98:13249.
56. Schielke, G. P., et al. 1998. Reduced ischemic brain injury in interleukin-1 beta converting enzyme-deficient mice. *J. Cereb. Blood Flow Metab.* 18:180.
57. Ona, V. O., et al. 1999. Inhibition of caspase-1 slows disease progression in a mouse model of Huntinton's disease. *Nature* 399:263.
58. Rodriguez, I., et al. 1996. Systemic injection of tripeptide inhibits the intracellular activation of CPP-32-like proteases *in vivo* and fully protects mice against Fas-mediated fulminant liver destruction and death. *J. Exp. Med.* 184:2067.
59. Hiramatsu, N., et al. 1994. Immunohistochemical detection of Fas antigen in liver tissue of patients with chronic hepatitis C. *Hepatology* 19:1354.
60. Mita, E., et al. 1994. Role of Fas ligand in apoptosis induced by hepatitis C virus infection. *Biochem. Biophys. Res. Commun.* 204:468.
61. Okazaki, M., et al. 1996. Hepatic Fas antigen expression before and after interferon therapy in patients with chronic hepatitis C. *Dig. Dis. Sci.* 41:2453.
62. Ando, K., et al. 1994. Class I-restricted cytotoxic T lymphocytes are directly cytopathic for their target cells *in vivo. J. Immunol.* 152:3245.
63. Kondo, T., et al. 1997. Essential roles of the Fas ligand in the development of hepatitis. *Nature Med.* 3:409.
64. Iio, S., et al. 1998. Serum levels of soluble Fas antigen in chronic hepatitis C patients. *J. Hepatol.* 29:517.

65. Pianko, S., et al. 2001. Fas-mediated hepatocyte apoptosis is increased by hepatitis C virus infection and alcohol consumption, and may be associated with hepatic fibrosis: Mechanism of liver injury in chronic hepatitis C virus infection. *J. Viral Hepat.* 8:406.

66. Bantel, H., et al. 2001. Caspase activation correlates with the degree of inflammatory liver injury in chronic hepatitis C virus infection. *Hepatology* 34:758.

67. Jaeschke, H., et al. 1998. Activation of caspase 3 (CPP-32)-like proteases is essential for TNF-alpha-induced hepatic parenchymal cell apoptosis and neutrophil-mediated necrosis in a murine endotoxin shock model. *J. Immunol.* 160:3480.

68. Lawson, J. A., et al. 1998. Parenchymal cell apoptosis as a signal for sinusoidal sequesteration and transendothelial migration of neutrophils in murine models of endotoxin and Fas-antibody-induced liver injury. *Hepatology* 28:865.

69. Bantel, H., et al. 2004. Detection of apoptotic caspase activation in sera from patients with chronic HCV infection is associated with fibrotic liver injury. *Hepatology* 40:1078.

70. Faubion, W. A., et al. 1999. Toxic bile salts induce rodent hepatocyte apoptosis via direct activation of Fas. *J. Clin. Invest.* 103:137.

71. Higuchi, H., et al. 2002. Cholestasis increases tumor necrosis factor-related apoptosis-inducing ligand (TRAIL)-R2/DR5 expression and sensitizes the liver to TRAIL-mediated cytotoxicity. *J. Pharmacol. Exp. Ther.* 303:461.

72. Higuchi, H., et al. 2003. Bile acids stimulate cFLIP phosphorylation enhancing TRAIL-mediated apoptosis. *J. Biol. Chem.* 278:454.

73. Goldin, R. D., et al. 1993. Apoptotic bodies in a murine model of alcoholic liver disease: Reversibility of ethanol-induced changes. *J. Pathol.* 171:73.

74. Yacoub, L. K., et al. 1995. Apoptotic and bcl-2 protein expression in experimental alcoholic liver disease in the rat. *Alcohol Clin. Exp. Res.* 19:854.

75. Castilla, R., et al. 2004. Dual effect of ethanol on cell death in primary culture of human and rat hepatocytes. *Alcohol Alcoholism* 39:290.

76. Arteel, G. E. 2003. Oxidants and antioxidants in alcohol-induced liver disease. *Gastroenterology* 124:778.

77. Thurman, R. G. 1998. Mechanisms of hepatic toxicity. II. Alcoholic liver injury involves activation of Kupffer cells by endotoxin. *Am. J. Physiol.* 275:G605.

78. Uesugi, T., et al. 2001. Toll-like receptor 4 is involved in the mechanism of early alcohol-induced liver injury in mice. *Hepatology* 34:101.

79. Kono, H., et al. 2000. NADPH oxidase-derived free radicals are key oxidants in alcohol-induced liver disease. *J. Clin. Invest.* 106:867.

80. Yin, M., et al. 1999. Essential role of tumor necrosis factor α in alcohol-induced liver injury in mice. *Gastroenterology* 117:942.

81. Morimoto, M., et al. 1993. Role of cytochrome P4502E1 in alcoholic liver disease pathogenesis. *Alcohol* 10:459.

82. Morimoto, M., et al. 1995. Modulation of experimental alcohol-induced liver disease by cytochrome P450 2E1 inhibitors. *Hepatology* 21:1610.

83. Morgan, K., French, S. W., and Morgan, T. R. 2002. Production of a cytochrome P450 2E1 transgenic mouse and initial evaluation of alcoholic liver damage. *Hepatology* 36:122.

84. Kono, H. et al. 1999. CYP2E1 is not involved in early alcohol-induced liver injury. *Am. J. Physiol.* 277:G1259.

85. Dey, A., and Cederbaum, A. I. 2006. Alcohol and oxidative liver injury. *Hepatology* 43:S63.

86. Farrell, G. C. 2003. Non-alcoholic steatohepatitis: What is it, and why is it important in the Asia-Pacific region? *J. Gastroenterol. Hepatol.* 18:124.

87. Clark, J. M., Brancati, F. L., and Diehl, A. M. 2003. The prevalence and etiology of elevated aminotransferase levels in the United States. *Am. J. Gastroenterol.* 98:960.
88. McCullough, A. J. 2006. Pathophysiology of nonalcoholic steatohepatitis. *J. Clin. Gastroenterol.* 40:S17.
89. Ludwig, J., et al. 1980. Nonalcoholic steatohepatitis: Mayo Clinic experiences with a hitherto unnamed disease. *Mayo Clin. Proc.* 55:434.
90. Day, C. P., and James, O. F. W. 1998. Steatohepatitis: A tale of two "hits"? *Gastroenterology* 114:842.
91. Shepherd, R., and Kahn, B. B. 1999. Glucose transporters and insulin action. *N. Engl. J. Med.* 341:248.
92. Chitturi, A., et al. 2002. NASH and insulin resistance: Insulin hypersecretion and specific association with the insulin resistance syndrome. *Hepatology* 35:373.
93. Cohen, P., et al. 2002. Role of stearoyl-CoA desaturase-1 in leptin-mediated weight loss. *Science* 279:240.
94. Matsusue, K., et al. 2003. Liver-specific disruption of PPARgamma in leptin-deficient mice improves fatty liver but aggravates diabetic phenotypes. *J. Clin. Invest.* 111:737.
95. Pessayre, D., Fromenty, B., and Mansouri, A. 2004. Mitochondrial injury in steatohepatitis. *Eur. J. Gastroenterol. Hepatol.* 16:1095.
96. Berson, A., et al. 1998. Steatohepatitis-inducing drugs cause mitochondrial dysfunction and lipid peroxidation in rat hepatocytes. *Gastroenterology* 114:764.
97. Sanyal, A. J., et al. 2001. Nonalcoholic steatohepatitis: Association of insulin resistance and mitochondrial abnormalities. *Gastroenterology* 120:1183.
98. Chalasani, N., et al. 2004. Systemic levels of lipid peroxidation and its metabolic and dietary correlates in patients with nonalcoholic steatohepatitis. *Am. J. Gastroenterol.* 99:1497.
99. Passayre, D., and Fromenty, B. 2005. NASH: A mitochondrial disease. *J. Hepatol.* 42:928.
100. Weltman, M. D., et al. 1998. Hepatic cytochrome P450 2E1 is increased in patients with nonalcoholic steatohepatitis. *Hepatology* 27:128.
101. Brun, P., et al. 2007. Increased intestinal permeability in obese mice: New evidences in the pathogenesis of nonalcoholic steatohepatitis. *Am. J. Physiol. Gastrointest. Liver Physiol.* 292:G518.
102. Crespo, J., et al. 2001. Gene expression of tumor necrosis factor a and TNF-receptors, p55 and p75, in nonalcoholic steatohepatitis patients. *Hepatology* 34:1158.
103. Hui, J. M., et al. 2004. Beyond insulin resistance in NASH: TNF-α or adiponectin? *Hepatology* 40:46.
104. Kirkland, R. A., et al. 2002. A Bax-induced pro-oxidant state is critical for cytochrome c release during programmed neuronal death. *J. Neurosci.* 22:6480.
105. Gao, W., et al. 1998. Apoptosis of sinusoidal endothelial cells is a critical mechanism of preservation injury in rat liver transplantation. *Hepatology* 27:1652.
106. Kohli, V., et al. 1999. Endothelial cell and hepatocyte deaths occur by apoptosis after ischemia-reperfusion injury in the rat liver. *Transplantation* 67:1099.
107. Natori, S., et al. 1999. Apoptosis of sinusoidal endothelial cells occurs during liver preservation injury by a caspase-dependent mechanism. *Transplantation* 68:89.
108. Caldwell-Kentel, J. C., et al. 1991. Kupffer cell activation and endothelial cell damage after storage of rat livers: Effects of reperfusion. *Hepatology* 13:83.
109. Sndram, D., et al. 2001. Synergism between platelets and leukocytes in inducing endothelial cell apoptosis in the cold ischemic rat liver: Kupffer cell-mediated injury. *FASEB J.* 15:1230.
110. Cursio, R., et al. 1999. A caspase inhibitor fully protects rats against lethal normothermic liver ischemia by inhibition of liver apoptosis. *FASEB J.* 13:253.

111. Jaeschke, H., and Lemasters, J. J. 2003. Apoptosis versus oncotic necrosis in hepatic ischemia/reperfusion injury. *Gastroenterology* 125:1246.

112. Gujal, J. S., et al. 2001. Mechanism of cell death during warm hepatic ischemia-reperfusion in rats: Apoptosis or necrosis? *Hepatology* 33:397.

113. Friedman, S. L. 2003. Liver fibrosis—From bench to bedside. *J. Hepatol.* 38:S38.

114. Bataller, R., and Brenner, D. A. 2005. Liver fibrosis. *J. Clin. Invest.* 115:209.

115. Canbay, A., et al. 2003. Apoptotic body engulfment by a human stellate cell line is profibrogenic. *Lab. Invest.* 83:655.

116. Zhan, S. S., et al. 2006. Phagocytosis of apoptotic bodies by hepatic stellate cells induces NADHP oxidase and is associated with liver fibrosis *in vivo*. *Hepatology* 43:435.

117. Syed, M., et al. 2005. Caspase inhibition improves ischemia-reperfusion injury after lung transplantation. *Am. J. Transplant.* 5:292.

118. Miyoshi, H., et al. 1999. Hepatocyte apoptosis after bile duct ligation in the mouse involves Fas. *Gastroenterology* 117:669.

119. Canbay, A., et al. 2002. Fas enhances fibrogenesis in the bile duct ligated mouse: A link between apoptosis and fibrosis. *Gastroenterology* 123:1323.

120. Hoglen, N. C., et al. 2003. IDN-6556, the first anti-apoptotic caspase inhibitor: Pre-clinical efficacy and safety. *Hepatology* 38(Suppl.1):579A.

121. Valentino, K. L., et al. 2003. First clinical trial of a novel caspase inhibitor: Anti-apoptotic caspase inhibitor, IDN-6556, improves liver enzymes. *Int. J. Clin. Pharmacol. Ther.* 41:441.

122. Schiff, E. R., et al. 2004. Oral, IDN-656, an anti-apoptotic caspase inhibitor, lowers aminotransferases in HCV patients. *J. Hepatol.* 40(Suppl. 1):24.

123. Baskin-Bey, E. S., et al. 2007. Clinical trial of the pan-caspase inhibitor, IDN-6556, in human liver preservation injury. *Am. J. Transplant.* 7:218.

124. Leers, M. P., et al. 1999. Immunohistochemical detection and mapping of a cytokera-tin 18 neo-epitope exposed during early apoptosis. *J. Pathol.* 187:567.

125. Alvarado-Kristensson, M., et al. 2004. p38-MAPK signals survival by phosphoryla-tion of caspase-8 and caspase-3 in human neutrophils. *J. Exp. Med.* 199:449.

126. Rossi, A. G., et al. 2006. Cyclin-dependent kinase inhibitors enhance the resolution of inflammation by promoting inflammatory cell apoptosis. *Nature Med.* 12:1056.

127. Riley, N. A., et al. 2006. Granulocyte apoptosis and macrophage clearance of apop-totic cells as targets for pharmacological intervention in inflammatory diseases. *Anti-infl. Antiallergy Agents Med. Chem.* 5:3.

128. Park, M., et al. 2004. LB84318, a novel and small-molecule caspase inhibitor, potently protects hepatic damage in mouse models of liver injury. *J. Hepatol.* 40(Suppl. 1): 147.

129. Park, M., et al. 2004. LB84318, a novel and small-molecule caspase inhibitor, potently protects hepatic damage in mouse models of liver injury. *Hepatology* 40(Suppl. 4):694A.

130. Chang, H. K., et al. 2004. Discovery of a novel caspase inhibitor and protection of liver damage in mouse models. In *228th ACS National Meeting*, abstract 267.

131. Manns, M. P., et al. 2001. Peginterferon alfa-2b plus ribavirin compared with inter-feron alfa-2b plus ribavirin for initial treatment of chronic hepatitis C: A randomized trial. *Lancet* 358:958.

132. Fried, M. W., et al. 2002. Peginterferon alfa-2a plus ribavirin for chronic hepatitis C virus infection. *N. Engl. J. Med.* 347:975.

133. Fried, M. W. 2002. Side effects of therapy of hepatitis C and their management. *Hepatology* 36:S237.

134. Wieckowska, A., et al. 2006. *In vivo* assessment of liver cell apoptosis as a novel biomarker of disease severity in nonalcoholic fatty liver disease. *Hepatology* 44:27.

135. Malhi, H., et al. 2006. Free fatty acids induce JNK-dependent hepatocyte lipoapoptosis. *J. Biol. Chem.* 281:12093.
136. Jaeschke, H., Farhood, A., and Smith, C. W. 1990. Neutrophils contribute to ischemia/reperfusion injury in rat liver *in vivo*. *FASEB J.* 4:3355.
137. Jaeschke, H., et al. 1993. Functional inactivation of neutrophils with a Mac-1 (CD11b/CD18) monoclonal antibody protects against ischemia-reperfusion injury in rat liver. *Hepatology* 17:915.
138. Jaeschke, H. 2003. Molecular mechanisms of hepatic ischemia-reperfusion injury and preconditioning. *Am. J. Physiol. Gastrointest. Liver Physiol.* 284:G15.
139. Kobayashi, A., et al. 2001. Mac-1 (CD11b/CD18) and intercellular adhesion molecule-1 in ischemia-reperfusion injury of rat liver. *Am. J. Physiol. Gastrointest. Liver Physiol.* 281:G577.
140. Contreras, J. L., et al. 2004. Caspase-8 and caspase-3 small interfering RNA decreases ischemia/reperfusion injury to the liver in mice. *Surgery* 136:390.
141. Malhi, H., Gores, G. J., and Lemasters, J. J. 2006. Apoptosis and necrosis in the liver: A tale of two deaths? *Hepatology* 43:S31.
142. Norman, J., et al. 1997. Severity and mortality of experimental pancreatitis are dependent on interleukin-1 converting enzyme (ICE). *J. Interferon Cytokine Res.* 17:113.
143. Melnikov, V. Y., et al. 2001. Impaired IL-18 processing protects caspase-1-deficient mice from ischemic acute renal failure. *J. Clin. Invest.* 107:1145.
144. Xu, H., et al. 2001. Attenuation of hypoxia-ischemia-induced monocyte chemoattractant protein-1 expression in brain of neonatal mice deficient in interleukin-1 converting enzyme. *Mol. Brain Res.* 90:57.
145. Koedel, U., et al. 2002. Role of caspase-1 in experimental pneumococcal meningitis: Evidence from pharmacologic caspase inhibition and caspase-1-deficient mice. *Ann. Neurol.* 51:319.
146. Bröker, L. E., Kruyt, F. A. E., and Giaccone, G. 2005. Cell death independent of caspases: A review. *Clin. Cancer Res.* 11:3155.
147. Hetz, C. A., Torres, V., and Quest, A. F. G. 2005. Beyond apoptosis: Nonapoptotic cell death in physiology and disease. *Biochem. Cell Biol.* 83:579.
148. Kim, R., et al. 2006. Regulation and interplay of apoptotic and non-apoptotic cell death. *J. Pathol.* 208:319.
149. Kroemer, G., and Martin, S. J. 2005. Caspase-independent cell death. *Nature Med.* 11:725.
150. Hail, Jr., N., et al. 2006. Apoptosis effector mechanisms: A requiem performed in different keys. *Apoptosis* 11:889.
151. Vandenabeele, P., Berghe, T. V., and Festjens, N. 2006. Caspase inhibitors promote alternative cell death pathways. *Science STKE,* October 24, pe44.
152. Xiang, J., Chao, D. T., and Korsmeyer, S. J. 1996. BAX-induced cell death may not require interleukin 1β-converting enzyme-like proteases. *Proc. Natl. Acad. Sci. USA* 93:14559.
153. Pérez-Galán, P., et al. 2002. Role of caspases and apoptotic-inducing factor (AIF) in cladribine-induced apoptosis of B cell chronic lymphotic leukemia. *Leukemia* 16:2106.
154. Cauwels, A., et al. 2003. Caspase inhibition causes hyperacute tumor necrosis factor-induced shock via oxidative stress and phospholipase A2. *Nature Immunol.* 4:387.
155. Bidère, N., et al. 2003. Cathepsin D triggers Bax activation, resulting in selective apoptosis-inducing factor (AIF) relocation in T lymphocytes entering the early commitment phase to apoptosis. *J. Biol. Chem.* 278:31401.
156. Carter, B. Z., et al. 2003. Caspase-independent cell death in AML: Caspase inhibition *in vitro* with pan-caspase inhibitors or *in vivo* by XIAP or survivin does not affect cell survival or prognosis. *Blood* 102:4179.

157. Maianski, N. A., Roos, D., and Kuijpers, T. W. 2003. Tumor necrosis factor a induces a caspase-independent death pathway in human neutrophils. *Blood* 101:1987.

158. Konopleva, M., et al. 2004. The synthetic triterpenoid 2-cyano-3, 12-dioxooleana-1, 9-dien-28-oic acid induces caspase-dependent and -independent apoptosis in acute myelogenous leukemia. *Cancer Res.* 64:7927.

159. Kanzawa, T., et al. 2005. Arsenic trioxide induces autophagic cell death in malignant glioma cells by upregulation of mitochondrial cell death protein BNIP3. *Oncogene* 24:980.

160. Lum, J. J., et al. 2005. Growth factor regulation of autophagy and cell survival in the absence of apoptosis. *Cell* 120:237.

161. Elmore, S. P., et al. 2001. The mitochondrial permeability transition initiates autophagy in rat hepatocytes. *FASEB J.* 15:2286.

162. MacDonald, G., et al. 1999. Mitochondria-dependent and -independent regulation of granzyme B-induced apoptosis. *J. Exp. Med.* 189:131.

163. Gozuacik, D., and Kimchi, A. 2004. Autophagy as a cell death and tumor suppressor mechanism. *Oncogene* 23:2891.

164. Kisen, G. O., et al. 1993. Reduced autophagic activity in primary rat hepatocellular carcinoma and ascites hepatoma cells. *Carcinogenesis* 14:2501.

165. Liang, X. H., et al. 1999. Induction of autophagy and inhibition of tumorigenesis by beclin 1. *Nature* 402:672.

166. Edinger, A. L., and Thompson, C. B. 2003. Defective autophagy leads to cancer. *Cancer Cell* 4:422.

167. Paglin, S., et al. 2001. A novel response of cancer cells to radiation involves autophagy and formation of acidic vesicles. *Cancer Res.* 61:439.

168. Degenhardt, K., et al. 2006. Autophagy promotes tumor cell survival and restricts necrosis, inflammation, and tumorigenesis. *Cancer Cell* 10:51.

169. Cuervo, A. M. 2004. Autophagy: In sickness and in health. *Trends Cell Biol.* 14:70.

170. Shintani, T., and Klionsky, D. J. 2004. Autophagy in health and disease: A double-edged sword. *Science* 306:990.

171. Kondo, Y., et al. 2005. The role of autophagy in cancer development and response to therapy. *Nat. Rev. Cancer* 5:726.

172. Jin, S., and White, E. 2007. Role of autophagy in cancer. *Autophagy* 3:28.

173. Nakagawa, T., et al. 2000. Caspase-12 mediates endoplasmic-reticulum-specific apoptosis and cytotoxicity by amyloid-β. *Nature* 403:98.

174. Mathiasen, I. S., Lademann, U., and Jäättelä, M. 1999. Apoptosis induced by vitamin D compounds in breast cancer cells is inhibited by Bcl-2 but does not involve known caspase or p53. *Cancer Res.* 59:4898.

175. Mathiasen, I. S., et al. 2002. Calcium and calpain as key mediators of apoptosis-like death induced by vitamin D compounds in breast cancer cells. *J. Biol.Chem.* 277:30738.

176. Lorenzo, H. K., and Susin, S. A. 2004. Mitochondrial effectors in caspase-independent cell death. *FEBS Lett.* 557:14.

177. Kim, R., Emi, M., and Tanabe, K. 2006. Role of mitochondria as the gardens of cell death. *Cancer Chemother. Pharmacol.* 57:545.

178. Susin, S. A., et al. 1996. Bcl-2 inhibits the mitochondrial release of an apoptogenic protease. *J. Exp. Med.* 184:1331.

179. Susin, S. A., et al. 1999. Molecular characterization of mitochondrial apoptosis-inducing factor. *Nature* 397:441.

180. Suzuki, Y., et al. 2001. A serine protease, HtrA2, is released from the mitochondria and interacts with XIAP, inducing cell death. *Mol. Cell.* 8:613.

181. Hegde, R., et al. 2002. Identification of Omi/HtrA2 as a mitochondrial apoptotic serine protease that disrupts inhibitor of apoptosis protein-caspase interaction. *J. Biol. Chem.* 277:432.

182. Li, L. Y., and Wang, X. 2001. Endonuclease G is an apoptotic DNase when released from mitochondria. *Nature* 412:95.

183. Vercammen, D., et al. 1998. Inhibition of caspases increases the sensitivity of L929 calls to necrosis mediated by tumor necrosis factor. *J. Exp. Med.* 187:1477.

184. Holler, N., et al. 2000. Fas triggers an alternative, caspase-8-independent cell death pathway using the kinase RIP as effector molecule. *Nature Immunol.* 1:489.

185. Denecker, G., et al. 2001. Death receptor-induced apoptotic and necrotic cell death: Differential role of caspases and mitochondria. *Cell Death Differ.* 8:829.

186. Lin, Y., et al. 2004. Tumor necrosis factor-induced nonapoptotic cell death requires receptor-interacting protein-mediated cellular reactive oxygen species accumulation. *J. Biol. Chem.* 279:10822.

187. Balkwill, F., and Mantovani, A. 2001. Inflammation and cancer: Back to Virchow? *Lancet* 357:539.

188. Coussens, L. M., and Werb, Z. 2002. Inflammation and cancer. *Nature* 420:860.

189. Lu, H., et al. 2006. Inflammation, a key event in cancer development. *Mol. Cancer Res.* 4:221.

190. Saijo, Y., et al. 2002. Proinflammatory cytokine IL-1β promotes tumor growth of Lewis lung carcinoma by induction of angiogenesis factors: *In vivo* analysis of tumor-stromal interaction. *J. Immunol.* 169:469.

191. Voronov, E., et al. 2003. IL-1 is required for tumor invasiveness and angiogenesis. *Proc. Natl. Acad. Sci. USA* 100:2645.

192. Jung, Y. J., et al. 2003. IL-1β-mediated up-regulation of HIF-1α via an NFκB/COX-2 pathway identifies HIF-1 as a critical link between inflammation and oncogenesis. *FASEB J.* 17:2115.

193. Teitz, T., et al. 2000. Caspase 8 is deleted or silenced preferentially in childhood neuroblastomas with amplification of MYCN. *Nature Med.* 6:529.

194. Teitz, T., Lahti, J. M., and Kidd, V. J. 2001. Aggressive childhood neuroblastomas do not express caspase-8: An important component of programmed cell death. *J. Mol. Med.* 79:428.

195. Fulda, S., et al. 2001. Sensitization for death receptor- or drug-induced apoptosis by re-expression of caspase-8 through demethylation or gene transfer. *Oncogene* 20:5865.

196. Harada, K., et al. 2002. Deregulation of caspase-8 and 10 expression in pediatric tumors and cell lines. *Cancer Res.* 62:5897.

197. Zuzak, T. J., et al. 2002. Loss of caspase-8 mRNA expression is common in childhood primitive neuroectodermal brain tumour/medulloblastoma. *Eur. J. Cancer* 38:83.

198. Joseph, B., et al. 1999. Differences in expression of pro-caspases in small cell and non-small cell lung carcinoma. *Biochem. Biophys. Res. Commun.* 262:381.

199. Shivapurkar, N., et al. 2002. Differential inactivation of caspase-8 in lung cancers. *Cancer Biol. Ther.* 1:65.

200. Shivapurkar, N., et al. 2002. Loss of expression of death-inducing signaling complex (DISC) components in lung cancer cell lines and the influence of MYC amplification. *Oncogene* 21:8510.

201. Hopkins-Donaldson, S., et al. 2003. Silencing of death receptor and caspase-8 expression in small cell lung carcinoma cell lines and tumors by DNA methylation. *Cell Death Differ.* 10:356.

202. Fulda, S., et al. 2006. Loss of caspase-8 expression does not correlate with MYCN amplification, aggressive disease, or progress in neuroblastoma. *Cancer Res.* 66:10016.

203. Ricci-Vitiani, L., et al. 2004. Absence of caspase 8 and high expression of PED protect primitive neural cells from cell death. *J. Exp. Med.* 200:1257.

204. Estrov, Z., et al. 1998. Caspase 2 and caspase 3 protein levels as predictors of survival in acute myelogenous leukemia. *Blood* 92:3090.

205. Campos, L., et al. 1999. Expression of apoptosis-controlling proteins in acute leukemia cells. *Leuk. Lymphoma* 33:499.

206. Prokop, A., et al. 2000. Relapse in childhood acute lymphoblastic leukemia is associated with a decrease of the Bax/Bcl-2 ratio and loss of spontaneous caspase-3 processing *in vivo. Leukemia* 14:1606.

207. Svingen, P. A., et al. 2000. Evaluation of Apaf-1 and procaspase-2, -3, -7, -8, and -9 as potential prognostic markers in acute leukemia. *Blood* 96:3922.

208. Oliver, L., et al. 2002. Assessment of caspase activity as a possible prognostic factor in acute myeloid leukemia. *Br. J. Haematol.* 118:434.

209. Kaufmann, S. H., and Gores, G. J. 2000. Apoptosis in cancer: Cause and cure. *Bioessays* 22:1007.

210. Zöring, M., et al.2001. Apoptosis regulators and their role in tumorigenesis. *Biochim. Biophys. Acta* 1551:F1.

211. Zhivotovsky, B., and Orrenius, S. 2006. Carcinogenesis and apoptosis: Paradigms and paradoxes. *Carcinogenesis* 27:1939.

212. Schwerk, C., and Schulze-Osthoff, K. 2003. Non-apoptotic functions of caspases in cellular proliferation and differentiation. *Biochem. Pharmacol.* 66:1453.

213. Oliver, L., and Vallette, F. M. 2005. The role of caspases in cell death and differentiation. *Drug Resistance Updates* 8:163.

214. Siegel, R. M. 2006. Caspases at the crossroads of immune-cell life and death. *Nature Rev. Immunol.* 6:308.

215. Lamkanfi, M., et al. 2007. Caspases in cell survival, proliferation and differentiation. *Cell Death Differ.* 14:44.

216. Zermati, Y., et al. 2001. Caspase activation is required for terminal erythroid differentiation. *J. Exp. Med.* 193:247.

217. Carlile, G. W., Smith, D. H., and Wiedmann, M. 2004. Caspase-3 has a nonapoptotic function in erythroid maturation. *Blood* 103:4310.

218. Sordet, O., et al. 2002. Specific involvement of caspases in the differentiation of monocytes into macrophages. *Blood* 100:4446.

219. de Botton, S., et al. 2002. Platelet formation is the consequence of caspase activation within megakaryocytes. *Blood* 100:1310.

220. Kuida, K., et al. 1996. Decreased apoptosis in the brain and premature lethality in CPP-32-deficient mice. *Nature* 384:368.

221. Woo, M., et al. 1998. Essential contribution of caspase 3/CPP32 to apoptosis and its associated nuclear changes. *Genes Dev.* 12:806.

222. Zheng, T. S., et al. 1998. Caspase-3 controls both cytoplasmic and nuclear events associated with Fas-mediated apoptosis *in vivo. Proc. Natl. Acad. Sci. USA* 95:13618.

223. Woo, M., et al. 2003. Caspase-3 regulates cell cycle in B cells: A consequence of substrate specificity. *Nature Immunol.* 4:1016.

224. Kang, T. B., et al. 2004. Caspase-8 serves both apoptotic and nonapoptotic roles. *J. Immunol.* 173:2976.

225. Pellegrini, M., et al. 2005. FADD and caspase-8 are required for cytokine-induced proliferation of hemopoietic progenitor cells. *Blood* 106:1581.

226. Doerfler, P., et al. 2000. Caspase enzyme activity is not essential for apoptosis during thymocyte development. *J. Immunol.* 164:4071.

227. Zhang, J., et al. 2000. Spontaneous thymocyte apoptosis is regulated by a mitochondrion-mediated signaling pathway. *J. Immunol.* 165:2970.

228. Li, J., et al. 2005. Survival versus neglect: Redefining thymocyte subsets based on expression of NKG2D ligand(s) and MHC class I. *Eur. J. Immunol.* 35:439.

229. Villunger, A., et al. 2004. Negative selection of semimature CD4⁺HAS⁺ thymocytes requires the BH3-only protein Bim but is independent of death receptor signaling. *Proc. Natl. Acad. Sci. USA* 101:7052.

230. Alam, A., et al. 1999. Early activation of caspases during T lymphocyte stimulation results in selective substrate cleavage in nonapoptotic cells. *J. Exp. Med.* 190:1879.

231. Kennedy, N. J., et al. 1999. Caspase activation is required for T cell proliferation. *J. Exp. Med.* 190:1891.

232. Wang, J., et al. 1999. Inherited human caspase 10 mutations underlie defective lymphocyte and dendritic cell apoptosis in autoimmune lymphoproliferative syndrome type II. *Cell* 98:47.

233. Chun, H. J., et al. 2002. Pleiotropic defects in lymphocyte activation caused by caspase-8 mutations lead to human immunodeficiency. *Nature* 419:395.

234. Worth, A., Thrasher, A. J., and Gaspar, H. B. 2006. Autoimmune lymphoproliferative syndrome: Molecular basis of disease and clinical phenotype. *Br. J. Haematol.* 133:124.

235. Zhu, S., et al. 2006. Genetic alterations in caspase-10 may be causative or protective in autoimmune lymphoproliferative syndrome. *Hum. Genet.* 119:284.

236. Salmena, L., et al. 2003. Essential role for caspase 8 in T-cell homeostasis and T-cell-mediated immunity. *Genes Dev.* 17:883.

237. Salmena, L., and Haken, R. 2005. Caspase-8 deficiency in T cells leads to a lethal lymphoinfiltrative immune disorder. *J. Exp. Med.* 202:727.

238. Chaudhary, P. M., et al. 2000. Activation of the NF-κB pathway by caspase 8 and its homologs. *Oncogene* 19:4451.

239. Kreuz, S., et al. 2004. NFκB activation by Fas is mediated through FADD, caspase-8, and RIP and is inhibited by FLIP. *J. Cell Biol.* 166:369.

240. Lamkanfi, M., et al. 2005. A novel caspase-2 complex containing TRAF2 and RIP1. *J. Biol. Chem.* 280, 6923.

241. Su, H., et al. 2005. Requirement for caspase-8 in NF-κB activation by antigen receptor. *Science* 307:1465.

Index

T - #0196 - 221019 - C8 - 234/156/14 - PB - 9780367386573